# Self-Organization of the Human Mind and the Transition From Paleolithic to Behavioral Modernity

Yury N. Kovalyov
*Mykhailo Boichuk Kyiv State Academy of Decorative Applied Arts and Design, Ukraine*

Nver M. Mkhitaryan
*National Academy of Science of Ukraine, Ukraine*

Alexander Yu. Nitsyn
*Kharkov Polytechnic Institute, Ukraine*

A volume in the Advances in Religious and Cultural Studies (ARCS) Book Series

IGI Global
DISSEMINATOR OF KNOWLEDGE

Published in the United States of America by
IGI Global
Information Science Reference (an imprint of IGI Global)
701 E. Chocolate Avenue
Hershey PA, USA 17033
Tel: 717-533-8845
Fax: 717-533-8661
E-mail: cust@igi-global.com
Web site: http://www.igi-global.com



Library of Congress Cataloging-in-Publication Data

Names: Kovalyov, Yury N., 1961- author. | Mkhitaryan, Nver. M., 1960- author. | Nitsyn, Alexander Yu., 1951- author.
Title: Self-organization of the human mind and the transition from paleolithic to behavioral modernity / by Yury N. Kovalyov, Nver. M. Mkhitaryan, and Alexander Yu. Nitsyn.
Description: Hershey, PA : Information Science Reference, [2020] | Includes bibliographical references and index. | Summary: “”This book examines theoretical assumptions with specific cultural phenomena that are a manifestation of the evolution of mind”--Provided by publisher”--Provided by publisher.
Identifiers: LCCN 2019032913 (print) | LCCN 2019032914 (ebook) | ISBN 9781799817062 (hardcover) | ISBN 9781799817079 (paperback) | ISBN 9781799817086 (ebook)
Subjects: LCSH: Evolutionary psychology. | Human evolution. | Brain--Evolution.
Classification: LCC BF698.95 .K69 2020 (print) | LCC BF698.95 (ebook) | DDC 155--dc23
LC record available at https://lccn.loc.gov/2019032913
LC ebook record available at https://lccn.loc.gov/2019032914

This book is published in the IGI Global book series Advances in Religious and Cultural Studies (ARCS) (ISSN: 2475-675X; eISSN: 2475-6768)

British Cataloguing in Publication Data
A Cataloguing in Publication record for this book is available from the British Library.



For electronic access to this publication, please contact: eresources@igi-global.com.

# Advances in Religious and Cultural Studies (ARCS) Book Series

Nancy Erbe
California State University-Dominguez Hills, USA

ISSN:2475-675X
EISSN:2475-6768

## Mission

In the era of globalization, the diversity of the world and various cultures becomes apparent as cross-cultural interactions turn into a daily occurrence for individuals in all professions. Understanding these differences is necessary in order to promote effective partnerships and interactions between those from different religious and cultural backgrounds.

The **Advances in Religious and Cultural Studies (ARCS)** book series brings together a collection of scholarly publications on topics pertaining to religious beliefs, culture, population studies, and sociology. Books published within this series are ideal for professionals, theorists, researchers, and students seeking the latest research on collective human behavior in terms of religion, social structure, and cultural identity and practice.

## Coverage

- Globalization and Culture
- Sociology
- Cross-Cultural Interaction
- Gender
- Politics and Religion
- Impact of Religion on Society
- Social Stratification and Classes
- Cultural Identity
- Human Rights and Ethics
- Group Behavior

IGI Global is currently accepting manuscripts for publication within this series. To submit a proposal for a volume in this series, please contact our Acquisition Editors at acquisitions@igi-global.com or visit: https://www.igi-global.com/publish/.

The Advances in Religious and Cultural Studies (ARCS) Book Series (ISSN 2475-675X) is published by IGI Global, 701 E. Chocolate Avenue, Hershey, PA 17033-1240, USA, www.igi-global.com. This series is composed of titles available for purchase individually; each title is edited to be contextually exclusive from any other title within the series. For pricing and ordering information please visit https://www.igi-global.com/book-series/advances-religious-cultural-studies/84269. Postmaster: Send all address changes to above address. 

## Titles in this Series

*For a list of additional titles in this series, please visit:*
*https://www.igi-global.com/book-series/advances-religious-cultural-studies/84269*

***Developing Effective Communication Skills in Archaeology***
Enrico Proietti (Ministry for Cultural Heritage and Activities and for Touris, Italy)
Information Science Reference • © 2020 • 347pp • H/C (ISBN: 9781799810599) • US $195.00

***Understanding the Relationship Between Religion and Entrepreneurship***
Khaled Tamzini (IHEC of Sousse, Tunisia) and Anis Ben Salem (ISG of Sousse, Tunisia)
Business Science Reference • © 2020 • 298pp • H/C (ISBN: 9781799818021) • US $235.00

***Adapting Human Thinking and Moral Reasoning in Contemporary Society***
Hiroshi Yama (Osaka City University, Japan) and Veronique Salvano-Pardieu (University of Tours, France)
Information Science Reference • © 2020 • 330pp • H/C (ISBN: 9781799818113) • US $175.00

***Impacts of Globalization on Youth in the Developing World Emerging Research and Opportunities***
Obediah Dodo (Bindura University, Zimbabwe) and Vincent Mabvurira (University of Zimbabwe, Zimbabwe)
Information Science Reference • © 2020 • 145pp • H/C (ISBN: 9781799819585) • US $170.00

***Recent Social, Environmental, and Cultural Issues in East Asian Societies***
Mika Markus Merviö (Kibi International University, Japan)
Information Science Reference • © 2020 • 298pp • H/C (ISBN: 9781799818076) • US $195.00

***Projects That Fuel Women's Empowerment and Global Impact Emerging Research and Opportunities***
Cassandra Sligh Conway (South Carolina State University, USA)
Information Science Reference • © 2020 • 160pp • H/C (ISBN: 9781799817147) • US $135.00

***Anthropological Applications of Human Thermodynamic Concepts Emerging Research and Opportunities***
Abhijit Pandit (Amity University, India)
Information Science Reference • © 2020 • 135pp • H/C (ISBN: 9781522597360) • US $155.00

***New Aesthetic Thought, Methodology, and Structure of Systemic Philosophy***
Jie Wu (China System Philosophy Research Centre, Shenzhen University, China)
Information Science Reference • © 2020 • 301pp • H/C (ISBN: 9781799817024) • US $195.00

***Multidisciplinary Issues Surrounding African Diasporas***
Uche T. Onyebadi (Texas Christian University, USA)
Information Science Reference • © 2020 • 330pp • H/C (ISBN: 9781522550792) • US $185.00

IGI Global
DISSEMINATOR OF KNOWLEDGE

701 East Chocolate Avenue, Hershey, PA 17033, USA
Tel: 717-533-8845 x100 • Fax: 717-533-8661
E-Mail: cust@igi-global.com • www.igi-global.com

# Table of Contents

# Foreword

When we are looking at the drawings of primeval humans, our feeling is most accurately described as astonishment.

Trying to understand what the reason for this astonishment is, we will most likely pay attention to the strange, from present point of view, methods of transmitting the depth of space, or the complete lack of depth.

Why did the ancient artists convey the depth of space without using the perspective, which is so usual to us?

One can refer that they have not known about it - indeed, the linear (classical, Renaissance) perspective as a method of depicting space was described only in the Renaissance Age and understood as a special case of the method of two images only in the first half of XX century.

But this answer will be too superficial. In fact, besides the linear perspective, there is a natural (perceptual) perspective, corresponding to the peculiarities of visual perception even better than the linear perspective. Why, then, did she begin to apply – rather late, and is little used even just now? Really the ancient people did not understand the world as three-dimensional? How, then, did they understand it?

And these questions lead to a scientific problem. If the image of the space in the works of painting is a kind of window into the psychic world of humanity (as a separate work is, from the standpoint of psychology, a projective test of an individual), then a consistent analysis of the methods of transmitting space to different historical eras is a documentary evidence of the evolution of human mind.

But the inquisitive mind of a person does not stop there, and immediately asks other questions:

- What is mind in general?
- How to determine the quality of rationality?

Reading the relevant literature - and quite a few outstanding scientists from the time of Antiquity to nowadays have asked themselves these questions and tried to find answers - does not lead to a clear understanding. Different points of view differ too much.

There are also many questions related to the process of evolution. The efforts of archaeologists, anthropologist, and geneticists have generally led to the establishment of a chronology of mankind origin and resettlement (despite the fact, that some researchers dispute these achievements), but many inexplicable details remain. Why, for example, upright walk and release of hands happened much earlier than the hand began to be used to make tools for labor? Why for the very long time the human, anatomically indistinguishable from the modern type, did not create of material and cultural phenomena, which clearly separates him from the imperfect ancestors? Why in the post-Paleolithic era, the time of culture, religion, science, and technology development, when the possibilities of the human brain should have become more and more necessary, the size of the brain decreased, and the structure was simplified? Are the mind and the intellect connected with anatomical evolution?

Perhaps the answers to these questions are related to rethinking the point of view on human evolution. There is no doubt that man and mankind are complex open systems, and, therefore, possess their properties, including self-organization. Perhaps an approach to evolution from the standpoint of self-organization of complex systems will help explain its regularities and paradoxes? Such a hypothesis has the right to exist.

That is why the authors, who are experts in the field of modeling and optimizing complex systems, turned to a topic that was so unusual for them but interesting for all representatives of the human genus.

So, the purpose is to substantiate the view on evolution as a process of self-organization of an open system, its verification, and consideration of possible consequences.

In this context, the definition of methods of depicting space in different historical epochs is considered as one of the markers that make it possible to trace the evolution of the human mind. Understanding this connection is important not only for the history of painting or the theory of images but also for anthropology and psychology. In addition to the theoretical value, it is possible to indicate the applied aspects, among which we single out the creation of a comfortable living environment for the whole variety of psychological and social types, which are present in the modern humanity. Creating a comfortable environment is another area of authors professional interest.

The publication of this innovative and possibly debatable book will contribute to the emergence of new perspectives on human evolution, a better understanding of the patterns of changes in human mind and behavior and their social and cultural implications, as well as improving the quality of life of modern people as a result of the comfortable environment creation.

I believe that the book will be of interest to various readers - mathematicians, anthropologists, ethnographers, art historians, as well as those who want to buy or build a truly comfortable dwelling. Such different target audiences for one book are another advantage of the systematic approach used by the authors!

*M. I. Yakovlev*
*National Academy of Arts of Ukraine, Ukraine*

# Preface

The book, presented to your attention, tells about the human mind and behavior patterns self-organization from ancient times for the present.

The authors' position is: a correct understanding of the human mind and behavior is possible in the framework of the system approach.

The author's hypothesis is: that since human is a complex system, then one of the main factors of his mind and behavior evolution is the realization of the general laws of self-organization of complex, open, emergent systems. Testing this hypothesis is the main purpose of this work.

But it is impossible to tell about the human mind and behavior even in ancient times without discussion about the evolution of mankind.

What was the process of human origin and what is its chronological framework? Was human evolution random, or did it have some kind of scenario? What are the paradoxes and unresolved problems of evolution? What are mind and rationality and how are they related to the development of the human brain? Why, having reached the maximum of the mass and complexity of our closest ancestors, the Cro-Magnons, it has reduced and degraded just when the main achievements of our civilization appeared?

This and the same circumstances are determining the course of self-organization and should be clarified.

So, the review of facts, their interpretation, and the presentation of modern theories of anthropogenesis are presented in the first chapter, *Human evolution: facts and problems*, which, therefore, is the background to the whole book.

Next, it is necessary to describe the research methods used in this work.

In the second chapter, *Self-organization of complex systems*, the basics of complex systems modeling apparatus are presented.

First, it is necessary to show that a human, his mind and behavior correspond to the characteristics of complex systems.

Second, is considered the main problem of mathematical modeling of complex, open, emergent systems - the discrepancy between the properties of such systems and the axiomatic foundations of geometry (as the most visible part of mathematics).

The most fundamental decision, and it is the third task, is a creation of the axiomatic model with open systems properties - wave model of S- space. The questions of completeness and consistency, means of description and verification methods are not the subject of this book; they have been reviewed in other publications, cited in the second section.

The next tasks are the construction of the theory of self-organization of complex systems and the definition of self-organization scenarios and their invariants. The driving forces of self-organization are the immanent variability of S-space waves and solitons and external factors. For a simple self-organization scenario, the structure of the system (levels and quantities of elements), the numerical characteristics of the elements (potentials, quantities, and a number of others), the integrity conditions (symmetries, conservation laws, Fibonacci numbers, relations based on the "golden section" and other) can be determined. Since the axioms of the S-space wave model include an emergent property, systems corresponding to different scenarios of self-organization can include heterogeneous, non-reducible S-sets and S-elements.

The presence of qualitative and quantitative characteristics of different scenarios of self-organization allows us to compare them with the known facts of human origins, presented in the first chapter, choosing the scenario, that best corresponds to these facts.

This verification is carried out in the third chapter, *Self-organization: body and mind*, the next task of it is to build a «Human-environment» model.

Making sure of the usefulness of this hypothesis at this stage, it is possible to go to the next step - examination of the structure and evolution of the human mind inside of the «Human-environment» model.

The fourth section, *Self-Organization: evolution and behavior*, contains results, related to four interrelated problems:

1. Explanation of the human evolution and mind paradoxes, listed in the Human evolution: facts and problems chapter.

Following the theory of self-organization, the paradoxes of evolution, due to the scenario and the wave nature of the Human-Environment system are regularities. This does not contradict the theory of natural selection, nor its modern additions; regularities of self-organization appear as limitations and potentialities that make natural selection "teleological".

2. Human mind definition.

The structure of a human subjective space[1] is determined by the scenario (1S, 1O), describing the relations of the levels and channels of human interaction with the environment according to a «Human-Environment» model. The script determines the number of channels on each level and the specific contribution of each of them to the formation of a picture of the world perceived by a person. By relating this contribution to the features of intuition, ego, will, mind, emotions, which do not have clear definitions in psychology, it becomes possible to give such definitions within the framework of the script. The human consciousness of itself as a whole is determined by the activity of a single channel of the second level, and the nature of relations with the environment is determined by the activity of other channels. The consideration of such heterogeneous components within the framework of one system became possible due to the emergence in the S- space the axioms.

3. In this context, the question of the knowability of the world being discussed. The authors substantiate the following point of view: the world is knowable insofar as it, its structure and levels and the human subjective space, its levels and channels and channels are described by the same laws of self-organization of complex systems. Thus, the structure of levels and channels is a peculiar gestalt, which limits the limits of cognition and influences the "image of the world" in the human mind.

Since in various groups of people, in individuals, the ratio of individual channels differs from the calculated ones, giving a kind of "ideal" picture, one should speak about different psychological types of groups of people and psychological portraits of individuals, basing their classification on a simple principle - the dominance one of the channels above the others. This separation determines the variety of ways people interact with the environment, their social division, and behavior patterns. All these results are presented in table form. A correlation with the specific needs and motivations for each psycho-type was also established, as far as specific features of the age groups and characteristics due to temperament.

The change in the activity of the channels from epoch to epoch determines a kind of generalized mind, a "spirit of the epoch", manifested as features of the world outlook and culture of each historical period.

4. The self-organization scenario gives a forecast of such changes, but with what markers can they be traced by comparing the predicted changes with the actual ones? The choice of two markers is substantiated - the way of imaging the space in painting and the way of representing the structure of the Universe in the myth of the World Tree. Last but not least, this choice is due to its global spread throughout the entire habitat of humanity. The presence of these markers defines the chronological framework of the study of the evolution of a human mind: from the Upper Paleolithic, when space constructions in painting and the myth of the World Tree arose, to present.

In the following sections painting of the Paleolithic and Neolithic, the epoch of pre-civilization, the Ancient World, Antiquity, the Middle Ages, and the Renaissance, as well as modernity is considered. For the same periods, the plot and expressive means inherent in the myth of the World Tree are tracked.

Let's note the following feature of the chronological scale used in the work. Since the subject of the study is the self-organization of modern man's mind, the authors found it convenient not to separate historical events across the border of the Christian era, but to keep a continuous account until 2019.

In the fifth section, *Paleolithic: unity*, the research is conducted as follows:

1. Examples of Paleolithic paintings are given, its features are determined, in particular, of space images. The results are tabulated.
2. The available data of the World Tree images, as well as ethnographic data on the world outlook of the tribes in the primitive state, based on which the reconstruction of the early version of the myth of the World Tree is made, is considered. Features of this version are tabulated.
3. Then markers analysis takes place, based on which the Paleolithic mind is reconstructed. The results are presented in the form of a generalized psychological portrait of a Paleolithic person (which is possible, since then there was still no expressed social division), showing the characteristic for the epoch correlations of the activity of the channels of subjective space. The same circumstance was taken into account in the process of the behavior patterns reconstructing, for which the results of the Self-organization: evolution and behavior chapter, were also applied.

The comparison of the predicted changes in the human mind and changes, in fact, are considered.

The same research sequence is used in the following chapters.

In the sixth chapter, *Neolithic: separation and war*, Neolithic painting is considered (samples are presented, features are identified and tabulated, compared with Paleolithic painting), as well as myth of the World Tree and its relation with the cult of ancestors (plot and the meaning of the myth reconstruction is carried out according to the available images and on the basis of subsequent representations supposedly of Neolithic origin, features are tabulated and compared with Paleolithic).

Then markers analysis is performed, during which changes in the activity of channels in the Neolithic era are revealed. The result of this analysis is the reconstruction of the Neolithic mind and behavior patterns.

The seventh chapter, *Eneolith: The Search for Balance*, begins with a justification of the choice of Eneolithic cultures, which could be a source of facts for research. The need for this is since by that time the homogeneity of the Neolithic culture was significantly impaired. Quite numerous centers of civilization, which differ significantly in their cultural development, appeared. To decide whether the evolution of mind was universal, or split into several local scenarios, the widest possible coverage of cultures is necessary; on the other hand, to consider them all in one book is not possible. Therefore, unrelated cultures were chosen, representing different paths of civilization development: settled and nomadic. Other conditions were the long existence of culture and its bright, outstanding character.

Further analysis of the artifacts was performed separately for each of the selected cultures: Cucuten-Tripolye, Harappa and the Central part of Eurasia cultures. There are no cultures of Mesoamerica and South America in this list that are undoubtedly well known to the American reader. As in the previous chapter, the features of Eneolithic painting, including plots and characters, were determined and compared with Neolithic painting. When considering the Eneolithic version of the World Tree myth (especially well represented by Harappa culture artifacts), special attention was paid to its connection with the Great Goddess, also known as Mother Goddess and Mother-Earth.

After the marker's analysis, conclusions about the activity of the channels in the Eneolithic era were made, which made it possible to reconstruct the Eneolithic mind and Behavior patterns.

A comparison of the predicted changes in the human mind and changes, in fact, is considered. Notes, that despite the different conditions and ways of the formation of cultures, the evolution of mind has remained global.

The next, eighth chapter, *The Ancient World: Ego, Archaic, Immortality*, opens up a discussion of the mental evolution in historical time. Its main features are the emergence of full-fledged and indisputable writing (for the previous period, the availability of writing was a subject of discussion), states, expressed social division, developed religious doctrines. Just as for the previous era, there are several cultures represented by a huge number of artifacts. For this reason, for the analysis, a civilization of ancient Egypt was chosen, which existed for a very long time and retained its originality up to the spread of Hellenism and, especially, Christianity; if necessary, a description and analysis of artifacts of synchronous cultures of the Middle East is provided.

Egyptian painting is represented not only by sacral and secular frescoes and paintings but by signs of writing, as well as reliefs, which can also be judged on the image of space. Moreover, there is a curious tendency: Neolithic features are preserved in sacred painting, while in secular painting a variety of techniques for displaying the depth of space are developed. As for the myth of the World Tree, it is present, but it goes to the periphery of the public consciousness - only one side of it, the path to the light through the world of the dead, causes unrelenting interest. But it is also in a different paradigm, which presented in the Book of the Dead and the Pyramid Texts.

This interest in the image of three-dimensional space, as well as the fact that the myth of the World Tree has gone to the background, provides rich material for the marker's analysis and conclusions about the activity of certain channels.

Because of the pronounced social separation (which is due to the predominance of a channel of interaction with the environment), it would be incorrect to compile a generalized psychological portrait outside the social context, so the mind and behavior patterns for main classes of the ancient Egyptians were reconstructed. For definiteness, the period of the highest prosperity of Egypt is taken - the time of the earliest dynasties of the New Kingdom.

Is interest in the depiction of three-dimensional space in painting a sign of a transition to a rational or even scientific understanding of the world? – This question is discussing in the conclusion.

In the ninth chapter, *Antiquity: the path to rationality*, the ancient world is considered, the main feature of which was the tendency towards ever-increasing rationalism, which led to the emergence of science in our usual sense. As well as for the previous period, a large number of monuments, artifacts, and written sources raises the problem of justification; so, selection criteria are given special attention.

The development of space imaging is considered. It is in many respects similar to the ancient Egyptian one: to replace the two-dimensional images made in the characteristic yet Neolithic technique, methods of transmitting the depth of space come. And they are different from Egyptian - by the late Republican time (according to written sources, much earlier), methods of perceptual perspective, aerial perspective and chiaroscuro are used. The names of various projections are even recorded in the Vitruvius Ten Books on Architecture. The connection between these methods and the development of Euclidean geometry seems undoubted, but not directly, but as "parallel" consequences of the same evolutionary process.

Very interesting changes have undergone the myth of the World Tree. Almost forgotten, seemingly gone from public discourse (for example, in the Etruscan tombs the canonical images of the World Tree are illustrative rather than sacral), it is suddenly reborn in Christianity. Jesus Christ sacrifices himself by crucifixion over Adam's grave (that is, on the World Tree), dies, descends into hell, defeats his masters and frees the souls of the righteous, is disciples in the flesh and ascends to heaven, revealing the status of the Living God - perfect "Shaman's journey", once again brought together all the episodes characteristic of ancient eras. And this is with the taboo of such a plot in Judaism - "hanging on a tree, damn it!", And with such colorful details as the dog-headed Saint Christopher, very similar to Anubis.

Of course, these changes, along with changes in the technique of image space, provide rich material for markers analysis and, accordingly, ancient human mind and behavior patterns.

The tenth chapter, *The Middle Ages: a failed return*, is dedicated to the Middle Ages. Not forgotten in the Dark Ages, developing, despite a series of almost continuous wars and other disasters, in Europe, in the Islamic world, and in China, the technique of depicting three-dimensional space by means of perceptual and aerial perspective, as well as chiaroscuro, reached almost perfect before the beginning of the Renaissance.

The myth of the World Tree, "preserved" by biblical ideas and the Gospel story of the crucifixion of Jesus, nevertheless also develops - in paganism (Odin and Yggdrasil), in China (the concept of San-Tsai, when human himself plays the role of the World Tree), in hermetic societies, in particular, among the Kabbalists (the Sefirot tree and Adam Kadmon).

All this again testifies to the global nature of the mind evolution, gives a lot of facts for the analysis and reconstructions of consciousness and behavior patterns. To compile a psychological portrait, this time an outstanding historical person was taken - Alienor of Aquitaine.

In the eleventh chapter, *Renaissance: the legalization offeelings*, describes the emergence and features of the scientific method of displaying spatial relations - a linear perspective that arose in Italy, and the perceptual perspective improved by the masters of the Northern Renaissance.

The myth of the World Tree finally passes into the form of natural-philosophical speculation.

All this is happening in the background of significant changes in the public consciousness, which can be traced from other sources.

A comparison of all these facts makes it possible to reconstruct the state of mind and patterns of behavior and compare them with the predicted ones.

In the twelfth chapter, *the Modern human mind*, the connection of the methods of depicting space characteristics of modern painting with the postmodern consciousness of our age, as well as the worldview, social and behavioral consequences of the current state of affairs are considerate. The state of mind is determined after analyzing the markers, just as it was done in the previous chapters. The result is generalized conclusions about the activity of the channels, as well as the expressive means inherent in each of them in all the considered historical epochs. These results confirm the main hypothesis of this research.

However, practical implications of the theory are becoming relevant, in particular, when designing a comfortable living environment. For this, it is necessary to move from generalizations for individual social groups to the compilation of a psychological portrait of a person.

Simple tests are given to determine the characteristics of the main channels, as well as the method of bringing the results of various tests to a commensurate scale and using them to build a psychological portrait of a person according to a systematic approach.

The final, thirteenth section – *Practice as a criterion of truth: the choice of a comfortable dwelling* – uses this portrait to optimize the part of the environment closest to a person - his habitat. Consistently considered, illustrated by examples and supplied with calculations, the following parts: comfort/discomfort conditions and design aims; design aims and dwelling specifications; dwelling for contemplators; dwelling for egoists; dwelling for fighters; dwelling for enterprising researchers; dwelling for artistic persons; dwelling for balanced persons.

Improved smart house technologies will make it possible to use the dwelling as a protector, adviser, and doctor, and a proper understanding of intra-family goals and conflicts can use the dwelling as a peacemaker.

The authors expect that this book could be interesting to a cross-disciplinary audience, such as scientists, researchers, practitioners, teachers, philosophers, architects, and education experts.

## ENDNOTE

1 This term, introduced following the system approach, replaces the categories of "mental organization", "psyche" and so on, that do not have precise definitions.

# Chapter 1
# Human Evolution:
## Facts and Problems

## ABSTRACT

*The main events and circumstances of human evolution are considered: classification of hominids, first descriptions, localization, chronology; artifacts characterizing their material and cultural activities; modern reconstruction of lifestyle and resettlement; and modern theories explaining the structural features of hominids and the processes of their occurrence. The manifestations of intelligent activity are discussed, in particular, their dependence from the structure of the body, the size, and complexity of the brain, for which comparisons with various animals are made. Particular attention is paid to unresolved or controversial issues. This material is necessary to assess the possibilities of the self-organization of complex systems theory (second chapter): if it adequately models the characteristics of a human's origin, then it can be used to understand the evolution of human mind and in the subsequent period, up to the current state.*

## BACKGROUND

In recent decades, a lot of paleontological and archaeological data have been accumulated (Leakey, 1979; Clarke & Tobias, 1995; Brunet et al., 2002; Brown et al., 2004; Linden, 2007; Johanson & Wong, 2009; Reid et al., 2011; Walter, 2013; Ryabinin, 2015; Drobyshevsky, 2017; Shannon & McPherron, 2017; Arnold, 2018).

A new dating method (potassium-argon, carbon), in addition to traditional ones (Kuzmin, 2017), made it possible to clarify the age of bones and artifacts. Particularly actively used methods based on genetic analysis (Cavalli-Sforza & Feldman, 2003; Borinskaya & Yankovsky, 2006; Krause et al. 2010; Reich et al., 2010; Wong, 2014; Fernando et al., 2016;), which may shed light on the course of human evolution.

Classification of findings underlying modern theories of human origin, according to (Khrisanfova & Perevozchikov, 2005; Markov, 2011; 2012; Drobyshevsky, 2017; Antropogenez, n/d) and is presented in Table 1. Modern reconstruction of the appearance and lifestyle of hominids made based on these findings delivers the following picture (Antropogenez, n/d). Table 1 will make it possible to take a general

DOI: 10.4018/978-1-7998-1706-2.ch001

look at the process of human origin and its chronological framework. Therefore, it is from her that the presentation of the main material will begin.

It should be remembered that if the findings themselves are undeniable, and the determination of their age is relatively reliable, reconstruction of the appearance, behavior, culture of human ancestors and predecessors, as well as the course of evolution, is usually presumptive. This is due to the lack and fragmentation of the findings, as well as the contradictions of different theories (Levins, 1968; Ungar, 2006; Palmer, 2007; Klyosov &, Tyunyaev, 2009; Falk, 2011; Walter, 2013; Boyd, 2017; Arnold, 2018; Primarashni, 2019).

The same applies to the key aspects of the evolution of human anatomy (Niemitz, 2002; Kwang, 2015), brain (Park et al., 2007; Leonard, Snodgrass & Robertson, 2007), mind, and culture (Vishnyatsky, 2010; Burlak, 2011; Wynn & Coolidge, 20012; Kevin, 2017).

A separate issue is the problem of the brain, the mind and consciousness correlation. There are many sources on this topic (Edelman & Mountcastle, 1978; Bohm, 1990; Stapp, 2001; Monserrat, 2016; Pally, 2018; Chung, 2018; Oakley & Plotkin (Eds.), 2018). A Human Brain Project coordinating numerous studies in this field (Human Brain Project, n/d).

## GENERAL VIEW: CLASSIFICATION, FIRST DESCRIPTIONS, LOCALIZATION, CHRONOLOGY

So, the process of human evolution has taken at least 7 million years; finds of bones, up to the appearance of the genus Homo, are concentrated in Africa; at the same time there were several types of hominids; the formation of mind and consciousness went at different speeds and accelerated since the advent of Neanthropes.

Let's considerate each period separately, highlighting finds and dating, anatomy features, habitats, and behavioral features, as well as commenting on the role in the origin of human and the problems that arise. Naturally, later periods, as more studied, will be discussed in more detail (Khrisanfova & Perevozchikov, 2005; Markov, 2011; 2012; Drobyshevsky, 2017; Antropogenez, n/d).

## EARLY AUSTRALOPITHECUS

Remains of Early Australopithecus have been found in Chad, Kenya, Ethiopia and dated to a period between 7 and 3.9 million years ago.

They were slightly more than a meter high. The structure of the skulls, femurs, and pelvis gives reason to believe that the early Australopithecine had already walked upright. On the other hand, the hand bones show that they retained the ability to climb trees or even had been four-handed beings, resting on the phalanges of fingers when walking. The structure of the teeth of Early Australopithecus in between monkeys and humans. The brain size was the same as that of a chimpanzee

Habitats are forests and forest-steppe. Virtually nothing is known about their lifestyle. Perhaps, they were an intermediate link between ape and human.

*Table 1. A generalized diagram of the basic hominid's family taxon's*

| Taxon's | First Descriptions | Region | Time, Million Years Ago |
|---|---|---|---|
| Early Australopithecus | Sahelanthropus tachadensis (Brunet et al., 2002) | Chad | 6-7 |
| | Orrorin tugenensis (Senut et al., 2001) | Kenya | 5.6-6.2 |
| | Ardipithecus ramidus (White et al., 1995) = Homo antiquus praegens (Ferguson, 1989) = Australopithecus praegens | Ethiopia | 3.9-5.5 |
| | Australopithecus anamensis (Leakey et al., 1995) | Kenya | 3.9-4.2 |
| Gracile Australopithecus | Australopithecus afarensis (Taieb et al., 1978) (probably 2 types) = Australopithecus africanus Tanzanians (Tobaios, 1980) = Australopithecus aethiopicus (Tobias, 1980) = Homo antiquus (Ferguson, 1984) = Meganthropus Africanus (Weinert, 1950) = Praeanthropus Africanus (Hennig, 1948; sensu Senuyrek, 1955) = Paranthropus Africanus (Weinert, 1950; sensu Olson, 1981) | Kenya, Tanzania, Ethiopia | 3.0-3.9 (4.4) |
| | = Kenyanthropus platyops (Leakey et al., 2001) | Kenya, Ethiopia | 3.2-3.5 |
| | Australopithecus bahrelghazali (Brunet et al., 1996) | Chad | 3.0-3.5 |
| | Australopithecus africanus (Dart, 1925) (probably 2 types) = Australopithecus prometheus (Dart, 1948) = Plesianthropus transvaalensis (Broom, 1936) = Homo transvaalensis (Mayr, 1950) | South and East Africa | 2.4-3.5 |
| | Australopithecus garhi (Asfaw et al., 1999) | Ethiopia | 2.5 |
| Massive Australopithecus | Paranthropus aethiopicus (Arambourg et Coppens, 1968) = Australopithecus walkeri (Ferguson, 1989) = Paraustralopithecus aethiopicus (Arambourg et Coppens, 1968) | Kenya, Ethiopia | 2.3-2.6 |
| | Paranthropus (Leakey, 1959) = Zijanthropus (Leakey, 1959) | East Africa | 1.1-2.3 (2.5) |
| | Paranthropus robustus (Broom, 1939) (probably 2 kinds) = Paranthropus crassidens (Broom, 1948) | South Africa | 0.9-2.5 |
| Early Homo | Homo (or Australopithecus) habilis (Leakey et al., 1964) = Australopithecus habilis = Homo erectus habilis (Campbell, 1965) = "Prezinjanthropus" = Homo microcranous (Ferguson, 1995) = Telanthropus capensis (Broom, 1947) | Eastern and Southern Africa | 1.5-2.3-2.6 |
| | Homo (or Pithecanthropus) rudolfensis (Alexeev, 1978) = Kenyanthropus rudolfensis | East Africa | 1.5-2.3 (2.5) |
| Homo genus | Homo (or Pithecanthropus) ergaster (Groves et Mazak, 1975) = Homo Kenyans (Zeitoun) = Homo okotensis (Zeitoun) = Homo georgicus (Gabunia, Vekua, de Lumley et Lordkipanidze, 2002) = Telanthropus capensis (Broom, 1947) | Africa, Georgia | 1.4-1.7 |
| Archantropes | Homo (or Pithecanthropus) erectus (Dubois, 1894) = Sinanthropus pekinensis (Black, 1927) = Pithecanthropus leakeyi (Heberer, 1963) = Meganthropus palaeojavanicus (Koenigswald, 1941) = Atlanthropus mauritanicus (Arambourg, 1955) and others. | Africa, Europe, Asia | 0.4-1.4 |
| | Homo (or Pithecanthropus) heidelbergensis (Schoetensack, 1908) = Cyphanthropus rhodesiensis (Woodward, 1921) = Javanthropus soloensis (Oppenoorth, 1932) and others. | Africa, Europe, Asia | 800-500 thousand years |
| Paleoanthropes | Homo neanderthalensis (King, 1864) = Homo sapiens krapinensis (Goryanovic-Kramberger, 1902) = Homo mousteriensis (Klaatsch et Hauser, 1909) = Palaeoanthropus palestinensis (Keith et Mc Cown, 1939) and others. homo sapienssubsp. Denisova (?) homo floresiensis (Brown et al., 2004) (?) | Africa, Europe, Asia | 500-130 (40) thousand years |
| Neanthropes | Homo sapiens sapiens (Linnaeus, 1758) = Neoantropus fossilis = Homo Africanthropus helmei (Dreyer, 1935) = Homo aurignacensis hauseri (Klaatsch et Hauser, 1912), and others. | whole planet | 0-40 (200) thousand years |

## GRACILE AUSTRALOPITHECUS

A large number of Gracile Australopithecines remains was found in Kenya, Tanzania, Ethiopia. The most known findings come from Afar area, including the skeleton of a female nicknamed Lucy (Johanson & Wong, 2009). The remains dated to a period between 4.4 and 2.4 million years ago.

They had an apelike skull, combined with an almost modern structure of legs and pelvis, elongated hands and fingers adapted to climbing trees. We can confidently assert that the gracile Australopithecus were creatures walking upright, about 1-1.5 meters tall. The gait was different from the human gait by shorter steps, and the hip joint did is not unbend completely when walking. The size and shape of the Australopithecus brain were like that of a monkey.

During the daytime, Australopithecus roamed about the savannah or forests, along with the banks of rivers and lakes, and in the evening, they climbed trees, as the modern chimpanzees do. Australopithecus lived in small herds or families and were able to move to fairly long distances. They ate mostly plant foods. Isolated findings of stone tools and antelopes' bones crushed by them do not yet give grounds to speak about the systematized tool production.

The most likely human ancestors are Afar australopithecines.

## MASSIVE AUSTRALOPITHECUS AND EARLY HOMO

Massive Australopithecus, known from findings in Kenya and Ethiopia (Omo, Olduvai, Koobi Fora), as well as in South Africa (2.6-0.9 million years ago).

Massive Australopithecus was a large (up to 70 kg) herbivorous creature, as evidenced by huge jaws and large bony crests, serving to attach the masticatory muscles. Bone structure gives reason to believe that they had a bipedal gait. Habitat is dense thickets along the banks of rivers and lakes, like that of modern gorillas. It is believed that the limited habitat and a narrow specialization caused their extinction. Massive australopithecines probably were able to make stone tools.

The questions important to understand the evolution progress are debatable. For example, some researchers believe that Australopithecus is a dead end. The appearance of Australopithecus can be judged by performed reconstructions, which are slightly different according to various groups of researchers. For example, let's refer to the reconstructions performed in the Laboratory of Anthropological Reconstruction of the Institute of Ethnology and Anthropology of the Russian Academy of Sciences, the advantage of which is the use of a single reliable method developed by M.M. Gerasimov on all patterns. Relevant photos are available at (Antropogenez, n/d). The dominance of ape-like facial features and skull can be ascertained.

Sometimes all hominids are combined into one genus, Homo erectus, to which a modern man belongs, i.e. it is included in the early Homo genus from Table 1. A more traditional approach is to allocate a special kind of Pithecanthropus for the ancient representatives of Homo, excluding the Early Homo.

Early Homo (2.7-1.5 million years ago) were found in East and South Africa.

The oldest stone tools of the so-called Olduvai type (Figure 1) are known from localities in Ethiopia, namely Gona, Shungura, and Hadar, and are dated as 2.5-2.7 million years old. The first tools were the pebbles split in half. Later, Early Homo began beating a few chips from stones, making out a sharp cutting edge.

*Figure 1. Olduvai type tools:1 - chopper; 2, 3 - choppings; 4, 5, 8 – tools on flakes; 6, 7 - discoid nuclei (Biofile, n/d)*

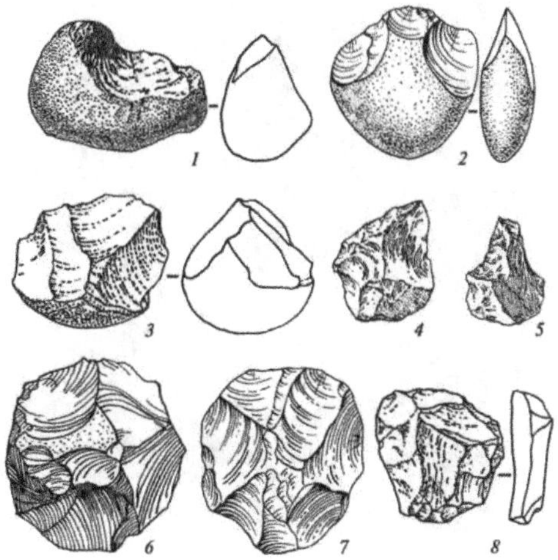

There appeared new kinds of hominids, i.e. small bipedal creatures with a relatively large brain and the monkey-like face. Large protruding jaws and a wide nose made them look like modern chimpanzees.

An important difference from Australopithecus was a big developed brain and a brush adapted to the manufacture of guns. Together with progressive symptoms, Early Homo retained many primitive features, including the structure of the hand and brain, in particular, a large size of jaws, thick skull walls, and pronounced frontal and occipital relief, because of which some researchers consider that they are a progressive kind of Gracile Australopithecines.

Early Homo shifted from plant food to the carnivorousness, probably picking up the remains of predators' prey (the traces of stone tools on bones, on top of lions' and hyenas' teeth marks, were found).

## ARCHANTHROPINES

Archanthropines represented by finds Africa, Europe, Asia (1.4-0.4 million years ago).

Approximately 1-1.5 million years ago (that is not earlier than the first Archanthropuses appeared), the tools have been improved so much that they have been attributed to a new, Acheulian culture (Figure 2).

A typical tool of Acheulian culture is a heavy ax, with a rough cutting edge. Besides stone tools, archanthropuses also made wooden spears, rarely found in the peat bogs of Europe.

About 1.4 million years ago Early Homo lost almost all similarities with Australopithecus, having become the people, almost identical to modern ones by skeleton structure (their hand has acquired a completely modern form). However, their skull remained primitive. Huge jaws, overhanging brow ridges, sloping forehead and the nape extending back are the characteristic features of Archanthropus. Archanthropines appearance reconstructions (Antropogenez, n/d) show the preservation of a quite archaic, apelike face.

The walls of the skull in some individuals reached 2 cm thick; the brain volume was 700-1100 $cm^3$. Archanthropines' brain continued to change progressively: the frontal, parietal and temporal lobes, the role of which is described in more detail below, was developing. The frontal lobe, responsible for social-

*Figure 2. Acheulean type tools: 1-4 - axes; 7, 8 - jibs; 5, 6 – ax production (Civilka. Humanity. Ashel, n/d)*

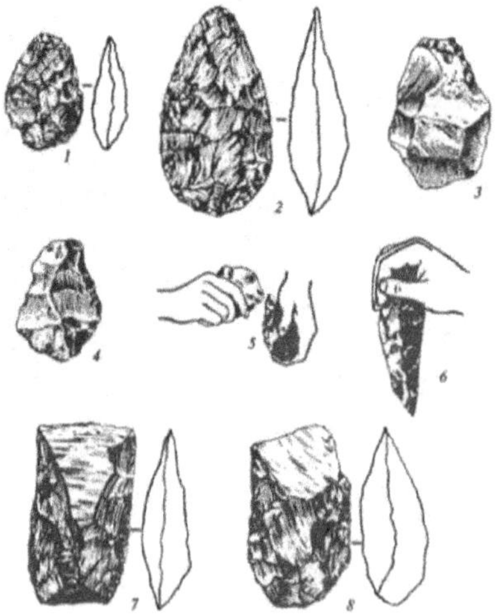

ity, communication, and the manifestation of altruistic behavior, was growing most actively in human ancestors. The men's height of later Archanthropines, Homo heidelbergensis, ranged between 155-175 cm, and the weight was about 65 kg. The modern features include brain volume, i.e. 1100-1400 cc, and the dental arch shape.

Archanthropuses colonized Europe and Asia, up to Indonesia. At that time, the people began to hunt large dangerous animals, which required team coordination and complex communication. Perhaps, at this time the speech emerged, which can be seen in brain casts. Then the people met fire (estimated 400 thousand years ago). Colonization of subtropical and temperate climates caused their need to protect themselves from temperature extremes, and the people began to settle in the caves and rocky coves. Cave settlements of that time are known in Spain, Caucasus, and China. The remains of primitive huts were found in Spain. Archanthropuses laid the foundations of human society.

Archanthropines included specialized groups, which left to posterity, while the others continued to evolve. Sometimes several species are allocated among them; there existed at least two main branches, eastern and western. It is also unclear what kind of Archanthropuses were a direct ancestor of modern humans, although they all are recognized as belonging to the Homo genus. Currently, paleoanthropologists are prone to think about the complex nature of evolution. Some population overtook the others, and migrations or isolations played a significant role in the uneven dissemination of progressive features. The progressive features could occur not comprehensively, but in various groups, uniting due to the mixing (retiform evolution). A typical example is the Heidelberg man. Homo Heidelbergensis has spread from Africa to Europe and Asia 500-800 thousand years ago. For a long time, European and African populations were isolated from each other. Heidelberg man could be the last common ancestor of modern humans and the Neanderthals. The Neanderthals have separated from this variety 300 thousand years and Homo Sapiens - 200-100 thousand years ago.

## PALEOANTHROPES

Remains of Paleoanthropes were found in Africa, Europe, Asia and have age 500–35 thousand years.

While early Paleoanthropes used Acheulean stone processing technique, approximately 200 thousand years ago (and 300 thousand years after their emergence) they switched to a more sophisticated Mousterian technique (Figure 3). The typical tools of that era are the sharp-pointed weapon and the scraper. The cultural difference of territorial groups was intensified. In Asia, the primitive stone processing methods have been kept for a long time. In Europe, Mousterian technology has reached its peak and was specialized noticeably.

The skeleton of early Paleanthropes was almost indistinguishable from the skeleton of a modern man, except for the head. They had a huge frontal bone and high face, broad nose, a heavy jaw with a sloping chin and a sloping forehead. The nape of many Paleanthropes was strongly extruded backward.

Late Paleoanthropes, particularly Neanderthals, also had some archaic features, for example, a very broad face coming forward, with beveled cheekbones.

Many signs of European Neanderthals could have arisen under the influence of harsh conditions of the Ice Age. The build of Neanderthals was stocky, with short legs, a barrel-like chest, and very broad shoulders. They were very strong men, accustomed to great physical stress. Interesting analogies to their morphological type can be found among modern Arctic peoples (broad shoulders, barrel-like chest, and

*Figure 3. Mousterian type tools: 1 – sub-prismatic nucleos; 2 - discoidal nucleos; 3 - scrubber; 4, 5, 7 - sharp-pointed weapon and its use; 6 - biface; 8 - scraper; 9 - cutter; 10 – edge (Civilka. Humanity. Mustie, n/d)*

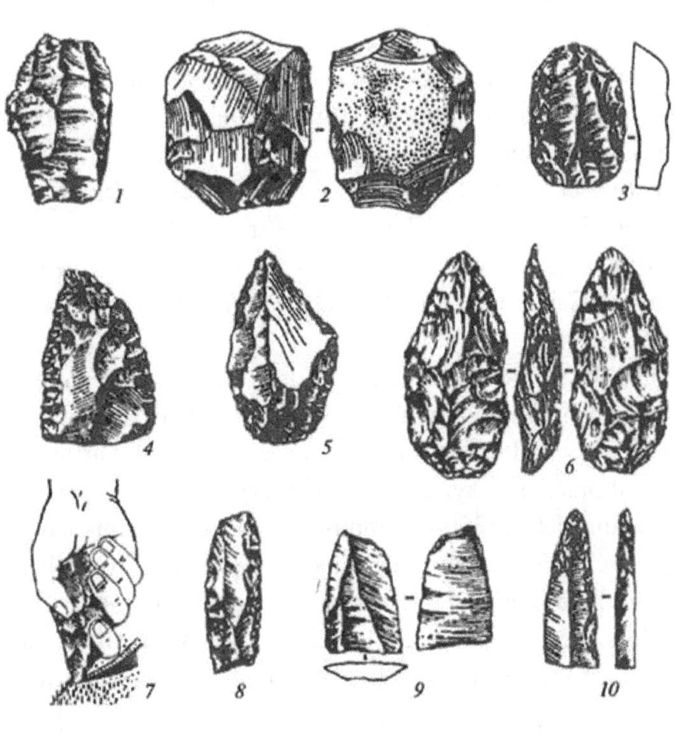

a stocky figure). However, Neanderthals' biological specialization to the cold has gone much further, and their differences were quite significant.

Some populations of Africa and the Middle East synchronous to the European Neanderthals were more like modern humans. Many researchers even classify them as the modern variety. People from Klazies River in South Africa had a chin boss, a rounded neck, and a high skull. The size and shape of the brain of these people are almost indistinguishable from the modern one.

The Middle East and Central Asia were populated by two types of people. One was more similar to the Neanderthals of Europe, while the other – to the progressive Paleanthropes from Africa. Specifically, the culture of all these people was very similar.

In general, one can state that biological evolution continued towards the reduction of the skull massiveness and increase (up to 1600-1700 $cm^3$, which is greater than that of modern people) and complications of the brain structure. Brain volume grew faster than the skull structure was developing and its shape was changing.

Paleoanthropes were setting the planet actively. Once in new conditions, the people learned to cope with natural difficulties: they invented the clothes, improved housing, inhabited the caves, driving large carnivores, i.e. bears, lions and hyenas out of there, and improved the hunting methods, as evidenced by numerous remains of bones at encampments.

There is evidence of cannibalism among paleanthropes, such as skulls with broken bases, incised and burned human bones in the caves of Sima de Los Uesos (Spain), Krapina (Yugoslavia), Steinheim (Germany), Bodo (Ethiopia), Klazies River (South Africa) and many other places. In the Monte Circeo (Italy), a human skull with the broken base was lying in a circle of large stones, which is the evidence of ritual cannibalism.

Social relationships among paleanthropes have become much more complicated. In the cave of Shanidar (Iraq), a skeleton of a man, who suffered from a whole set of serious diseases, was discovered. He was unable to walk and to earn his living but has reached the age of 40. He was fed, cared for, and buried after death. In another tomb from the same cave, a high concentration of pollen of the mountain flowers put into the cave was found. The skeleton of an old man, who suffered from severe forms of arthritis, has been found in La Chapelle-aux-Seine (France), in a cave, in which the people never lived, i.e. this cave was used as a burial place.

The earliest burials date back to about 100,000 years ago. Presumably, at the same time, the ideas about the afterlife appeared. Sometimes the burials have traces of rituals and are arranged in a quite sophisticated manner. For example, the caches with skulls of cave bears were found in the caves of Germany, Yugoslavia and the Caucasus. At Teshik-Tash cave in Uzbekistan, the child was buried in the circle of the mountain goat horns.

Neanderthal burials had the following features: the dead were always put to one side in a crouched position; only one person was buried in a tomb; the Neanderthals never put any things with the dead. Quite often, the Neanderthals buried several dead in a cave, with graves arranged in a certain order.

Perhaps due to the ritual practice, a symbolic activity originated. For example, parallel cuts on the bones of an elephant in Bilzingsleben (300-400 thousand years old), scratched parallel bars and arrows painted on the bone fragments in Oldisleben (Germany). The evidence from a later period is the drawing from the Gorham cave (Gibraltar, Figure 4), also done by the Neanderthals.

About 80 thousand years ago, in the North, South and East Africa, there appear the beads made of shells. Probably, the Neanderthal consciousness has lacked imagination and was subject-oriented. Judging from the drawings, it has changed a little over hundreds of thousands of years.

*Figure 4. A sample of the Neanderthals' rock pattern. Graham Cave, Gibraltar, not later than 39 thousand years ago. (Stewart Finlayson. Gibraltar Museum, n/d)*

Because the larynx of the Neanderthals was very different from that of the modern humans, it is likely that their speech was much more primitive as regards the sound composition and grammar, included facial expressions and gestures, denoted objects and actions, and expressed the moods and intentions, rather than concepts and features with a high degree of abstraction (Burlak, 2011).

Some researchers believe that Neanderthals are the ancestors of modern humans, while the others treat them as a dead-end, completely extinct branch, and some others speak of a certain percentage of the genetic code inherited from them by Neanthropines as a result of mixed marriages (Klyosov & Tyunyaev, 2009; Krause et al., 2010). In particular, it is believed that for an individual, known as "Denisova 11", 1.2% of the DNA fragments correspond to the modern man, 38.6% to the Neanderthal and 42.3% to the Denisovan (Reich et al., 2010; Popmech. Science.News-437992-7, n/d). Svante Pääbo, the head of the laboratory of the Max Planck Institute for the Science of Human History (Leipzig, Germany), assesses the share of the Neanderthal's genes in the genome of a modern human as 10-15%, while the share of the Denisovan's genesis 1.5-5%.

## NEOANTHROPES

In several African localities dating back 200 to 100 thousand years ago, the bones of the people who did not have a strongly extruded nape, a large superciliary roller and at the same time had a very large brain and prominent chin was found. Similar findings were also made in the Middle East, in Es Skhul and Kafzeh caves (100 thousand years). According to its anatomic structures, they are consistent with the people of the modern type.

This correlates with the molecular biology data, according to which the direct ancestors of a modern man (figuratively called the mitochondrial Eve and the Y-chromosomal Adam; they have not been a married couple), existed, respectively, for 230-140 thousand years with the maximum probability in the range of 200-180 thousand years and 156-120 thousand years ago with the maximum likelihood of 138,000 years ago (there are the other dates, which differ significantly from the above).

The population of Europe, classified as the modern variety, but larder, then modern people (brain was 1,800-1,900 $cm^3$), which lived in the Upper Paleolithic era (40 -10 thousand years ago), is called the Cro-Magnons. This name as a generic term is often used for synchronous populations in other parts of the world, which are much less known.

In different parts of Europe, the skeletons with differing height, weight, and other characteristics were found, which gives grounds to speak about racial differences, are not identical, however, to the differences between present-day races (for comparison, the reconstructions from the burials in Arena Candida (Italy) and Sungir (Russia), (Antropogenez, n/d).

During the Paleolithic, the tool manufacturing technology based on specifically prepared blanks, and have much more sophisticated shapes than Mousterian counterparts, has improved significantly, and the tool range has expanded (Figure 5). Cro-Magnons widely used animal bones to make the tools. A bow and arrows appeared. There is a time gap between the appearance of more advanced brain and hands, and their potential implementation when making tools (100-150 thousand years).

Since the most significant changes of a modern man in comparison with its predecessors are related to the brain structure (Park et al., 2007), let's dwell on this in more detail.

The brain of a modern human (Figure 6) contains, according to various estimates, 86-100 billion neurons and consists of several parts.

*Figure 5. Upper paleolithic tools (40-10 thousand years): 1-3 - retouched microplates; 4, 5 - scrapers; 6, 7 - lugs; 8, 9 - tips; 10 - nuclei; 1-13 - cutters; 14, 5 - toothed and notched tools; 16 - borer (Civilka, n/d)*

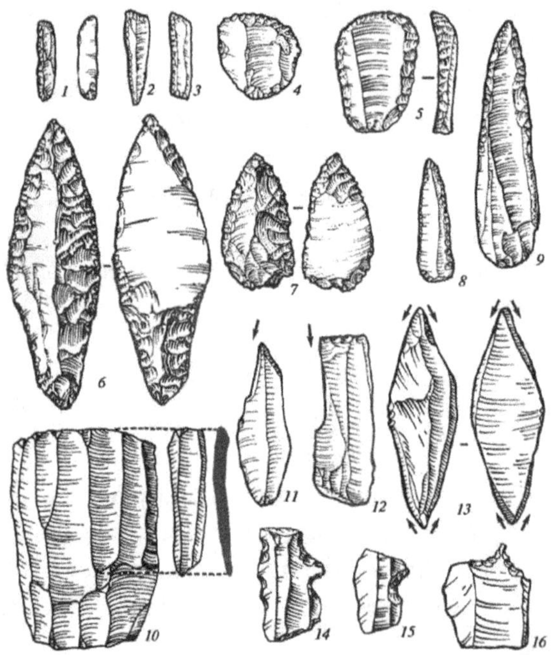

*Figure 6. A modern man's brain (Mbradio, n/d)*

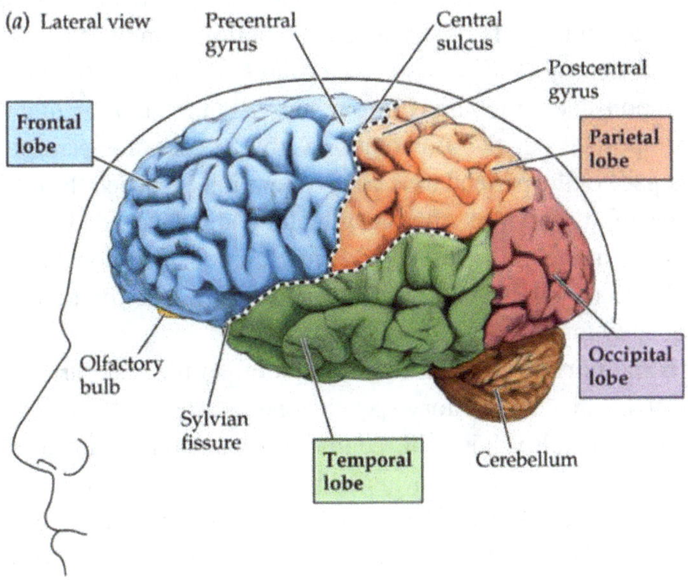


The brain is commonly described as composed of three main parts: the occipital, middle and frontal brain. Historically, the occipital and middle brains are considered older. They are responsible for the body's vital internal functions, such as maintaining blood flow and breathing. The frontal brain is responsible for the human form of communication with the outside world (thinking, memory, speech). Some researchers believe that it is this part of the brain that is most important for the human mind and consciousness (Chung, 2018).

In the cortex, one can distinguish four lobes: occipital, parietal, temporal and frontal; each lobe has its corresponding pair.

The frontal lobe **is thought to be responsible for thinking, consciousness, and the ability to communicate.** The length of the frontal lobe during the hominid evolution changed twice; it increased during the transition from Australopithecus to Homo erectus and began to decline since the beginning of the Upper Paleolithic. The width of the frontal lobe peaked with Neanderthals, and then also began to decline. An important area of the frontal lobe evolution was the lower lobe region, which increased to the greatest extent from Australopithecus to Homo erectus. Within the lower lobe region, the orbiting part was changing actively, having reached the maximum manifestation in the Neanderthals in the form of form hillocks, which further flattened due to the growth of the neighboring regions. At the rear portion of the lower frontal gyrus, there is Broca's area, the motor speech area; and the precentral gyrus, i.e. a zone of conscious motor activity. It is believed that the development of the frontal lobe made the altruistic actions, such as care for the helpless kindred and their burial, possible. Finally, the greatest expression of the ability to communicate was the domestication of animals, the first of which was a dog 13-17 thousand years ago; therefore, a man extended its sociability beyond its species (let's note that a modern man has lost the ability to domesticate new species of animals). The evolution of the frontal lobe in the Upper Paleolithic is correlated with the jewelry appearance and development of mystical ideas.

The parietal lobe is mainly responsible for sensitivity and spatial orientation. The anterior portion of the lower parietal lobule is responsible for hand sensitivity, determination of weight, shape, humidity, temperature and surface topography of objects by touch; coordination of sensitivity and motor activity, i.e. labor, and ensures self-control. The back section ensures spatial orientation and labor actions. The upper parietal lobe is responsible for the body layout and vision integration into the movements of the whole body. The most important stage of its evolution is the process of expansion in the area of the supramarginal gyrus. The pace of the parietal lobe width alteration exceeded the rate of increase in length and height; the processes of primary growth in length and width succeeded each other. It is important to note the variable predominance of the frontal and parietal lobe evolution pace. The parietal lobe development is correlated with the manufacture and use of tools, the construction of dwellings, as well as considerable human capacity for resettlement in new areas.

The temporal lobe **is primarily responsible for hearing.** At the rear part of the superior temporal gyrus, there is the Wernicke area, an auditory speech zone. Changes in Broca's and Wernicke's areas were directly related to speech ability. It is believed that speaking ability can be determined by the brain structure. On this basis, it can be assumed the availability of speech rudiments yet at the earliest members of the Homo genus, and the speech ability of Neanderthals seems almost undoubted. Probably, the emergence of musical abilities is related to the evolution of the temporal lobe, too; ancient authentic musical instruments made of mammoth bones are known from Mezin (Ukraine) and are dated about 20 thousand years ago. The temporal lobe also evolved unevenly. There were two main proportion options: extended in the back and elongated in the bottom (Australopithecines, Homo erectus, European Neanderthals) and extended in the front and elongated on top (Homo heidelbergensis, early Middle Eastern Neanderthals, Upper Paleolithic, and modern humans). The size and proportions of the temporal fossa, very wide and shallow in some groups (massive Australopithecines, European Neanderthals) and deep and narrow in others (early Middle Eastern Neanderthal, Upper Palaeolithic people) have been changing significantly.

**The occipital lobe is mainly responsible for vision, and the cerebellum – for the movement coordination.** It is believed that the evolution of the occipital lobe, together with the frontal lobe, allowed reproducing visual images. The dimensions of the occipital lobe evolved in a non-synchronized manner: from Australopithecus to the earlier Homo, the width increased sharply, while maintaining the length; further, the length started increasing gradually, but until Homo erectus, the growth in width had been prevailing; thereafter, the rate of increase in length began to prevail; from Neanderthals to Homo sapiens, the width starts to decrease, while maintaining the length. From the Upper Paleolithic to the present day, the pace of width change, but now in terms of its reduction, gained superiority over the lobe shortening pace. In the background of all these changes, the dominance of the upper lobe portion above the bottom one has been increasing steadily. Cerebellum has been increasing from the earliest stages until the Upper Paleolithic, with the hemisphere becoming relatively longer and narrower.

From the Upper Paleolithic, the trend has changed fundamentally, and to the present day, all dimensions have decreased, while the hemisphere has relatively extended. Thus, a conclusion about a certain degradation of the human brain regarding both size (from 1,800-1,900 $cm^3$ to the modern 1,350-1,400 $cm^3$) and structure is possible. Thus, the most striking manifestations of human culture were created at the brain degradation stage, which the period of its highest development, 100 or even 200 thousand years before the Upper Paleolithic, has not left any evidence of creativity superior to that of the Neanderthals. It is believed that a modern man uses only 7-8% of its brain capacity.

*Figure 7. Presumably, the oldest human painting of a modern type. Blombos Cave (Haaland, 2011)*

Human settlements of the Upper Paleolithic usually were the regularly visited hunting encampments. Here the houses were built, the community life evolved, the rituals were performed, and the dead were buried. The Cro-Magnons put guns, spears, stone knives, and numerous decorations to the graves with the dead. The tomb was filled with red ocher and sometimes covered with mammoth bones on top, indicating the availability of the developed ideas about the afterlife and related rituals.

The excellent examples of rock art are preserved in European caves; the figurines of people and animals made of the bones and limestone were found at encampments in Europe and Asia. The handles of knives were decorated with carvings. Clothing was decorated with beads and painted with ocher. Art also had a magic meaning. The images of animals were accompanied by signs of arrows and spears to facilitate the upcoming hunt. Judging by the footprints of teenagers in the clay in front of the cave paintings, here they were made hunters.

An important question is about the dating of the primitive artworks. The age of the Upper Paleolithic frescoes in European caves is estimated at 35-30 thousand years. There are opinions based on the findings in the other parts of the world and new interpretations that the art emerged 50 or even 60 thousand years ago. The oldest painting was found in Namibia (Figure 7). Its age, previously defined as 26-28 thousand years old, is now estimated at 59 thousand years old.

Therefore, for at least several tens of thousands of years, the art of anatomically modern humans did not differ from Neanderthals paintings, and only after some 25 thousand years, it reached the level characteristic of rock art in the caves of Lascaux, Chauvet, and Altamira.

It is believed that the man of a modern type colonized the planet in several migration waves from Africa (Figure 8). This is indicated by some biological features of the Upper Paleolithic people, i.e. long limbs, tall, elongated body proportions, big jaws, and elongated cranium are characteristic of modern tropical populations.

The same conclusion is indicated by archaeological findings and genetic studies (Figure 9). It is believed that America was populated later, during the existence of the land on the site of the Bering Strait.

*Figure 8. Approximate dating of mankind migration (Borinskaya & Yankovsky, 2006)*

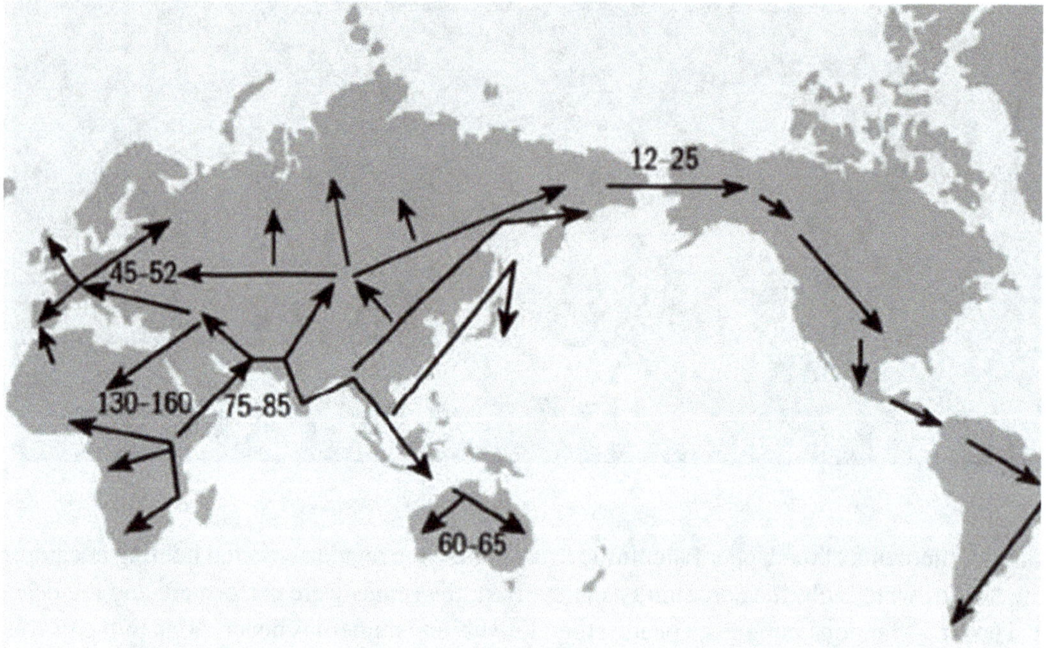


*Figure 9. Human migration according to genetic research (Treegene, n/d)*

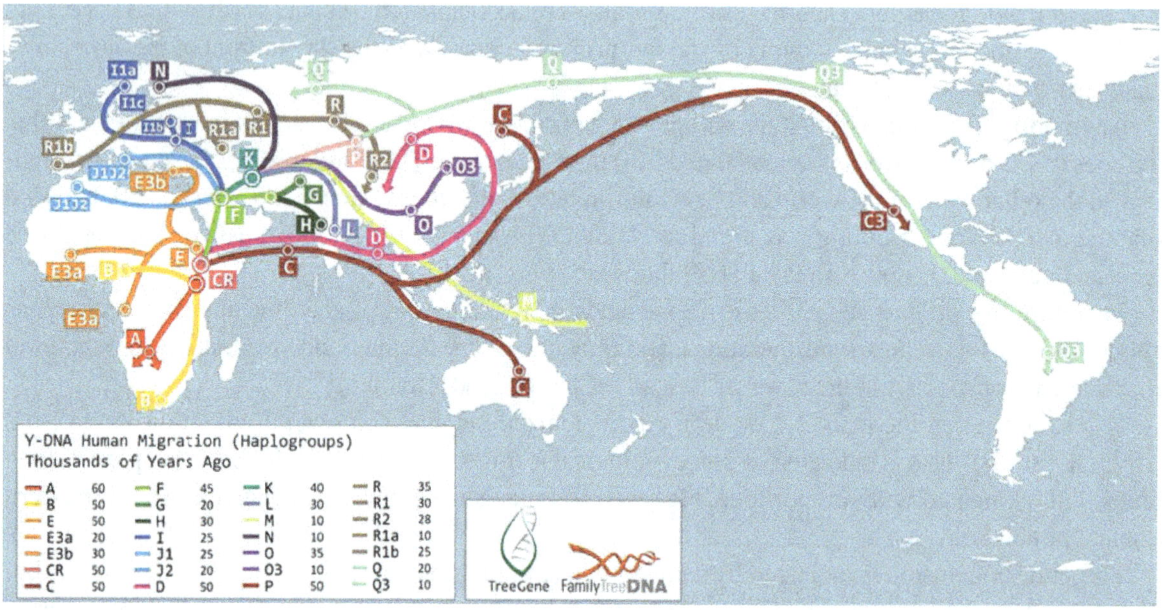

## RESUME

Now let's summarize the results of human evolution.

First, let's compare human and animals' brains:

1. Clear predominance over the spinal cord: the spinal cord weight in humans is 2% of the brain weighs; in gorillas - 5-6%; in other mammals - 22-48%.
2. A high square brain pointer value (the product of the absolute and relative brain weight): in humans, it is 32.0; in apes - 7.35; in mammals – 0.19-1.14 (but in dolphins - 6.27).
3. New brain predominance over the ancient one.
4. Developed frontal lobe: the human frontal lobe accounts for 30% of the surface of cerebral hemispheres; in apes - 16%; and in the lower monkeys - 8-12%.
5. Subcortex predominance over the cortex, i.e. 54% of the brain volume.
6. Pronounced brain folding; the furrows and convolutions are asymmetrical and much more pronounced than in animals.
7. The asymmetry of brain hemispheres' functions (the right hemisphere in most people is responsible for the figurative and creative work, while the left is associated with abstract thinking, symbolic activities, and rationality).
8. Distribution of functions by different brain areas (although the brain works as a whole and many higher functions are supported by coordinated operation of various parts).
9. Anatomical background of the second signal system (speech), which is reflected in the development of the surface layers of the cerebral cortex.
10. The volume of short-term memory, measured by the number of objects, which can be retained in memory, is 7 in humans, 3 in chimpanzees, and 2 in the majority of animals, allowing a man to think recursively.
11. A greater ductility and lesser dependence on genes, giving advantages in training and specialization.
12. For these changes had to pay a considerable price. In the adult human, the brain metabolism share in the total energy needs of the organism is 9% during sleep and 20-25% during the intense intellectual work, which is significantly more than in primates (8-10%), and mammals (3-5%). Still, the brain weight is 1.5-2% of the bodyweight (Leonard, Snodgrass & Robertson, 2007; Weekend, n/d).

Second, let's track the general changes. Throughout the evolution of hominids:

1. Like in all mammals, their body structure pattern does not change: one head, one body, two symmetrical halves and two pairs of limbs, consisting of three parts and ending with five fingers.
2. The topology of individual body systems common to all mammals does not change.
3. The body size and weight are changing from early Australopithecs to Cro-Magnons, mainly upwards; at the same time, for certain species (e.g., Homo floresiensis, the body size decreases).
4. At a relatively stable pace (1-3), the proportions of the body parts are changing, the "correctness" of various symmetries is broken (location of fingers and toes, right and left hemispheres asymmetry, hand functions), and the complexity of network components of separate body systems is increasing.

5. Very minor genetic changes occur (1-3% of human genetic differences from apes); probably, the difference between a man and its direct ancestors is even smaller.
6. However, the SRGAP2 gene and two of its replications considered responsible for the brain growing cannot be found in primates.[1]
7. The role of hybridization in human origin is, apparently, quite substantial.
8. The most important changes are walking upright, release and improvement of hands; binocular vision, expressed vision prevalence over the other senses, a change in the larynx structure, loss of hair.
9. An even more important change is the greater complexity of the brain structure, its increased volume, and power consumption (signs 1-12 noted above), followed by their degradation since the Upper Paleolithic.
10. There is a correlation between anatomical changes and subsequent external manifestations of smart activity. At the same time, there is a considerable time lag, from hundreds of thousands of years between the release of the hands and production of the first tools to tens of thousands of years between the formation of the modern brain and the emergence of symbolic activity.
11. However, ultimately, there is formed a homogeneous commonality of modern people, the genetic differences among which are 0.1%.
12. More and more new species emerged and disappeared for still unclear reasons. Why, for example, gorillas, having a limited range, were preserved, while the seemingly more advanced Australopithecus with similar habitat is now extinct?

Let's comment on the general changes.

Now let's assess these changes in terms of their eventual explanations within the natural selection theory. It is obvious that characteristics of 1-5, available over a long period, are quite consistent with this theory and are explained by the adaptation of hominids to climate change and taking of new ecological niches during migration outside the primary habitat. The emergence of mutations responsible for the increase in brain size (6) could be random and is not an argument against the natural selection theory. However, the reasons for an activity of the relevant gene over a long period, the energy supply of the relevant changes, their utility, as well as the subsequent tendency to reduction of the brain size need to be explained. The signs of hybridization of different species of hominids (7) contradict the fact that different species cannot deliver the reproducible offspring. There may be at least two explanations for this fact: perhaps because of the small amount and fragmentation of findings, the population differences are mistaken for the species differences; perhaps some genetic commonality is explained by a common ancestor, and not species hybridization. The paradoxical nature of changes 8-10 is that their positive effect appears after many generations and initially is useless or even harmful to the survival of the species. For example, walking upright and an increase in brain volume required additional energy resources and, consequently, an increase in the required amount of food. Loss of hair gave rise to the problem of overheating and hypothermia. Impaired hearing and smell deteriorated the security, etc., whereas the production of tools and clothing, as well as speech development, giving the evolutionary advantage, occurred much later.

All these comments suggest that natural selection, as noted by H. Spencer, cannot be the driving force of evolution, since "intermediate links" not only fail to provide advantages for survival but are often harmful; further, there is a problem of their lack among the findings. These arguments are supplemented by modern scholars with the fact that the evolution term, calculated based on the random mutation occurrence and their selection, is much longer than the one determined using the modern dating methods. Therefore, the researchers are trying to suggest other evolutionary mechanisms. For example, Johanson and Wong (2009) uses the following hypothesis to explain the upright posture.

There are two approaches to the species reproduction: the production of a huge number of eggs with the minimum energy spending for each of them (R-strategy); or the production of a small number of eggs (kids), requiring a lot of energy (K-strategy). An extreme example of K-strategy is large anthropoids (gorillas, chimpanzees, and orangutan), giving birth to a cub once every five or six years, making this strategy ineffective. Bipedalism is a way to reduce the orientation towards the K-strategy; a set of changes accompanying the same (terrestrial life, paired connections, leading to the family formation, a reduced mobility of females with cubs - the potential for emergence of a "home", the beginning of the use and making of tools, brain increase, constant sexuality, taking care of a few cubs) gave evolutionary advantages from the very beginning:

"Hominids have mastered the two-legged walking when they lived in the forest, and not after moving to Savannah, although that is where they subsequently lived. They came to the savannah as two-legged creatures. The transition to upright walking occurred about four million years ago. And it was done without the aid of tools. Primarily sexual and social factors were at the heart of change in the locomotion pattern. They probably came to Savannah because the forest was becoming too crowded. They started reproducing will, and they did not have enough living space. Eventually, hominids migrated to the savannah and after a couple of million years spread over a vast area.

The tools might have contributed to this process. Did the hominids use the tools from the very beginning? Probably no more than do chimpanzees. However, as the bipedal gait developed, the hominids gained a better opportunity to transfer, hold and throw items than before. In between four million and, say, one-half million years, the use of tools has developed from an irrelevant factor into the main driving force of the hominids' evolution. These were the monkeys-innovators who have invented new forms of social and sexual behavior, switched to the upright posture and, consequently, significantly increased the pace of their species reproduction as compared to the other anthropoids. They were lucky enough, so the upright posture and release of hands from the movement function led to the further development of tools and culture, an increase in the brain and, ultimately, the appearance of living beings, such as humans. Still, all these are the side effects of a single process, which was to improve the reproduction strategy. The appearance of a man was made possible due to innovations in the K-strategy, and tools improved it. They have led to the emergence of Homo erectus, but not of the Homo" (Johanson & Wong, 2009).

Let's compare hominids with the birds two-legged "by definition":

- Bipedalism is not associated with any particular habitat: condors spend a large part of life in flight, penguins - in the water, ostriches - on the ground, and the chickens have been domesticated long ago. Still, bipedalism is the most characteristic of penguins, the Antarctic living conditions of which do not resemble the forests and savannas of Africa in any aspect.
- Paired connections occur ("swan fidelity"), but not always, e.g., the chickens are not monogamous.
- K-strategy and the potential for the nest-home occurrence do exist, but with certain restrictions – birds quickly "get on the wing" and begin an independent life without the need for parental care. And, for example, the cuckoo limit the care for its offspring by tossing her eggs to the others' nests.
- There is seasonal sexuality, except for poultry, where it is constant, and taking care of a few birds (limited by available food resources).

- No tool production, even by ostriches with "free wings." Moreover, these wings become rudimental, and do not develop into the hands.
- No increase in the brain, even with crows, which are considered the most "intelligent." Figuratively speaking, the crows would have compared the increase in the size, weight, and power of their brain with cancer, leading to guaranteed death.

Thus, the set of changes associated with bipedalism is by no means mandatory.

Let's go back to the hominids and consider the arguments of D. Johanson in terms of genetic modifications accompanying the set of changes described above. SRGAP2 gene and its two replications are responsible for the brain volume increase; where the genes responsible for upright posture mutated synchronously, then the formation of a set of changes would have a genetic basis.

However, it is not the case. The transition to upright walking is possible by shifting the center of gravity back and down, to the sacral spine and the spine, allowing to keep the balance and carry the head on the shoulders. According to A. Fisher (Gazeta. Science, n/d), these changes are associated with mutation of a homeotic gene responsible for the formation of repetitive elements in the embryo. Our research on the remains of the oldest known bipedal mammal, Morotopithecus bishopi, related to eventual human ancestors, showed that this mutation occurred at least 20 million years ago.

Therefore, bipedalism is in no way related to an increase in brain volume.

Other views on the origin of bipedalism are presented in (Niemitz, 2002; Kwang, 2015)

Another argument against is the homogeneity of mankind genetic composition (Palmer, 2007). Why did the other kinds of hominids having the same set of progressive changes and implementing the K-strategy failed to survive? Maybe it was an accident, caused by some relatively recent disaster. Then this objection would be null and void. However, the Homo variety has long ago migrated to the vast expanses of Africa and Eurasia, and only a global catastrophe could lead to its extinction, such as a rapid climate change due to the eruption of a super volcano or fall of large meteorites, which would inevitably lead to the extinction, comparable to the extinction of the dinosaurs or the Devonian extinction. No trace of such recent and global catastrophe has been revealed yet.

Therefore, some evolutionary features of a man as a biological creature was explained within the natural selection theory, but of paradoxical features also presented.

## CONSCIOUSNESS, MIND, BRAIN: PRELIMINARY DISCUSSION

So, the long process of evolution led to the emergence of Cro-Magnons - people of a modern type, undoubtedly possessing rationality.

And here begins the most difficult and interesting. What is "rationality"? Often say "human culture", "human civilization", implying that the author of these phenomena is a "human" with "human consciousness", "soul", "spirituality", and "reason". However, attempts to define these categories are now causing no less controversy than in ancient times. Let`s discuss this problem preliminary; author`s definitions will be done in chapter 3.

At the center of the debate is the problem of consciousness phenomenology. Here is how J. Montserrat defines it (Monserrat, 2016):

1) "The unitary character of *consciousness*. Our consciousness is noticed as a system that integrates with unity the different sensory modalities (vision, audition, proprioception…) which are projected to the psychical subject that coordinates them and sets responses.
2) The *holistic* character of consciousness. Our experience of consciousness is wide open: we feel the openness of the external space through vision, the unitary extension of our sensations through our own body, or the wide unity of our internal experience when we close our eyes and follow the stream of our thoughts.
3) The *indeterminate nature* of the responses of a conscious subject. Subjects notice themselves as open to a multitude of possible responses and consider themselves as the cause of these responses. Responses may be driven by programmed automatisms, but subjects strategically exercise their control and feel that their lives are played out without an absolute determinism, with free indeterminate options (which does not mean unconditioned). This phenomenology of our own experience of indetermination (*free will*) creates the basic persuasion that gives sense to our personal and social life".

Another side of a problem is the relationship between consciousness and mind with the brain.

It has been proposed numerous solutions to these problems, which are reduced to two basic approaches (review's done by Monserrat, 2016; Chung, 2018)[2]:

1) Consciousness and mind are consequences of the activity of a highly organized human brain, the especially cerebral cortex. An example of this approach realization is a Human Brain Project, coordinating numerous studies in the fields of multi-scale scaffold theory and models for the brain, cognitive neuroscience, brain-inspired computing, data analytics and robotics, medicine and, importantly, ethics.
2) Consciousness and mind are consequences of the general laws of nature, and the structure of the brain affects only the distinctness of their manifestation. This approach is presented, for example, by Kaznacheev V.P. and Spirin E.A., whose book is eloquently called the "Cosmoplanetary Phenomenon of Human" (Kaznacheev & Spirin, 1991).

It may seem that these approaches are complementary, but practically they are difficult to combine it due to methodological differences.

Therefore, before determining the position of the authors (chapters 2-4), let's try to pre-evaluate them and choose the most preferable one using the following way.

Let's consider a set of interactions of a living being with the environment based on the systematic principle, and then let's check to see whether the findings contradict the facts, both for humans and other creatures, trying to find similarities and differences.

To understand evolution, it is also important to understand whether the body anatomically identical to a modern human type is a prerequisite for the emergence of consciousness? Or is the "calibration" and "adjustment" of the nervous system specific to humans and absent in animals a more important factor? What are the general features of the human "consciousness", "mind", "spirit", which could be tracked to declare that they are present in humans and certainly lacking in animals?

Well, the interaction with the environment is impossible without a sense of a certain unity with it. If the creature and the environment are completely alien to each other, how can the creature take the substances, energy, and knowledge required for it?

The interaction is impossible without a sense of separation from the environment. The creature may feel itself only in relation to something external, which is not a creature; no "object" means no "subject", no interaction and no knowledge.

The interaction is shifting the boundary between the creature and the environment towards the environment (affecting the creature), or towards the creature (environmental reaction). To generate the action and to assess the reaction as favorable or unfavorable for the creature, the relevant structures are required.

The sequence of actions and reactions is developed in space and time. To distinguish the sequence, it is required to distinguish between the past, present and future, as well as the relationship "to the left - to the right", "front-back", "top-bottom". Spatial and temporal relationships and the associated senses of "inside", "outside", "on the boundary" shall also be identified by a separate structure.

There are more specific differences associated with features of the habitat, i.e. water, land, air, and the inner, as well as the boundary environment of the creature, which must be distinguished by senses. Here, there is no clear definition, and the development of senses is determined by environmental conditions.

The next level is the distinction between homogeneous environmental impact, i.e. colors, sounds, scents, etc. Here again, it is appropriate to recall Miller's numbers (7-9) and ascertain both the tendency of the number of colors or similar objects to these values and the lack of the same certainty for fragrances or tactile sensations.

Starting with the second level, we need to have the organs affecting the environment and developing sequentially at all subsequent levels. The creature with such coordination will get advantages when adapting to the environment.

Let's check the described system on the human being.

A sense of unity with the environment takes place and is described in different words, such as foresight, prediction, meditation and the like, even love can be seen as a strong sense of unity. The most general term here is intuition, while the other sensations can be regarded as different degrees of its manifestation.

The feeling of separation from the environment and one's unity is expressed in our ego, in the presence of homeostasis, in the instinct of self-preservation. It can be stated that it is here the "consciousness" appears.

The generation of human activity on the environment is determined by the will, and assessment of the reactions – by the reason.

Distinguishing between spatiotemporal relationships with environment, as well as the classification of effects, both external, internal and boundary, falls within nous (practical reason), whether this happens instinctively or by generation of abstract geometric, physical and other similar systems, which is inevitably linked with the description means, in particular, with natural or abstract language, as well as the verification, means.

Human senses are external (vision, hearing, smell, touch, taste), internal (hunger, thirst, and the like) and boundary (a sense of balance and the like) receptors.

Finally, a person can distinguish between homogeneous objects, sounds, and colors (7-9), gradations of taste, and so on (where there is no such certainty). Thus, the system of human interaction with the environment is described by a scheme 1-1-2-3-5-(7-9) including six levels and the channels of interaction described above.

A similar scheme, except for the sixth level, describes the structure of the human body (head, trunk, developed organs of movement and labor, consisting of three parts and ending with five fingers, 1-1-2-3-5).

A man mastered not only the surface of the earth but also partly water and air elements within a short time, using the *creation of an artificial habitat* for this purpose, comfortable for the members of its species. So far, the boundaries of human expansion are invisible (e.g., relatively recently, the people began to explore the space).

Now let's give some examples of interaction systems of living beings, assessing them in terms of the possibility of expanding the range and the number of populations and their survival over a long time, as well as the potential for consciousness emergence.

Let's start with relatively simple creatures, **mollusks**. Mollusks appear in the oceans and seas since the Paleozoic era, i.e. approximately 540 million years ago. At the same time, they had mastered the land. Mollusks heyday is considered to be the Ordovician period (488-443 million years ago), and their mass extinction (as well as that of many other species) occurred on the boundary of the Permian and Triassic, about 250 million years ago. However, even now, they are quite common in the water and, to a lesser extent, on land.

Since the time of the Ordovician period, the most perfect group of mollusks is the class of cephalopods, including octopus (Figure 10), having the highest intelligence among shellfish. Let's consider them closer.

The body of cephalopods is symmetrical.

It is usually divided by an interception between the trunk and the bid head, while the leg is modified into a funnel located on the ventral side, i.e. a muscular conical tube (siphon) and long, muscular tentacles located around the mouth.

Octopuses have eight tentacles, while cuttlefish and squids - ten. The inner side of tentacles is covered with numerous large discoid suckers.

The brain does not have a remarkable size. Thus, the octopuses' structure is described by pattern 1-1-1-2-4.

Octopuses have well-developed senses. The eyes have reached the highest perfection: they are not only very large (occupying a large part of the head), but also have a sophisticated structure. The octopus eye structure is not essentially different from the human eye; there are the beginnings of binocular vision. The visual angle of the convex eyes is approaching 360°. Further, light-sensitive cells allowing to define the general direction of the light are scattered about the octopus' skin. The octopus taste buds are located on suckers. Octopus can detect infrasound.

*Figure 10. Octopus structure: 1 - horny mandible; 2 - brain; 3 - siphon; 4 - liver; 5 - pancreas; 6 - stomach; 7 - gown; 8 - gonad; 9 - kidney; 10 - heart; 11 – gills; 12 - ink sac (Fakty-ob-osminogah, n/d)*

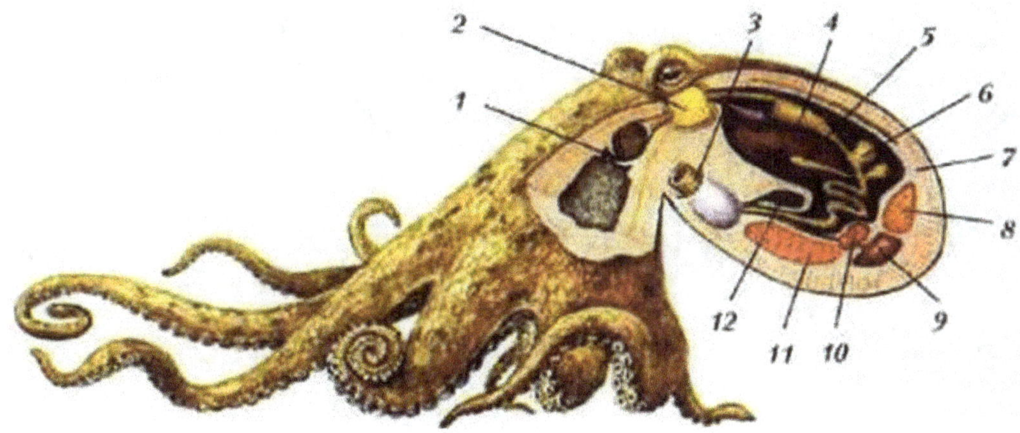

As regards the environmental interaction channels, let's note the sense of belonging to the environment (manifested as instinctive behavior), a sense of one's integrity and separation from the environment (manifested as the survival instinct), the rudiments of will and reason (influence on the environment and assessment of reaction), the rudiments of intelligence (the ability to learn, memorize, solve the problems related to orientation in space-time) and four external senses, so such a system can be described as 1-1-2-3-4.

Let's also note:

1) Mismatch, in contrast to human, of body and environmental interaction system structure schemes.
2) The evolutionary success of the octopus' organization (habitat expansion, reasonableness rudiment development).
3) Depletion of the evolution potential at some point (a lack of intellectual development, a lack of new behavior patterns, a lack of environment changes).

*Figure 11. The Ant structure: 1 - brain (approximately 500 thousand neurons); 2 - intestines; 3 - stalk - a narrow part of the body connecting the chest and the belly and adding flexibility to the movement; 4 - metapleural gland (producing antibiotics ant protecting from bacteria); 5 - nervous system made of three thoracic and abdominal moreganglia; 6 - small struma; 7 - chitinexoskeleton; 8 - tip (not in all varieties); 9 - Dufour's gland (generating pheromones serving as signals for ants walking along the track); 10 – a leg made of five parts; 11 - poison gland; 12 - stomach; 13 - eye (some ant species see items located hundreds of meters away, while the others are almost blind); 14 – jaw gland (generating pheromones of alarm during the predator attack or appearance of strangers); 15 – a spur on the leg to clean the antennas; 16 - claws (two on each leg); 17 - antennas, by means of which the ant receives information of the smell, taste, chemical composition, as well as the texture of various subjects and share it with fellows; 18 - retropharyngeal gland; 19 - mandibles to capture and chew the food, as well as weapons, (Antclub. Muravi. Stroenie-muravya, n/d).*

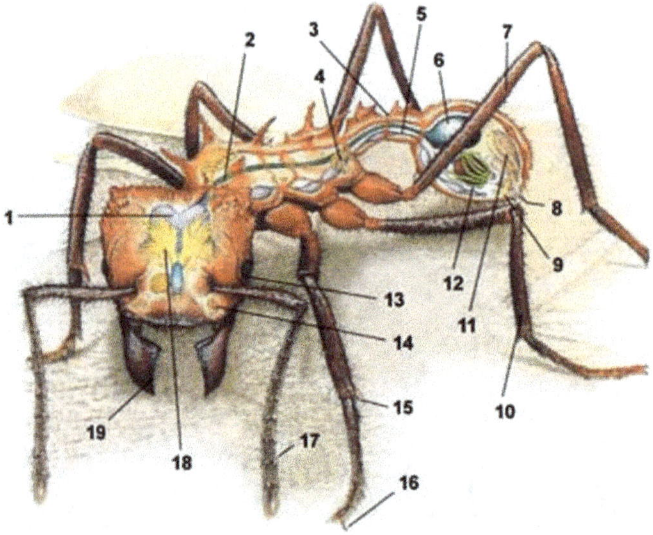

As another example, let's consider the insects.

The finds of the first **insects** date back to the Devonian, 400 million years ago. The insects master the land, water (larval stage) and air quickly enough. To date, having survived all extinctions, they constitute the largest group of living beings in terms of the number of species and individuals therein. The most advanced family of insects are ants (Figure 11), known from the Cretaceous period, about 100 million years ago. Figure 11 shows the Ant structure.

The body structure corresponds to the following scheme 1-1-1-1-2-3-5-2.

The brain is rather small and contains about 500 thousand neurons.

The ant senses are vision, smell, taste, touch, hearing not playing a significant role, as well as the fields of bristles, which are the receptors of gravity and orientation.

The system of environment interaction channels is characterized by a sense of belonging to the environment, a sense of one's integrity and separateness from the environment, the rudiments of will and reason, the rudiments of intellect and six external senses described by the scheme of 1-1-2-3-6 (or 5.5, if take into account the underdevelopment of the hearing).

Let's note the following:

1) The inconsistency of the body structure and channels organization scheme.
2) Evolutionary success manifested by the habitat extension range and a marked environment transformation. There is a fierce competition between different types of ants; the authors witnessed the displacement of red wood ants building nests by black ants digging the underground burrows occurring in Ukraine over the past decades.
3) The presence of a sophisticated social organization divided into the "castes", quite different by appearance, lifestyle and behavior, and the ability to build highly organized housing (Figure 12). For example, Formica rufa lives in large families in anthills. In a mature anthill (with the dome diameter of about one meter), there are 600,000-700,000 individuals, while in the largest ones (with the diameter up to 1.8 meters) - up to 2 million. If talking about the man, the presence of a complex society and housing would unambiguously be interpreted as signs of intelligence; as regards the ants, given the simplicity of their brains structure, the scientists prefer to abstain from such an interpretation.
4) The presence of a kind of "language" based on the "ritual" behaviors, generation of chemicals and tactile sensations, allowing top convey information to individuals and groups.
5) The ability to perform simple arithmetic operations, which is quite developed, as compared to the other insects and animals (Reznikova & Ryabko, 2011).
6) A limited evolutionary potential.

Let's consider the **reptiles**. Reptiles appear in the Devonian period and successfully survive the Permian-Triassic extinction. They flourished in the Mesozoic era when they assimilated land, water, and air space; at the end of the Cretaceous period (about 65 million years ago), there was a mass extinction. Currently, the reptiles inhabit the land or are amphibians in areas with a warm and temperate climate. Let's consider the reptiles' structure based on the example of a crocodile (Figure 13).

The overall crocodile's body structure is described by the following scheme: 1-1-2-3-5.

Reptiles have six major senses: vision, smell, thermal sensitivity, hearing (somewhat rudimentary), and touch.

*Figure 12. The anthill structure and the ant caste representatives (Fb. Article 410425, n/d)*

*Figure 13. Crocodile (Kozlov, n/d)*

The structure of environmental interaction channels is described by the following scheme: 1-1-2-3-6 (or 5.5, given the hearing underdevelopment), including the sense of belonging to the environment, the sense of one's integrity and separation from the environment, the rudiments of the reason and will, the rudiments of the intelligence and six external senses.

Let's note the following:

1) The proximity of the body structure patterns and channels organization to both reptiles and human beings.
2) Evolutionary success in the past and evolutionary end in the present.
3) The absence of any significant sign of environment transformation.
4) A reputation of not too clever creatures, which is consistent with the size of the brain, and the absence of a complex language.

*Figure 14. Bottlenose dolphin skeleton (Horek, n/d)*

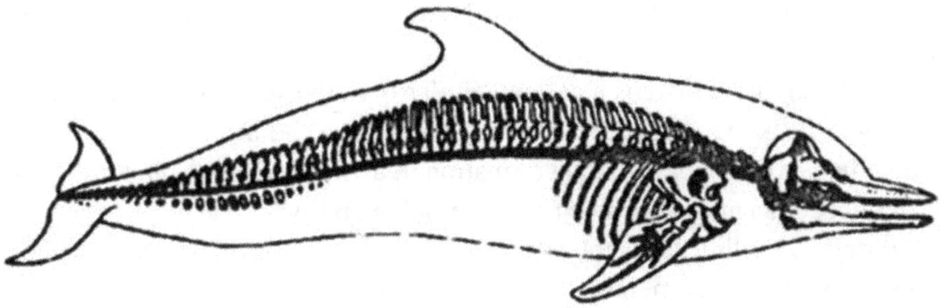

Finally, let's consider the **dolphins**. This example is interesting because, on the one hand, there is a narrow specialization of their limbs adapted to the aquatic environment, and on the other, the Dolphins have a reputation of highly intelligent beings. The ancestors of dolphins became fully aquatic animals, according to various estimates, 45-38 million years ago, before our time. It is uncertain when the dolphins of modern species appeared. The skeleton of the bottlenose dolphins is shown in Figure 14.

Given the rudimentation of the rear limbs, the body plan should be described as follows 1-1-2-3-5.

The senses include hearing (very well-developed), echolocation, touch, taste, and vision. The sense of smell is considered to be lost.

In general, the system of environment interaction channels is characterized by a sense of belonging to the environment, a sense of one's integrity and separateness from the environment, the rudiments of will and reason, the rudiments of intelligence and five external senses, described by scheme 1-1-2-3-5.

Let's also note:

1) An inconsistency of the body structure and channels organization schemes.
2) The presence of a quite large and complex brain, close to the primates' brain characteristics by many indicators.
3) Evolutionary success, manifested by the development of new habitats.
4) The presence of the social organization, social behavior, and communication mean. However, there are no signs of a substantial habitat transformation.
5) The absence of evidence to speak of the evolutionary potential exhaustion.

Analysis of the above cases leads to the following conclusions:

- All "abstract" levels of interaction with the environment, as the required existence preconditions, are available in all considered living creatures, despite the distance between them on the evolutionary track.
- *Ego* and self-awareness, as well as awareness of one's place in the world, *(mind, consciousness)*, appear on the second and subsequent levels of the environment interaction system.
- *Reason and will* appear on the third, and *nous* – on the fourth level. Still, the intelligence development degree is quite different and depends on the species position on the evolutionary track; the intelligence development degree is quite a different event within the same species.
- The composition of senses is inconstant and depends on the environment.

- The evidence of the ability to distinguish between sensations is insufficient to make firm conclusions; judging by circumstantial evidence, it is believed that for some sense, it tends to the Miller's numbers.
- For reptiles and mammals, there is a certain similarity in the structure of the body and action organs, reaching its peak in humans; it is absent for mollusks and insects.
- A sign of intelligence is the habitat transformation in a direction favorable for the species, as well as a well-developed language and the abstract thinking ability.
- A sign of conscious intelligence is the efforts to transform the environment, based on the abstract thinking ability, summarization, long-term preservation and conveyance of the experience to the other individuals and generations. We will consider the symptoms and markers of conscious intelligence in more details starting with chapter 3, *Self-organization: body and mind.*
- The differences between humans and animals are expressed in the better conformity of the environment interaction system to the structure of the body and action organs, as well as different configurations and proportions of the body and organs as compared to the other mammals, primates and hominids, in particular, a closer connection between consciousness, reason, will and intelligence, creating conditions for a conscious intelligence.
- Another important statement is the availability of signs of a cyclical existence of species, i.e. rapid appearance, which cannot always be tracked (a lack of intermediate forms), expansion and occupation of new habitats and even new environments, a continued existence without significant changes, a rapid extinction as the end of the life cycle, although provoked by external factors, which cannot be resisted by the exhausted species potential.

Thus, the above examples show that consciousness and mind appear within the framework of self-organization of the system of interaction of a living being with the environment as special levels of the system. The structural features of living organisms, in particular, the brain, affect the specifics of the manifestations of consciousness and mind, but they are not the cause of their occurrence.

Now let's imagine a robot controlled by artificial intelligence based on, for example, a quantum computer (Stapp, 2001) located in a certain environment.

His artificial intelligence must also have a level of communication with the environment; a level of separation from the environment; a level of generation of impacts and their estimates; a level of orientation in space and time; a level of sensory perceptions. Their implementation can be carried out in a way similar or different from the above examples - but the levels themselves must be present. And this makes it possible to discuss it's (or his?) consciousness and mind in the sense reviewed above.

This point of view is closer to the second of the above-mentioned approaches.

## CONCLUSION

The considered facts made it possible to determine the chronological framework of the study. By the time of the Upper Paleolithic, human evolution reaches its peak. The anatomy of the body and the structure of the brain already correspond to the modern type. The presence of speech, symbolic and material activity, the success of the survival and settlement of almost the entire earth in harsh climatic conditions show that this is already a truly intelligent creature. So, the evolution of the mind can be explored from the Upper Paleolithic to the present day.

Preliminary considerations show that since Homo Sapiens has already appeared with a formed system of interaction with the environment, which has structures of consciousness and mind, its evolution consists in their adjustment during the life cycle, which is consistent with other levels and channels of the human-environment system. That is, these changes should be researched in the framework of a systematic approach.

However, a simple declaration of a systematic approach is not enough. A useful theory must be a general enough to connect the evolution of the human mind and consciousness with the general laws of nature, but also concrete enough to pass the test using modern scientific data and historical facts.

The above features of anthropogenesis can just be used to verify the adequacy of the description of evolutionary processes in the theory of self-organization of complex systems, which will be described in the next chapter.

## FUTURE RESEARCH DIRECTIONS

Thus, the nearest goal is to develop a system approach apparatus so that it:

- Allowed to explain the patterns of anthropogenesis.
- Made it possible to determine the priorities and weights of the channels of human interaction with the environment, qualitative and quantitative relationships.
- Linked the characteristics of perception with the characteristics of the behavior of the individual and social groups.
- Allowed to predict the scenario of the evolution of human consciousness and behavior in different historical eras.
- Made it possible to determine adequate markers of such changes, sufficient to verify the proposed evolutionary scenario.

## REFERENCES

Antclub. Muravi. Stroenie-muravya. (n.d.). Retrieved from http://antclub.ru/muravi/stroenie-muravya

Antropogenez. (n.d.). Retrieved from http://antropogenez.ru

Appenzeller, T. (2012). Human migrations: Eastern odyssey. Humans had spread across Asia by 50,000 years ago. Everything else about our original exodus from Africa is up for debate. *Nature*, *485*(7396). PMID:22552074

Arnold, O. R. (2018). People and animals: myths and reality. Alpina non-fiction.

Beyin, A. (2011). Upper Pleistocene Human Dispersals out of Africa: A Review of the Current State of the Debate. *International Journal of Evolutionary Biology*, *2011*(615094), 1–17. doi:10.4061/2011/615094 PMID:21716744

Biofile. (n.d.). Retrieved from http://biofile.ru/his/24488.html

Bohm, D. (1990). A New Theory of the relationship of Mind to Matter. *Philosophical Psychology*, *3*(2-3), 271–286. doi:10.1080/09515089008573004

Borinskaya, S., & Yankovsky, N. (2006). People and genes: the threads of fate, Vek 2.

Boyd, R. (2017). *A Different Kind of Animal: How Culture Transformed Our Species*. Princeton University Press. doi:10.2307/j.ctvc7799z

Brown, P., Sutikna, T., Morwood, M. J., Soejono, R. P., & Saptomo, E. WDue, R. (2004). A new small-bodied hominin from the Late Pleistocene of Flores, Indonesia. *Nature*, *431*(7012), 1055–1061. doi:10.1038/nature02999 PMID:15514638

Brunet, M., Guy, F., Pilbeam, D., Mackaye, H. T., Likius, A., Ahounta, D., ... Zollikofer, C. (2002). A new hominid from the Upper Miocene of Chad, Central Africa. *Nature*, *418*(6894), 145–151. doi:10.1038/nature00879 PMID:12110880

Burlak, S. V. (2011). *The origin of language. Facts, research, hypothesis*. CORPUS.

Cavalli-Sforza, L. L., & Feldman, M. W. (2003). The application of molecular genetic approaches to the study of human evolution. *Nature Genetics*, *33*(S3), 266–275. doi:10.1038/ng1113 PMID:12610536

Chip, W. (2013). Last Ape Standing: The Seven-Million-Year Story of How and Why We Survived. Walker Books, 2013.

Chung, S. J. (2018). A review of the relationship among self, mind, and brain in functional magnetic resonance imaging study: Tree-pattern image of the semantic map in human brain viewed from the Ultron-Logotron Theory. *Open Journal of Philosophy*, *8*(04), 408–427. doi:10.4236/ojpp.2018.84028

Civilka. Humanity. Ashel. (n.d.). Retrieved from http://civilka.ru/humanity/ashel.html

Civilka. Humanity. Mustie. (n.d.). Retrieved from http://civilka.ru/humanity/mustie.html

Civilka. Humanity. Verhniypaleolit. (n.d.). Retrieved from http://civilka.ru/humanity/verhniypaleolit.html

Clarke, R. J., & Tobias, P. V. (1995). Sterkfontein Member 2 Foot Bones of the Oldest South African Hominid. *Science*, *269*(5223), 521–524. doi:10.1126cience.7624772 PMID:7624772

Drobyshevsky, S. V. (2017). Non-missing link. Book 1, 2. Monkeys and all-all-all.- Corpus, 2017.

Edelman, G. M., & Mountcastle, V. B. (1978). *The Mindful Brain: Cortical Organization and the Group-Selective Theory of Higher Brain Function*. Cambridge, MA: MIT Press.

Fakty-ob-osminogah. (n.d.). Retrieved from https://www.voprosy-kak-i-pochemu.ru/interesnye-fakty-ob-osminogax/

Falk, D. (2011). The Fossil Chronicles: How Two Controversial Discoveries Changed Our View of Human Evolution. University of California Press Fb. Article 410425. (n.d.). Retrieved from http://fb.ru/article/410425/muraveynik-ustroystvo-etapyi-postroyki-foto-muraveynik-iznutri-delenie-na-kastyi-i-interesnyie-faktyi-iz-jizni-muravev

Ferring, R., Oms, O., Agustí, J., Berna, F., Nioradze, M., Shelia, T., ... & Lordkipanidze, D. (2011). Earliest human occupations at Dmanisi (Georgian Caucasus) dated to 1.85–1.78 Ma. Proceedings of the National Academy of Sciences, 108(26), 10432-10436.

Finlayson, S. (n.d.). A sample of the Neanderthals' rock pattern. Graham Cave, Gibraltar, not later than 39 thousand years ago. Retrieved from Gibraltar Museum.

Gower, P. (2019) Groundbreaking study reveals southern Africa as birthplace of modern Homo sapiens, who migrated due to climate shifts. Retrieved from up.ac.za. University of Pretoria.

Haaland, M. (2011). Presumably, the oldest human painting of a modern type. Blombos Cave, 2011. Retrieved from https://vokrugsveta.ua/science/najden-drevnejshij-v-mire-risunok-13-09-2018

Horek. (n.d.). Retrieved from http://horek.spb.ru/skelet-delfina-foto.html

Human Brain Project. (n.d.). Retrieved from https://www.humanbrainproject.eu/en/

Johanson, D. C., & Wong, K. (2009). *Lucy's Legacy: The Quest for Human Origins*. Crown.

Kaznacheev, V. P., & Spirin, E. A. (1991). *Cosmoplanetary Phenomenon of Human: The Problem of Integrated Study*. Novosibirsk, Russia: Nauka.

Kevin, N. L. (2017). *Darwin's Unfinished Symphony: How Culture Made the Human Mind*. Princeton University Press.

Khrisanfova, E. N., & Perevozchikov, I. V. (2005). Anthropology, 4th ed.: Moscow, Russia: Publishing house Moscow University, Science, 2005.

Klyosov, A. A., & Tyunyaev, A. A. (2009). *The origin of man according to archeology, anthropology and DNA genealogy*. Boston, MA.

Kozlov, C. (n.d.). Retrieved from https://www.e-reading.club/chapter.php/102033/2/Filip%27echev%2C_Kozlov_-_Krokodily.html

Krause, J., Fu, Q., Good, J. M., Viola, B., Shunkov, M. V., Derevianko, A. P., & Pääbo, S. (2010). The complete mitochondrial genome of an unknown hominin from southern Siberia. *Nature*, *464*(7290), 894–897. doi:10.1038/nature08976 PMID:20336068

Kuzmin, Y. F. (2017). *Geoarchaeology: natural science methods in archaeological research*. Tomsk, Russia: TSU Publishing House.

Kwang Hyun, K. (2015). Origins of Bipedalism. *Brazilian Archives of Biology and Technology*, *58*(6), 929–934. doi:10.1590/S1516-89132015060399

Leakey, M. D. (1979). *Olduvai Gorge: My search for early man*. Collins.

Leonard, W. R., Snodgrass, J. J., & Robertson, M. L. (2007). Effects of brain evolution on human nutrition and metabolism. Annu Rev Nutr. 2007. – Vol. 27. – pp. 311-327. doi:10.1146/annurev.nutr.27.061406.093659

Levins, R. (1968). Evolution in Changing Environments: Some Theoretical Explorations. Princeton University Press.

Linden, A. (2007). Reclassification of Homo floresiensis as Homo species indeterminate to accurately reflect a lack of existing knowledge. *Totem: The University of Western Ontario Journal of Anthropology, 15*(1), 5–20.

Markov, A. V. (2011-2012). Human evolution. Book 1, 2. Moscow, Russia. *Astrel: Corpus, 2011*, 2012.

Mbradio. (n.d.). Retrieved from https://mbradio.ru/publication/1356/

McPherron, S. P. (2017). The Age of the Homo sapiens fossils from Jebel Irhoud (Morocco) and the origins of the Middle Stone Age. *Nature, 8*(June), doi:10.1038/nature22335

Mendez, F. L., Poznik, G. D., Castellano, S., & Bustamante, C. D. (2016). The Divergence of Neandertal and Modern Human Y Chromosomes. *The American Journal of Human Genetics*, 2016. Gazeta. Science. (n.d.). Retrieved from https://www.gazeta.ru/science/2007/10/11_a_2232959.shtml?updated

Monserrat, J. (2016). How the brain and neural networks explain human reality. *Pensamiento, 72*(273), pp. 1043-1070.

Niemitz, C. A. (2002). Theory on the Evolution of the Habitual Orthograde Human Bipedalism - The "Amphibisce Generalistheorie". *Anthropologischer Anzeiger, 60*(1), 3–66. doi:10.1127/anthranz/60/2002/3 PMID:12058577

Pally, R. (2018). *The mind-brain relationship* (n.d.). London, UK: Routledge. doi:10.4324/9780429482465

Palmer, D. (2007). *The Origins of Man: an illustrated history of human evolution*. New Holland Publishers.

Park, M. S., Nguyen, A. D., Aryan, H. E., U, H. S., Levy, M. L., & Semendeferi, K. (2007). Evolution of the human brain: Changing brain size and the fossil record. *Neurosurgery, 60*(3), 555–562. doi:10.1227/01.NEU.0000249284.54137.32 PMID:17327801

Popmech. Science News-437992-7. (n.d.). Retrieved from https://www.popmech.ru/science/news-437992-7-faktov-o-gibride-neandertalca-i-denisovskogo-cheloveka/#part0

Potts, R. (1998). Environmental Hypotheses of Hominin Evolution. *American Journal of Physical Anthropology*, 107, 93–136. doi:10.1002/(SICI)1096-8644(1998)107:27

Reich, D., Green, R. E., Kircher, M., Krause, J., Patterson, N., Durand, E. Y., ... Pääbo, S. (2010). Genetic history of an archaic hominin group from Denisova Cave in Siberia. *Nature, 468*(7327), 1053–1060. doi:10.1038/nature09710 PMID:21179161

Reznikova, Z., & Ryabko, B. (2011). Numerical competence in animals, with insight from ants. *Behaviour, 148*(4), 405–434. doi:10.1163/000579511X568562

Rose, K. D. (1994). The earliest primates. Evolutionary Anthropology: Issues. *News Review (Melbourne), 3*(5), 159–173.

Ryabinin, I. (2015). *Origin of man*. Baltic Book Company.

Sperry, R. W. (2018). Consciousness, free will, and personal identity 1. In D. A. Oakley, & H. C. Plotkin (Eds.), *Brain, behavior, and evolution* (pp. 219–228). London, UK: Routledge. doi:10.4324/9781315149523-10

Stapp, H. (2001). Quantum Theory and the Role of Mind in Nature. *Foundations of Physics*, *31*(10), 1465–1499. doi:10.1023/A:1012682413597

Teilhard de Chardin, P. (1956). *Oeuvres, t.1: Le phénomène humain*. Paris, France: Les Éditions du Seuil.

Thomas, W., & Coolidge, F. L. (2012). *How to Think Like a Neandertal*. Oxford University Press.

Treegene. (n.d.). Retrieved from https://treegene.kz/uslugi/ychromosome-heritage/attachment/567

Vishnyatsky, L. B. (2010). Neanderthals: the story of a failed humanity. Publishing house "Nestor-Istoria".

Weekend. (n.d.). Retrieved from https://weekend.rambler.ru/read/38910641-skolko-kaloriy-szhigaet-mozg-i-kak-eto-rasschitat/

Wong, K. (2014). Tiny Genetic Differences between Humans and Other Primates Pervade the Genome. *Scientific American*.

## ENDNOTES

1 Intelligence formation is thought to be driven by at least 52 genes; SRGAP2 gene and two of its replications refer only to an increase in brain size.

2 The theological point of view in the evolutionary process context presented by Teilhard de Chardin (Teilhard de Chardin, 1956).

# Chapter 2
# Self-Organization of Complex Systems


## ABSTRACT

*The previous chapter suggested it is possible and productive to apply a systematic approach to the study of the evolution of human and his mind. It is time to move on to research methods. A human is a complex open system, and thus, application of a systematic approach is correct. It is necessary to check how well-known methods are suitable for modeling complex systems as part of a systematic approach. For this, the axiomatic basis of the methods was compared with the known properties of complex systems. Such verification leads to the conclusion that it is necessary to create such a modeling apparatus that will most closely correspond to these properties. The chapter describes the components of such an apparatus (axiomatic basis and self-organization scenarios).*


## BACKGROUND

Let's look at previous studies, grouping them by the questions in this chapter.

Researches have shown that the human body and its subsystems are complex systems date back to the post-medieval period and associated with the names M. Servetus and W. Harvey.

The achievements of their followers made it possible to rely on reliable anatomical and physiological foundations when discussing a human as a complex system (Georgieva et al., 1981; Luria, 1985; Gegenfurtner, & Sharpe (Eds.), 1999; Ivanitskiy, 2008; Carter, 1998; 2014).

One of the founders of neurophysiology was Pavlov (1949) and Anokhin (1978). Modern brain research (including using MRI) has allowed us to go much further. In particular, a detailed map of the brain was compiled (Carter, 1998), correspondences between certain abilities and areas of the brain were found (Miller & Cohen, 2001; Fuster, 2003), the state of neural networks for various types of brain activity was recorded Human Brain Project (Human Brain Project, n/d,), brain rhythms were determined for various states of wakefulness and sleep (Zidermane, 1988; Libet, 2004), and data on brain physiology were substantially replenished.

DOI: 10.4018/978-1-7998-1706-2.ch002

Numerous works have been devoted to the search for the relationship of consciousness, mind, emotions with various areas of the brain and states of its activity (Edelman, 1989; Pribram, 1991; Abbott, Sejnowski, (Eds.), 1999; Damasio, 1999; Beauregard, (Ed.), 2003; Chalupa, & Werner, (Eds.), 2004; Hameroff, 2006; Harris, & Jenkin, M. (Eds.) 2007; Monserrat, 2001; 2007).

These results provide enough material to substantiate the idea of a human in general and his brain, in particular, as complex open systems.

The idea of the unity of man and the Universe has philosophical and mathematical components.

The origins of this doctrine are in ancient times. Let's remember the myths of Purusha or Pan-gu, from whose bodies the Universe was created; very developed representations of Taoism and, especially, Sankhya (Radhakrishnan, 1956-1957), describing a complex picture of the structures of consciousness and the Universe, their mutual connections and synchronizations. Ancient natural philosophy shows that such ideas were truly global. Antiquity includes the concept of the unity of microcosm and macrocosm; ideas about the origin of all things from the primary elements of water (Thales) and fire (Heraclitus), Protagoras thesis about man, as the measure of all things.

Modern researchers sometimes return to these ideas, of course, already at the level of modern knowledge. For example, M. Todeschini's (Todeschini, 1949) theory of the role of the fluid in explaining the motion of celestial bodies, gravity, manifestations of life and nervous activity is consistent with Thales's ideas, the Big Bang theory is consistent with Heraclitus's point of view, and the anthropic principle of cosmology (Kuzhekina, Voronina & Lukoyanova, 2016) correlates with Protagoras thesis. Some scientists find a similarity between quantum physics and Taoism (Capra, 1999) or Confucian philosophy (Chung, 2014)[1].

The founder of the theory of complex systems – a kind of reincarnation of ancient natural philosophy of philosophy – was a biologist L. von Bertalanfi (Bertalanffy, 1950).

The development of mathematical methods for modeling complex systems is associated with the names of M. Mesarovich, and Y. Takahara (Mesarovic, Mano & Takahara, 1973; Mesarovic, Takahara, 1978). Note, that the classical mathematical paradigm, largely based on the logic of Aristotle, was used.

Subsequently, numerous modelling and optimization methods were applied to complex technical and human-machine systems, natural biological systems, social systems (Urmantsev, 1974, 1988; Lomov, 1979; Voronin et al., 1995; Korobko, 1997; Olsen, 2006). Let us single out works of S. Kauffman, (Kauffman, 1993; 1995) and works, where a systematic approach is combined with quantum physics (Wolff, Hameroff, 2001; Schwartz, Stapp & Beauregard, 2005; Penrose, Hameroff, 2011). All of them were also in the framework of classical mathematics.

However, a comparison of the properties of geometric and mathematical objects fixed in axioms, with the properties of complex systems shows their almost complete opposite. A difference was noted by researchers (Orlando, 2015). For this reason, classical methods of modelling and optimization of complex systems need empirical corrections and are, in fact, theoretical and experimental.

Understanding this circumstance led the authors to develop an axiomatic wave model, and subsequently – the theory of self-organization of complex systems (Kovalyov, 1996). This device was subsequently used to model and optimize complex technological and human-machine systems, create a comfortable home, optimize the living environment, and study natural languages (Mkhitaryan, Badeyan & Kovalyov, 2004; Kovalyov, Grebenyuk, 2007; Mkhitaryan, 2016).

Now this theory will be used to research the structure and evolution of human mind. The use of the new mathematical apparatus is the main difference between this work and previous studies

## HUMAN AS A COMPLEX SYSTEM

To start, let's begin of some facts which describe the complexity of the human organism (Georgieva et al., 1981; Luria, 1985; Gegenfurtner, & Sharpe, (Eds.),1999; Ivanitskiy, 2008; Carter, 1998; 2014; Polesn_inters_anatomy, n/d; Aif-archive161698, n/d).

Human body consists of about one quadrillion of cells ($10^{15}$) which make groups as tissues, organs, organ system, body parts.

In particular, there are the following organ systems:

- Supporting-motor (bones, joints, muscles, ligaments).
- Excitatory.
- Integumentary (skin and mucous membranes).
- Cardio-vascular and blood.
- Respiratory.
- Digestive.
- Urinary.
- Immune.
- Endocrine.
- Reproductive.

Let's give some illustrations.

*Supporting-motor system*: Skeleton consists of 206 bones, 85 paired and 36 unpaired ones. Shin's bone is the longest and it is about 27% of human height. The smallest bones of the human organism are auditory ossicles, their weight doesn't exceed 0,05 g. Shoulder bone is not connected with any bones of the body as it is fixed with 15 muscles. The total amount of muscles in the human body is more than 600.

*Excitatory system*: It includes central (cerebral axis) and peripheral (nerves and nerve knots) ones; according to functions it can be defined as somatic one which regulates muscles and autonomic one which coordinates cardio-vascular, digestive and excretory system, as well as internal and external secretion glands. The weight of the male brain comprises on the average 1375 g, and female one - 1275 g, individual deviations - from 900 up to 2000 g. The weight of the brain doesn't influence intelligence. More than 50% of neurons are in cerebral hemispheres. A total area of brain cortex varies from 1468 up to 1670 $cm^2$.

Approx. 4 (according to other data – 7-8) percent of available brain cells function, the rest are on reserve. Beginning from the age 30, every day 30-50 thousand of human nerve cells perish. Main dimensions are decreased and the shape of the brain is changed. Male brain reaches its max. weight at the age of 20-29 years, female one – at the age 15-19 years.

In cranial nerves, 2 600 000 nerve fibers enter the brain and 140 000 exits it. About 50% of output fibers convey orders to eyeball muscles and control fast eye movements. The rest of the nerves control facial gestures, chewing, swallowing and activity of internal organs. Two million of input nerve fibers are optic ones. 740-750 ml of blood runs through the brain within one minute. Length of male spinal cord is about 45 cm, female one - 40-42 cm. The speed of nerve impulse may reach 120 m/sec.

*Cardio-vascular system and blood*: Heart of adult human pumps about 10 000 litres of blood per day. Resting heart rate is 60-80 beats per minute, the female heart rate is 6-8 beats faster per minute than the male one. During heavy physical activity heart rate may speed up to 200 and more beats per minute.

Within the life cycle, the bone marrow makes 650 kg of red blood cells and 1 ton of leucocytes. Each red blood cell contains approx. 270 million hemoglobin molecules. A life time of mature blood cells: red blood cells – up to 120 full days, thrombocytes - 10-14 full days, lymphocytes - 2 full days, neutrophilic leucocytes - 8-10 hours.

The total length of a blood vessel in human organism comprises approx. 100 thousand km.

About 100 g of protein is produced within 1 hour in the human organism. Energy reserves of human whose weight is 70 kg: ca. 15 kg of fats (fat tissue), 6 kg of proteins (mainly in skeletal muscles) and only 0,9 kg of carbohydrates (in form of glycogen).

The critical weight of organs and tissues which is enough for survival: liver – 15%, kidneys – 25%, red blood cells – 35%, lungs – 45%, volume of plasma circulation – 70%. Main biological systems of human have a triple safety factor, and in terminal conditions when it is a matter of life or death of the organism these reserves may double or more but for short term only.

Now let's recollect composition of tissues for short.

Human consists of 60% of water: fat tissues - 20% of water, bones - 25%, liver - 70%, muscles - 75%, blood - 80% and brain - 85% of their weight. The rest 40% of weight is spread as follows: proteins - 19%, fats and lipoids - 15%, mineral substances - 5%, carbohydrates - 1%.

The following elements i.e. oxygen, carbon, hydrogen and nitrogen which create the body play the most vital role. A body of an adult human contains ca. 70 kg of them. Calcium and phosphorus comprise about 2 kg, they compose bones and make them resistible. Potassium, sulfur, sodium, and chlorine comprise only dozens of grams. Iron weighs ca. 6 g as it composes hemoglobin.

Some data on exchange indices with an environment (Actistudy.info-vashhestvo, n\d).

In a quiescent state and lying, human consumes 400-500 l of oxygen within full day making 12-20 breath and out-breath per minute.

Average full day need of human in water is 2-2,5 l; this need is covered with food, water as well as oxidation processes in an organism which also produces water.

Food ration has to cater to the needs of human in energy (200-5000 kcal for adult human depending on type of activity), as well as other substances including proteins (obligatory – essential amino acid – arginine, valine, histidine, isoleucine, leucine, lysine, methionine, threonine, tryptophan and phenylalanine), 80-100 g, fats, 80-100 g, carbohydrates, 450-550 g, mineral substances, 14-21 g, and vitamins, totally 0,1 g.

A human organism can't exist without water and food.

Information exchange should be added to the substances and energy exchange with the environment.

From above illustrations, it's obvious that human is a complex system[2]; to confirm such understanding, let's compare features of human organism with the most essential and known properties of complex systems.

1. Complex systems are open systems and they emerge in the *dissipative area at the border of different environment* making a base for consumption of substances, energy, and information.

It can be seen see from the above exchange indices human is an open system. Within human evolution environment has changed its properties multiple times; the same is fair for artificial life environment. It is an information environment which changes most rapidly.

2. Complex systems are *emergent* (integral), i.e. are not limited by their components having very diversified properties, which don't belong to particular components or their total. This feature counters additivity of multitudes completely defined by their components.

Much as it's not the total amount of particular cells or a multitude of organism systems which makes living human as they differ in a quality manner. Such properties of human as homeostasis (parameter constancy of internal environment) or ego may describe a human only in general.

3. Complex systems are *inhomogeneous*; their components may considerably vary from each other, change their composition, quantity, and parameters, therefore saving some *remote similarity*.

Human organs system may substantially differ from each other as for the composition of substances and their functions. Nevertheless, all of them consist of cells and molecules and atoms. Organs systems are formed when the fetus develops from the egg cell. During their life, a human may lose eyesight, sense of hearing, limb, etc.

4. A complex system is *well arranged* and *self-regulated.*

Besides homeostasis, the human organism has multiple means to arrange its functions and self-regulate them. For example, the heart has autoregulation mechanism; besides, its operation is affected by some sections of the spinal cord and spinal bulb, sympathetic and vagus nerves, activity of supramental capsules, emotional condition, temperature, chemical substances, mechanical influence, bright light, loud noise, pain. Several organism systems concurrently react to any external factor, for example, a reaction to aggression condition of an excitatory system, supporting-motor system, cardiovascular system, and endocrine system will change.

5. The behavior of a complex system is notable for the *nonlinear effect* and *target setting*.

Human behavior is *nonlinear too*. For example, stress which develops for long results in a nervous breakdown and sharp change in behavior. Target setting shows up, for example, when we study, run a business or carry out research and similar activities.

6. At the same time, the behavior of a complex system is notable for *the cycle of operation*, and its lifetime is limited (*finiteness*).

Human has their life cycle and their lifetime is limited. In the course of a lifetime, organism follows different rhythms, beginning with annual cycles and ending with those which last fractions of a second (for example, brain activity).

7. The elements of a complex system *depending on the system and each other, immanently changeable, non-additive ones*.

Above example of heart, regulation confirms the property.

8. Elements of complex system *immanently extended, dependable, changeable, they may be complex systems as well.*

Cells being elements of the human organism have all the above-mentioned properties.

9. Interaction of the systems, components, and elements among themselves and with the environment is *nonlinear*; there is *cut-off of nonlinear effect.*

The samples are as follows: different cases of self-regulation, smooth change of brain activity which results in sleep, peace, physical activity, sensitivity threshold of sense organs, etc.

Thus, a human has all 10 attributes of a complex system, and any modeling of their evolution, mind or behavior should stem from this fundamental fact.

## THE MAIN PROBLEM OF THE COMPLEX SYSTEMS MODELING

Let's consider available modeling and optimization methods of a complex system from this point of view. Depending on the system type, they substantially differ from each other.

If we research artificial technical system, such properties, as emergence, inhomogeneity and nonlinear effect are minimized. In this case, the mathematical tool corresponds to the research object, and if you define different system parameters upon necessary experiment – its model, as a rule, is functional enough.

However, in the study of natural biological it is impossible to neglect any of its 10 properties, and this leads to a contradiction with the axiomatic properties of mathematical objects, operations, means of description and verification. As a result, some structures and functions of the system cannot be formalized, and the models themselves need constant experimental corrections.

For visual clarity, let's compare features of complex systems 1-10 with properties of geometrical objects (innated in axiomatic statements). It's possible as geometry being a part of mathematics is built on similar principles (Table. 1).

*Table 1. Comparison of complex systems and space, sets, elements, operations properties*

| Complex Systems | Space, Sets, Elements, Operations |
|---|---|
| **Systems** appear in the dissipative region on the border of environment. They are open, heterogeneous, emergent ones with a variable number of parameters at different evolution stages. They can be self-organizing and self-regulative. Their behavior may be non-linear and teleological. Time of existence is limited. | **Space** is absolute, homogeneous, isotropic, with a fixed dimension, no self-organization, and self-regulation. The existence time is unlimited |
| **components** are dependent on a system, and from each other, inherently variable, not additive | **Sets are** independent of space and from each of other, inherently immutable, additive |
| **elements** are inherently long, dependent, variables, may be complex systems | **elements** (points) are not extended, independent, variable, simple |
| **interoperability** of systems, components, and elements with each other and with the environment is nonlinear, the nonlinearity threshold exists, there is a synergy | **functions** allow bundle. **operations** are linear |
| **measurement** (a special case of interaction) changes value, measurable, and coordinate system | **the measure** does not change the measurable value and coordinate system |

As available modeling methods of complex systems (Voronin et al., 1995) are mathematical ones, we have to admit that their axiomatic basic principles also differ from properties of complex systems. Here you have the fundamental reason for difficulties arising when we apply them.

This implies the main issue of complex systems modeling – we have to create a suitable tool which corresponds to properties of complex systems.

But how to do that? To understand, lets recall the history of the issue.

By the beginning of the 20th century, D. Gilbert developed, as he believed, a complete and consistent system of axioms of Euclidean geometry (Gilbert, 1948). The same approach was extended by him to all mathematics Gilbert & Berneis, 1982).

By that time, other geometries were also known - Lobachevsky and Riemann, obtained by replacing the postulate of parallel lines, with alternative axioms. F. Klein proposed another approach to the creation of non-Euclidean geometries (Klein, 1956). As a result, for example, affine and projective geometries were obtained. The refusal to fix the metric properties of geometric objects led to another alternative - topology.

Sometimes later, the role of logic as a means of verifying mathematical statements, as well as the role of the description language, became completely clear (Kleene, 1957; 1973).

Such a paradigm seemed complete and even beautiful; it can rightly be called classical.

However, already in the thirties of the 20th century, the picture completely changed. K. Gödel formulated two theorems that received the general name of "incompleteness," (Gödel, 1931) the essence of which was that any axiomatic system can be supplemented by an infinite number of true, but irreducible statements. After the works of Tarski (1931; 1936), the difference between the class of deduced and true statements became clear, and there was no longer any hope for a complete and consistent axiomatics.

Moreover, all these results were based on the laws of logic formulated by Aristotle (Kleene, 1973). The right column of table 1 contains the properties of geometric objects and operations; all these properties are determined by the conditions of applicability of logical judgments. But the use of logic does not save from paradoxes, known both in antiquity (for example, the Epaminondas paradox), and new ones connected with set-theoretic relations (for example, the Cantor paradox).

A legitimate question is: why is logic one of the forms of human thinking, that is, a subjective factor, accepted as a means of verifying spatial relations - an objective factor, as it is believed?

The correct answer opens the way to building a new class of geometries that are not based on the applicability of logical verifications and are neither Euclidean nor non-Euclidean.

Various researchers have proposed different ways of verifying statements, both in the general philosophical and mathematical sense (Lossky, 1992; Kleene & Wesley, 1978).

Returning to the problem of modeling complex systems, we clarify it as follows.

The most critical decision is to make geometrical system $G=(E,O,R)$, where $E$ – non-empty set of geometrical objects, $O$ – circuit $o_i$: $E^{ni} \rightarrow E(i \in I)$ and $R$ – total of relations $r_j \subseteq E^{mj}$ $(j \in J)$, defined by axiomatic statements, suitable properties of open complex systems.

To avoid mistakes, we have to build a geometrical system and create verification methods and special language (Kovalyov, 1996; Mkhitaryan, Badeyan & Kovalyov, 2004; Kovalyov & Grebenyuk, 2007); as this book is not dedicated to mathematics, they are not presented here.

*Figure 1. Universal, object, subject and limit S-space*

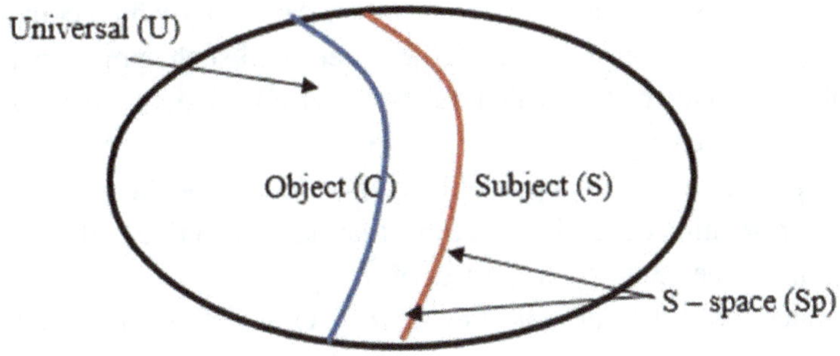


## AXIOMATIC S -SPACE MODEL

It is necessary to introduce concepts and definitions that will be used in the description of S-space (Figure 1).

Let's represent Universal (U) which contains everything in the form of an ellipse and divide it into two parts which we call Object (O) and Subject (S). These names don't fix meaning and depending on the situation they may be interpreted differently (as variables $x$ and $y$ in mathematics). As Universal includes everything hence border between Object and Subject which reflects their differences cannot be introduced from outside, and whereas nothing can be added to Universal, this border doesn't bring in anything new.

*Figure 2. Wave nature of S -space, potential, modalities of existence*

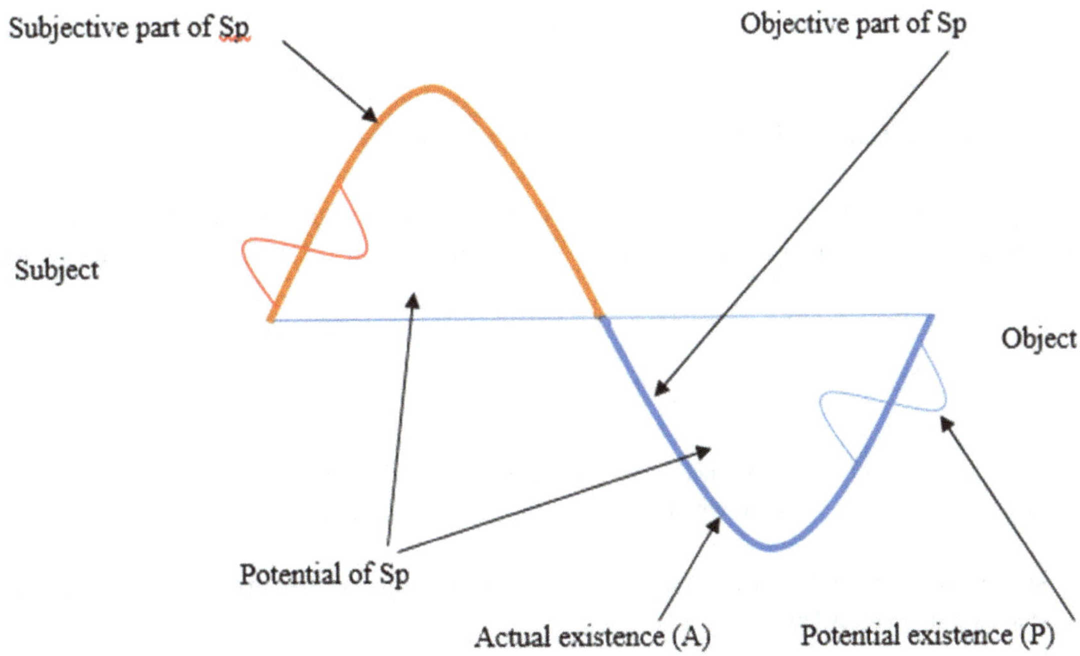

Let's attribute mark «+» to differences between Object and Subject, and differences between Subject and Object as having the opposite meaning mark «-», then the border which contains both differences will be equal «0», i.e. both conditions will be fulfilled. Let's call such symmetrical border exposed to the influence of Subject and Object and which exists because they both exist - *S-space*. Appropriate abstraction of S-space is a wave (Figure 2).

A wave of S-space "grasps" a particular part of Universal; let's name this part *potential* (π) of S-space and give it different interpretation depending on the situation. Subjective and objective parts of Sp are symmetrical and it is one of the integrity properties of Sp.

Further, let's introduce modalities of *actual existence* (***A***), which we understand as the unification of Sp, or S- sets or S- elements and *potential existence* (***P***), a chance for them to emerge (unlike non-existence). Fig. 2 shows that big wave of Sp exists actually and small waves - potentially – they emerge due to existence first, of Sp and second, an influence of Subject and Object.

Interaction of S and O – *observation* (***H***) – changes existence modality of Sp and causes its *bundle* (P) into actually existing S- sets and S-elements. An absence of ***H*** triggers a backward process to P – *convolution* of Sp.

Let's understand *S-space* (°) as a carrier of sets and elements; *S- sets* (□) – as potentially or existing structure which coordinates potentially or actually existing and relatively homogenous S-elements.

*S-elements* include: *wave* (∪) – potentially or actually existing element which is marked by immanent alterability, extension, relative instability, ability to interaction, condition modalities, capability to transfer in soliton wave and *soliton wave*[3] (•) – a relatively stable wave which can transfer in ∪. Entry ∪ may also describe multiple waves; • – multiple soliton waves; ∪• – waves and soliton waves. These elements are chosen because waves and soliton have the above properties. Elements may have only one component (prototype – a sound wave of one frequency), or they may be created out of several perpendicular components (as an electromagnetic wave which contains perpendicular electric and magnetic elements). Accordingly, it is necessary to distinguish between single-component and multicomponent waves and solitons.

S-space is described by four groups of axiomatic statements.

## AXIOMATIC STATEMENTS OF EXISTENCE 1. 1 -1. 5

(1-st group) codify ratio of existence modalities (***A, P***) of space, sets and elements as well as rules of their mutual transfer depending on the dissolution of U and interaction of C и O, i.e. "from up to down".

### 1.1. Observation Transfers S-space in the Condition of Actual Existence:

$$S°A \tag{1}$$

Axiomatic statement assigns existence actuality of Sp as a result of S and O interaction when U dissolves. Thus Sp is a primary notion which can't be converted to sets or elements.

### 1.2. S-space is a Carrier of Potentially Existing Sets Which Contains Potentially Existing Elements – Waves and Soliton Waves:

$$^{\circ}\supset(\square P(\cup\bullet P)) \quad (2)$$

It is stated that S-space is arranged "from up to down" and it is a carrier of potentially existing sets and elements. It means that space keeping its integrity is not converted to its components. Therewith there is no Cantor antinomy due to non-set-theoretic nature of relations space – sets – elements.

The relative homogeneity of S-space is secured as follows: 1. Two types of S- elements are relatively homogenous; 2. Sp has at least two (not considering Sp in general) organization levels which have a different attitude to Sp in general and react to external influence differently. There are no limits as for a number of levels.

Waves and solitons are the primary elements of Sp and cannot be reduced to simpler ones. So, at the elemental level, it is impossible to correlate a wave and a function (which implies a reduction), but only a wave and a point. It follows that the wave and discrete models of space do not reduce, but do not contradict one another, and are abstractions that are mutually complementary and together correspond to the particle-wave dualism of physical reality.

### 1.3. Observation of S-space Triggers its Bundle into Actually Existing S- sets:

$$H^{\circ}P\square A \quad (3)$$

The role of external influences – observations – is postulated as the causes of the evolution of Cn, which affects the change in the modality of the existence of C-sets. Thus, Cn is characterized as open and dual.

### 1.4. Observation of S- set Triggers its Bundle into Actually Existing S-elements:

$$H\square P\cup\bullet A \quad (4)$$

This statement defines the role of observation as a reason for the change of existence modality for S- elements.

### 1.5. Observation of S- elements Trigger Transfer from Waves to Soliton Waves and Vice Versa; its Stop – Disappearance of Elements:

$$H\cup\bullet A\rightarrow(\cup A\leftrightarrow\bullet A) \quad (5)$$

$$\cup\bullet A\rightarrow\cup\bullet P\downarrow$$

where ↔ – means transfer from waves into solitons and vice versa.

The transformation of waves into solitons and vice versa upon observation is postulated; their disappearance in the absence of observation makes convolution possible, since the transition of C-elements to the state of potential existence reverses the bundle. The irreversibility of evolution is not defined explicitly, but it is derived from these and the following axioms.

Axiomatic statements of existence have top priority.

Let's distinguish modalities of waves and solitons: *expansion* (╭) – emergence and expansion as they extend and absorb other S- elements and within cooperation with them; *weakening* (╮) – a decrease of length and destruction within interaction; *disappearance* (↓) – transfer from actual to potential existence. Entries of form ∪╮, •╯ mean wave (waves) within the condition of weakening and solitons (soliton) within the expansion.

## AXIOMATIC STATEMENT OF CONDITION 2.1 (2-ND GROUP)

It defines indicated special aspects of behavior of waves and it makes global link of Sp, its sets and elements on macro and micro levels as dependence of condition modality of elements from evolution path of Sp and vice versa; dependence of evolution direction of Sp from modality condition (╭, ╮, ↓) of elements.

2.1. **Bundle Progress of S-space Defines of S- sets Structure and Condition Modality of S- Elements – Expansion or Weakening; Condition of Elements Depicts S- sets Structure and Change Directivity of Sp:**

°P(□A((∀∪•A(∪•╯∨∪•╮)))) (6)

In abstract form, the axiomatic statement contains properties of integrity, organization, non-additivity, non-binarity. Its priority is lower than the axiomatic statement of the first group but higher than the following ones have.

Let's illustrate the execution of axiomatic statement for natural systems.

Let's view lake water as limit space between ground and air. The gust of the wind stirs water surface. Properties of the waves will be different in particular areas. It should be noted certain factors which predetermine these differences. First, these are an external influence – wind and bottom relief. Second, the general properties of the space (water density, size, and depth of the lake, etc.). The first group of factors predetermines the movement direction of the wave group; second one – their frequencies and amplitude. These properties are agreed. Here, coordination directivity is "from up to down" and its essence is defined as subordination and dependence. And the waves redirect different temperature layers and thus they change water density. Thus, they influence the properties of space in general. Thus, feedback is realized.

It should be noted another consistent pattern – irregular behavior of waves depending on their existence stage. Some of them express expansion which means an increase of amplitude, they embrace a bigger amount of water, etc.; other, quite the opposite, are exposed to regression – a decrease of amplitudes, confusion, disappearance.

Consequently, changes of elements are mutually agreed and they also depend on their condition.

## AXIOMATIC STATEMENTS OF INTERACTION 3.1-3.9 (3RD GROUP)

Let's distinguish: *parallelism* (||) – ratio of two or more waves when their interaction doesn't increase the number of components of the resultant wave as compared with initial ones, and *perpendicularity*

(⊥) – ratio of two or more waves when interaction increases the number of components of the resultant wave as compared with initial ones. Entry of the form ∪⊥ means perpendicular waves.

Let's consider the following interactions (operations):

- **Superposition Operation (*S*)**: Unification of two or more parallel waves which components are summed up or compensated.
- **Coincidence Operation (*C*)**: Unification of two or more perpendicular S- elements.
- **Interference Operation (*I*)**: Interaction of two or more waves and number of waves, their components and their condition are changed.
- **Diffraction Operation (*D*)**: Interaction of waves and solitons with a change in the number and state of waves and components (except for diffraction of solitons only).

Operations over S- sets, and S-elements are abstractions of different interaction types of physical waves.

Axiomatic statements which describe interaction are applied only with axiomatic statements of existence and condition.

Under superposition operation, parallel components of single or multi-component elements interact. As a result of an operation, their amplitudes are summed up. As axiomatic statement 1.1 is executed therefore above-mentioned operation doesn't breach conditions of non-binarity and non-additivity.

3.1. **Superposition Operation of Actually Existing Single Component Waves Turns Them into Single Component Waves with Linearly Changed Properties:**

$$S(\cup A\|) \rightarrow \cup' \qquad (7)$$

Depending on the condition of initial elements the amplitudes are summed up or weakened; in the last case, all interacting waves can transfer from actual to potential existence.

During coincidence operation, perpendicular elements interact and it results in their unification which contains all perpendicular components of initial elements. These unifications are not binary or additive.

3.2. **Coincidence Operation of Actually Existing Perpendicular S- elements Transfers Them into the Element Which Perpendicular Components Equal to the Total Amount of Perpendicular Components of Initial Waves and Soliton Waves:**

$$C(\cup \bullet A\perp) \rightarrow \cup' \bullet C \qquad (8)$$

There is Interference operation for waves consisting of parallel and perpendicular components.

3.3. **Interference Operation of Actually Existing Waves Turns Them into Waves with Changing Number of Waves and Components and non-linear Change of Properties:**

$$I(\cup A) \rightarrow \cup' \qquad (9)$$

Diffraction operation is understood as the interaction of waves and soliton waves. The latter is considered as obstacles which disrupt linear movements of waves. The result stipulated by the axiomatic statements of the first and second group has the following parameters changed: quantity of waves and components, the condition of waves and movement direction of soliton waves. It should be noted the particular case: if only soliton waves take part in diffraction operation, they have pushed away and they change movement direction i.e. they behave like particles.

**3.4. Diffraction Operation of Actually Existing Waves and Solitons Transforms Waves – into Waves with Changing Quantity of Waves and Components and Non-linear Change of Properties; Soliton Waves – into Solitons Which Retain Set of Components and Changed Movement Direction:**

$$D(\cup\bullet A)\rightarrow\cup' \qquad (10)$$

$\rightarrow\bullet$

We should mention that during diffraction operation properties of solitons don't change.

Diffraction operation is the most general type of operation in S- space; all others are considered as its special cases, which emerge depending on properties of components of interacting elements.

The property of immanent variability from Table 1 is also interpreted in this way: there should not be such combinations of operations that would translate C-elements to their original state, and such waves and solitons that could be defined as zero and single ones. That's why operations don't make groups except for the special case of diffraction operation of solitons only (Axiomatic statement 3.9.). The layers are closed as for operations with structures which build them.

**3.5. There is no Such Consequence of Two and More Superposition Operations Which Would Turn S-elements into Their Initial Condition:**

$$CC^{-1}(\cup\|A)\rightarrow\cup'' \qquad (11)$$

**3.6. There is no such consequence of two or more coincidence operations which would transfer S-elements into their initial state:**

$$CC^{-1}(\cup A\perp)\rightarrow\cup''C \qquad (12)$$

**3.7. There is no Such Consequence of Two or More Interference Operations Which Would Transform Waves into Their Initial State:**

$$II^{-1}(\cup A)\rightarrow\cup' \qquad (13)$$

**3.8. There is no Such Consequence of Two or More Diffraction Operations Which Would Turn Waves into Their Initial State:**

$$DD^{-1}(\cup\bullet A)\rightarrow\cup''\bullet \qquad (14)$$

3.9. **There is at Least One Consequence of Two Diffraction Operations Which Transforms Solitons into Their Initial State:**

$$DD^{-1}(\bullet A)\rightarrow\bullet A \tag{15}$$

Irreversibility of operations corresponds to the irreversibility of evolution path.

Based on axiomatic statements in groups 1-3 self-organization theory is built which is given below; they moderate acceptable ways to create S- sets within self-organization of S-space.

## AXIOMATIC STATEMENTS OF MEASURING 4.1 -4.2 (4$^{TH}$ GROUP)

Let's introduce the following notions.

*Coordinate system* (Cs) is called any set of soliton waves which belongs to different organization levels of Sp (layers); due to their interaction with other waves and solitons, it's possible to assess their properties. Let's define a total of such layers as *measurable space. The result* (Rs) of measuring should be interpreted as changes in coordinate system after interaction with a measurable element.

The total number of perpendicular components of all solitons Cs should be accepted as *the dimensionality* of the coordinate system. Sp is an absolute if compared to totals and elements, therefore Cs will describe only measurable subspace. Components Cs are not related with components of the measurable element and depending on its attribute to subspace, there are different grades of comparison of element and Cs.

Measuring is the observation of interaction results of element and Cs. Such a procedure is not always suitable. Wave and soliton react in a different way when they are measured: the wave is measured till disappearance (axiomatic statement 3.4.), and repeated measurements can't make results more specific; for soliton waves, only movement direction is changed. Cs is also changed.

Axiomatic statements of measuring prescribe interaction procedures with Cs for waves and soliton waves. They satisfy axiomatic statements of groups 1-3, and special aspects mentioned above.

4.1. **Observation of Diffraction Operation in the Coordinate System and an Actually Existing Wave of Measurable Space Results in Changing of Coordinate System – Measuring Result –and Changing of the Measured Wave:**

$$HD(Cs\wedge\cup A)\rightarrow(Cs'\wedge\cup'\wedge Rs) \tag{16}$$

4.2. **Observation of Diffraction Operation of the Coordinate System and Actually Existing Soliton in Measured Space Leads to Changes in the Coordinate System – Measuring Results – and Change of Direction of Soliton Wave:**

$$HD(Cs\wedge\bullet A)\rightarrow(Cs'\wedge\bullet'\wedge Rs) \tag{17}$$

Let's give the examples which show that there are special aspects which differ from measurements in a rectangular coordinate system.

Thus, when forecasting economic development, the necessary data is not only about production resources, but also the level of qualifications and education, the availability of minerals, climatic conditions, historical development trends, and the like, which are impossible to bring to one measure.

The influence of the observer on the measurement object for, for example, human-machine systems is common; most pronounced in extreme situations. Another example is W. Heisenberg's uncertainty principle.

The object does influence the observer as we may notice in operator's activity – the more experience operators gain the more different parameters they choose to control, a time necessary to assess the situation, algorithm of actions in typical situations, etc.

Thus, the wave model of S-space is built; all properties of complex systems are included in axiomatic statements according to Table 1.

## THEORY OF SELF-ORGANIZATION

It's necessary to check completeness and consistency of axiomatic statements and consider if external influence leads to particular organized nature (self-organization scenarios) of waves and solitons of S-space, or they will emerge and extinguish chaotically.

Verification of the completeness and consistency of the system of axioms traditionally comes down to proving the independence and consistency of the consequences of the axioms of a group from the axioms of other groups, as well as finding out whether it is necessary to add assumptions not fixed in the axioms. Considering incompleteness theorem and the incongruity of classes of true and conclusive expositions, such a check should be made in a limited form – for a defined type of tasks. It's also necessary to consider priorities in the system of axiomatic statements of wave model, and limits in using formal ways for description and verification. Thus, checking procedure will be defined as follows:

1) first, let's examine the consequences of axiomatic statements in the first group and check if they are consistent, independent and they don't require additional suppositions.
2) the axiom of the second group is added and their general consequences are considered. If they are also consistent, independent, etc., the axioms of the third and then fourth groups are added and the procedure is repeated.
3) the answer to the question – which consequences should be considered – is defined by the tasks of specific research.

According to the above provisions let's consider only those properties of Sp, which derive from axiomatic statements in the first group.

Axiomatic statements of existence qualitatively describe the structure of Sp and consequence of bundles, but they don't specify reasons and don't introduce quantitative properties. Let's find out these reasons and properties.

### Scenario (1S, 1O)

Let's assume that one C interacts with one O and mark such a self-organization scenario (1S, 1O) [4]. Following axiomatic statements 1.1.–1.5., Sp is independent limit space open for the external influence which emerged as a result of the disintegration of Universal. Its structure expresses an identity relation of S and O in U and their differences. In the first case, Sp is a link of S and O – the ternary bond, in the second one – a projection of S or O. Let's examine the consequences of identity S and O.

The existence of a ternary bond means that S and O are similar in all aspects apart from those that are expressed in Sp. That's why S and O are symmetrical as for Sp and Sp as a projection of S and O reflects this fact. Consequently, 1) S-space is symmetrical. As within bundle relations S – Sp – O don't change then 2) symmetry of Sp is retained within its bundle. As long as Sp emerged and added nothing to Universe, the total potential of its halves equals zero – it means that 3) S-space is skew-symmetric.

These observations give us the reasons to distinguish symmetrical halves of Sp and introduce similarity type for membership properties of sets and elements of Sp. Marks □⁺, ∪⁻ will be interpreted as structures that belong to subjective and objective parts of Sp respectively.

Let's refer:

*{P}* – total of acceptable bundles.
$p_i$ – current bundle from *{P}*.
*{x(*$p_i$*)}* – set of its properties;
$x_j$ – j property from *{x}*.

There is

**Exposition 2.1:** Only those bundles are acceptable which retain oblique symmetry of Sp:

$$\{P\}:x_j^{+}, x_{j-}\{x(pi)\} \tag{18}$$

To claim that S and O are identical it needs to secure the process opposite to bundle – decrease of the number of elements and sets as they disappear after we stopped observation – i.e. convolution (Axiomatic statement 1.5.).

As for P, for convolution it is correct:

**Exposition 2.2:** Oblique symmetry of Sp remains unchanged within convolution:

$$\{Co\}:x_k^{+},x_k^{-}\{x(f_i)\}, \tag{19}$$

where *{Co}* – total convolution; *c* – current convolution.

Link of $Co_k$ with respective *P*, under a condition, that observation is stopped on N-bundle, shown in the formula:

$$x_j(c_k)\leftrightarrow x_j(p^{-}_{i+1}), \tag{20}$$

where the symbol ↔ means a change of process direction.

Available symmetry as a feature of an integrated system is subject to numerous researches.

Now let's consider difference relations of C and O. In this case, a ternary bond disintegrates. Relations S - Sp and O - Sp are interpreted as **H** which makes S- sets and S- elements actual ones.

The bundle process has to reflect the fact of the disintegration of integral Universal into S and O. It is possible only when properties of S- elements and S- sets which reveal themselves on next stage of a bundle are attributed to respective properties of the current layer as they would, in turn, be attributed to the respective properties of the previous layer, so they emerge when harmonic relations are compiled. Consequently, the bundling process has to meet the following requirement:

**Exposition 2.3:** Accepted are only those bundles which properties of S- elements and S- sets on current stage are in harmonic relations with properties of respective elements and sets of the previous stage:

$$\{P\}:x_j(p_i)/x_j(p_{i-1} = \Phi, \tag{21}$$

where $\Phi = \dfrac{\sqrt{5}-1}{2} \approx 0,618...$ – quantitative property of "golden proportion"[5].

As long as Φ is the irrational number and quantity of sets and elements which emerge within bundle – is whole as well, it's correct that:

**Consequence 1:** Total elements and sets which emerge within the bundle are not limited.

**Consequence 2:** The relation of a number of elements and sets within the current stage to a number of elements and sets of the previous stage under an unlimited bundle tends to Φ.

Number Φ is also considered as integrity factor.

**Relations of Sets and Elements:** A number of sets and elements of the layer will be marked with *layer index* (analog to a cardinal number of sets), for example, entries $\square^{+}_{5}$, $\square^{+}_{5i}$ mean sets of subjective half which are attributed to the layer which contains five sets; the second index shows the number of elements for each set.

Let's link the index and layer number (bundle stages).

According to consequence 2 of exposition 2.3, a relation of indices for current and previous layers tends to Φ. The Fibonacci sequence obtained as follows complies with this consistent pattern.

Suppose n– current member of the line, and n– 1, n– 2 – previous ones. Then

$$n=n-1+ n- 2 \tag{22}$$

Substitution gives the first numbers in line as follows:

1, 1, 2, 3, 5, 8, 13, 21, 34, 55...

It's easy to come to exposition that relation of current quantity to the next one does tend to Φ as numbers rise. Then accordance between number and layer index will be attributed to known correlations:

1 2 3 4 5 6 7 8 9 10... m

1 1 2 3 5 8 13 21 34 55... n,

where $n \approx \frac{1}{\sqrt{5}}((\Phi+1)^m + (\Phi+1)^{-m})$ (23)

Let's describe the bundle–convolution scenario.

**Exposition 2.4:** In case if one S interacts with one O, only one bundle–convolution scenario is executed in which the quantity of elements in all layers beginning from the first one equals to quantities of the Fibonacci sequence.

Research of complex systems confirms that the Fibonacci sequence imposes the number of their elements on different organization levels.

## S- Set and S- Element

Axiomatic statement 1.4 establishes the existence of S- elements; at the same time, according to the consequence 1 of exposition 2.3, bundle process is limitless and there is no such final stage that contains only elements. Simultaneous execution of both conditions is only possible when the bundling process changes the status of some S- sets and S- elements. Considering exposition 2.4, have place

**Property 2.1:** Structures of the current bundle stage are S-elements as for one of the structures of the previous stages. At least one of the structures of the current stage is S- sets for the structures of the next stage.

**Consequence 1:** All the layers which emerge but the last one contain at least one soliton wave.

Thus, S- elements include all structures of the last layer which emerge and part of structures of preceding layers. There is the same property for the convolution:

**Property 2.2:** All structures of the current stage of the convolution are elements at least one structure of the next stage.

## Wave and Soliton

Following axiomatic statement 1.5, the soliton is created from the wave within its observation. As long as (property 2.2.) formation $\bullet^+$ will be obligatory accompanied by the formation of skew-symmetric one $\bullet^-$, the mechanism of how soliton waveforms should be as follows:

1) ***H*** triggers a mutual weakening of halves of skew-symmetric waves.
2) Halves which are left become soliton waves as they lose the capability to change according to the laws on periodical fluctuations.

## Scenarios of a Bundle

Considering the role of **H** in triggering bundle Sp and S- sets (axiomatic statements 1.2 and 1.4) and within transfer from wave to soliton wave (Axiomatic statement 1.5), as well as a duality of set and element established by property 2.1.2, let's assume the following bundle scenario.

Due to the influence of ***H*** actualization of Sp occurs (axiomatic statement 1.1)

Further on simultaneous (according to the property 2.1) transfer takes place:

$$H\{\cup^{+}{}_{1}\rightarrow\bullet^{+}{}_{1};\cup^{-1}\rightarrow\bullet^{-1}\} \quad (24)$$

There is no need to weaken halves of waves.

Following bundle of S- sets happen by the same scheme:

- One of the waves in the last layer turns into soliton wave.
- Waves on the "surface" of soliton wave emerge in such a way that their total amount would equal index of the respective layer.

## Quantitative Properties and Qualitative Bundles

*Potential* (π) of Sp will mean the property of capability to interaction and change, "captured" share of "space" Universe, stipulated by the disintegration ratio into S, O, Sp. Potential is intuitive, formally infinite property, such as, for example, energy, weight, charge in physics; it is reflected for each of S-elements in conditional units as for the potential of Sp according to the measurement results (axiomatic statements of the 4th group).

The notion of potential enables to interpret weakening of halves of waves when they turn into soliton waves as a potential decrease of Sp in whole, its dissipation in U. This implies irreversibility of Sp evolution in this case; it is reflected by

**Property 2.3:** Suppose the potential ∪ of layer n equals $\pi_n$. Then $\pi_n > \pi_n^*$, where $\pi_n^*$ – potential after $\{\cup\rightarrow\bullet, Pn_{\rightarrow m}C\text{B}_{m\rightarrow n'}\}$, m>n.

But as axiomatic statements don't define if ***H*** compensates this loss, thus it's theoretically possible bundle scenarios when potential decreases, retains and increases. This possibility will be considered when a summary classification is built. There several variants available it enables the modeling of different situations of interaction. Assume that observation does not compensate for potential losses.

To assess bundle let's define that part of the potential of objective or subjective half of Sp moves from the previous stage to the current one. It's correct for each S- set

**Property 2.4:** When S- set with potential $\pi$ stratifies each element gains potential $\pi_i$:

$$\pi_i = \frac{\pi\Phi}{i}, \quad (25)$$

where $i$ – index of the current layer.

Thus, *potential Sp (S- sets) is not equal to the total potentials of composing elements.* Consequently, (2.8) is a formal reflection of non-additivity and relative homogeneity of Sp and S- sets. This implies:

**Exposition 2.5:** All properties related to the potential of any layer differ from the properties of the next layers qualitatively. Quite the opposite, for the properties of the next layer of the current layer, may be used.

Therewith there is:

**Consequence 1:** Shannon assessment of information amount codified with using of S- elements, has to be carried out for each layer separately.

Exposition 2.5 is of great importance for modeling of complex systems.

Let's introduce two properties that depend on $\pi$. It is necessary to define the length of waves for current and previous bundle stages. As long as waves emerge when solitons are observed which in this case S- sets are, and their quantity is a whole number, they have to "pack up" integrally in soliton. Here take place

**Property 2.5:** Length of $l_k$ waves in the current stage as for the length of the wave in the previous stage is defined with the formula

$$l_k = \frac{l_{k-1}}{2i}, \tag{26}$$

where $k$– a number of the current stage.

Let's unite potential of the wave with length and amplitude A:

$$\pi=f(l/2,A), \tag{27}$$

where $f$ – a function of proportionality defined by the experiments.

The wave potential will be estimated as half of wave potential, for example, by the compression phase only.

## Conditions of Elements

Let's examine properties which stem from the general use of axiomatic statements of the first and second group in the assumption of interaction between one O and one S.

Due to condition modality, it is possible to describe the behavior of S- elements within a period between bundles, as well as define the evolution path of Sp. The conditions change following the following scheme:

1. Observation of soliton makes waves appear on its surface. Their oscillatory amplitude increases till

$$\sum_{i=1}^{I} \pi_i < \pi_k, \tag{28}$$

where $\pi_i$ – potentials of waves in the layer; $I$ –index of the layer; $\pi_k$ – the total theoretical potential of waves in the layer according to (25).

Such behavior of waves is defined as expansion ⎠. U⎠ begins to interact which doesn't destroy their structural independence and therewith length of Sp increases.

2. When $\sum\pi_i = \pi_k$, observation is shifted to one of the waves or consistently to several waves which results in the creating of solitons.
3. Waves which are not subject to observation decreases their amplitude and gradually disappear – they return to the state of potential existence. Such a transformation of waves is defined as weakening ⎞. Their structure is destroyed within an interaction.
4. When ***H*** of last layer elements ceases bundle stops and convolution of Sp are started. It means that differences between S and O diminish opposite to its increase when a bundle is in progress. That's why ⎠ is explained as simultaneous action of S and O, and ⎞ – cessation of their interaction.

This implies a time limit of *waves and solitons existence* and its dependence as interaction with other elements and available ***S***. Let's describe time rating and define layer existence time.

*Time rating*. It's convenient to use time rating unit $t$ as a time equal to one period of the wave. In such case take place:

**Property 2.6:** Each layer has its time rating unit equal to a period of the wave in this layer.

**Property 2.7:** Time rating unit in current layer $t_k$ is to time rating unit of the previous one as a length of waves in these layers:

$$\frac{t_k}{t_{k-1}} = \frac{l_k}{l_{k-1}}, \tag{29}$$

where $k$ – a number of the layer; $l_k$ and $l_{k-1}$ are related according to (2.8).

**Property 2.8:** It's possible to have such bundle scenarios for which the existence time of each layer expressed in relative measuring units is identical:

$$\forall k, \frac{T_k}{t_k} = const, \tag{30}$$

where $T_k$ – existence time $k$ layer.

## Duration of a Bundle

The above information implies ***an*** unlimited bundle in that meaning that there is no such layer that would contain elements that are not S- sets for the following ones as it would mean process end (consequence 1 of property 2.1). However, there is a limit as to time stipulated by the existence time of the first wave of Sp.

**Property 2.9:** Bundle and convolution processes have a total duration equal to the existence time of the first wave Sp.

This implies several consequences:

**Consequence 1:** Sp existence time is limited i.e. its evolution is equifinal one.
**Consequence 2:** The bundle process is only possible when the existence time of the current layer is shorter than the existence time of the previous one:

$$\forall k, T_k < T_{k-1} \tag{31}$$

**Consequence 3:** Suppose the time of convolution is equal to bundle time. Then time $T_{ok}$, which is left till the bundling process ends at the beginning of $k$ stage is equal to

$$T_{0k} = \frac{T_1}{2} - \sum_{j=2}^{k-1} T_j \tag{32}$$

Thus, time for S-space is described with two properties:
*"Time arrow"* which reflects irreversibility of evolution (property 2.9 and consequences);
*A bundle* which reflects relative homogeneity qualitatively different processes (properties 2.6–2.9).
Let's define features and those properties which stem from a general examination of axiomatic statements in 1–3 groups.

## Parallelism and Perpendicularity of Elements

When S interacts with several {O} Sp stratifies according to scenarios *{P}* in such manner that $P_i \rightarrow O_i$ (symbol "→" means "stipulated $O_i$»), $P_i \in \{P\}$, $O_i \in \{O\}$.
The following conditions are executed:

1) Qualitative differences $S \wedge O_i$, $S \wedge O_j$, $O_i$, $O_j \in \{O\}$, have to correspond to relatively homogeneous structures of their bundles.
2) Similar differences $S \wedge O_i$, $C \wedge O_j$, $O_i$, $O_j \in \{O\}$, have to correspond to homogeneous structures, which are made as a result of $P_i$, $P_j \in \{P\}$, which supposes quantitative comparison.

The first condition is interpreted as the interaction

$$\cup\bullet_i \text{ and } \cup\bullet \tag{33}$$

(33) is necessary and sufficient as it embraces all operations over relatively homogeneous elements (axiomatic statements 2.1.2–2.4, 2.6, 2.9.), which requires their perpendicularity.

The second condition is understood as an interaction

$$\cup\bullet_i \text{ and } \cup\bullet_j \|;\ \cup\bullet_i \in \square_k,\ \cup\bullet_j \in \square_k \tag{34}$$

(34) is necessary as an interaction of homogeneous elements – superposition operation – requires available parallel elements according to axiomatic statement 2.1. Sufficiency (2.17) implies the absence of superposition operation of parallel elements in one bundle i.e.:

- Elements of one layer.
- Elements of different layers.

Let's show that such elements virtually won't interact according to the superposition operation rule.

a. By property 2.5, the length of waves in any layer is the same as the one of soliton which is S- set of this layer and makes the whole number of waves. According to the exposition 2.1, the symmetry of Sp will retain within its bundle. That's why each layer is symmetrical and it's correct that

**Property 2.10**: The suppose layer contains only waves and doesn't contain solitons. Then it has a pattern of symmetry equal to its index.

As long as an interaction of waves in one layer would destroy symmetry, there is

**Consequence 1:** Waves of one layer don't interact until it keeps its symmetry.

b. Elements of different layers also don't undergo superposition operation as they have potentials which considerably differ. For example, potential $\cup_8$ is to potential $\cup_3$ as Φ/2*5*3 = 1/30 (2.8), and their length – as 1/2*5*2*8=1/160 (2.9), i.e. it differs more than once or twice and it results either in unification of "small" wave, or it avoids "big" wave, which leads to interaction as Interference operation according to axiomatic statements 2.2 and 2.4.

Now let's consider bundle scenarios that make parallel and perpendicular elements emerge.

## Formation of Parallel Structures Only

In this case for each with $P_i$ exposition, 2.4 and property 2.10 are executed. Only single component waves emerge; therewith quantity of waves and as a bundle is in progress symmetry pattern increases according to the Fibonacci series. The scenario describes

**Exposition 2.6:** Suppose for $□^+_{-i}$ only group of symmetry $S_i$ acts (not obligatory $S_i$=$S_n$ acting for $□^+_{-n}$). Then within a bundle, only parallel structures emerge and therewith symmetry pattern is *m=i.*

## Formation of Perpendicular Structures Only

In this case, multi-component waves and soliton waves emerge. The scenario of such bundles describes

**Exposition 2.7:** Suppose for $□^+_{-i}$ within formation $U_k \in □_i$, k=1,...i, own symmetry group $S_{ik}$ acts (not necessarily $S_{ik}$= $S_{jl}$, which acts for $□^+_{-j}$ for $U_{jl}$). Then within a bundle, only perpendicular structures emerge, therewith a number of symmetry groups n for the layer with index i: n=i.

## Formation of Parallel and Perpendicular Structures

Exposition 2.6 and 2.7 are executed. Their interrelation establishes

**Exposition 2.8:** Suppose for $P_i$ M group symmetry exists and for $S_{ik}$ $S_{ik}$=$S_{jl}$ is executed. Then the formation of perpendicular structures will stop, when the quantity of realized groups of symmetries for each layer L=M, and further on only parallel structures will be able to form.

As dimensionality of S- sets equals the number of perpendicular components of structures, therefore, there is

**Consequence 1:** Within bundle which follows the scenario described in exposition 2.8, dimensionality r of S- sets decreases from M to 1: $r_k$=$r_k$– 1– i, and has reached 1, it remains unchanged.

There are diversified symmetries for S- sets, and S- elements which make it possible to state such provisions.

1. By analogy with the Noether theorem, it can be assumed that *the presence of particular groups of symmetry predetermines respective conservation principles*, for example, group of parallel transfer for a time corresponds to energy conservation, shift and movement of space – momentum conservation, etc. In this case, the properties of the layer and the symmetry group can be determined, since the results of observations are used and the invariants of the organization of the system are preserved.
2. Due to a bundle of space and time (properties 2.6-2.9), these laws are limited by:
   **Exposition 2.9:** For $□^+_-$ any layer which has M symmetry groups is subject to M* conservation principles; their influence ceases when they transfer to the previous or next layer.
3. Property 2.10 and exposition 2.6-2.9 for an organization of Sp are similar to Pauli exclusion principle for quantum systems: feasibility of interaction and its result are defined not only by properties of the elements but also necessity or absence of necessity to realize symmetry group and pattern change stipulated by the bundle progress of Sp (axiomatic statement 2.1).

Thus, the role of symmetry in the organization and evolution of Sp is suitable for its role in the organization and evolution of natural systems.

## Properties of Operations

4 operations of arithmetic are reduced to additive binary addition; set-theoretic ones – to unification and crosscut. By contrast to this, 4 operations of the wave model of Sp are special cases of non-additive and non-binary diffraction operation. Let's mark their multitude as {**DICS**}. Each operation will be described according to the following plan:

1. Areas of use.
2. Special cases (given in tables).
3. Role in structure formation, bundle, and convolution.
4. Additional remarks.

It is necessary to highlight that special cases of particular operations are similar to some extent to the interaction of their physical prototypes. Nevertheless, this analogy doesn't embrace formulae to define quantitative properties.

The set $\{x(pi,_j)\}$ will include:

- Potential $\pi$.
- Length $l$; amplitude $A$.
- Frequency $o$.
- Phase $F$.
- Period $T$.
- Coefficients.
- Measures of nonlinearity.

$$e = \frac{A}{M} \leq e^*, \quad d \leq d^* \tag{35}$$

where $e$ and $d$ describes nonlinearity on amplitude and movement respectively; $e^*$ and $d^*$ – limit values of $e$ and $d$ beginning from which it can't be to ignore nonlinearity; $M$ – property of the non-excited state of S- sets;

- relation of length:

$$s=l_k/l_m; \tag{36}$$

- relation of potentials:

$$p=\pi_k/\pi_m; \tag{37}$$

- modality $\rfloor$, $\rceil$, $\downarrow$, and attribute of $\|$and $\perp$.

Some of these properties are related to each other according to (2.9–2.12); it is necessary to recollect that not all of them may be defined for all layers.

**Superposition operation** (axiomatic statements 3.1, 3.5).

**Condition 2.1:** Superposition operation is defined for single component waves only if the following conditions are executed

$\cup_{ik}, \cup_{lm} \|, \cup \rightarrow P_i, P_j$

$l_k < l_m, s \in \{\sim 0.2 \div \sim 1'\}$

$\pi_k < \pi_m, p \in \{\sim 0.2 \div \sim 1'\}$ (38)

$d < d^*$

$e < e^*$,

where the symbol ‘ means an exception.

1.2. Superposition operation breaks the symmetry of the layer. There is

**Property 2.11:** If the symmetry of the layer is broken the following occurs:

- Waves move on the surface of their soliton waves.
- Waves of one layer interact and potentials are re-distributed till symmetry is restored but with lesser order.

1. Special cases are given in table 2.

*Table 2. Special cases of superposition operation*

| $\{x(p_i)\}, \{x(p_j)\}$ | Description |
|---|---|
| $\cup_k\rfloor$ and $\cup_m\rceil$ | saved: $F_k, o_k$; not saved: $F_m$, $o_m$; changed: $\pi_m$, $A_m$ |
| $\cup_k\rfloor$ and $\cup_m\rfloor$ | saved: $F_k$, $F_m$, $o_k$, $o_m$; re-distributed: $\pi_k, \pi_m$, $A_k$, $A_m$ |
| $\cup_k\rceil$ and $\cup_m\rceil$ | not saved: $F_k$, $F_m$, $o_k$, $o_m$; decreased: $\pi_k$, $\pi_m$, $A_{k, Am}$ |
| $\cup_k\rfloor$ and $\cup_m\rfloor$ and $\cup_k\rceil$ and $\cup\rceil_m$ | standing waves: $F_k - F_m = T/2$, $A_k + A_m = 2A$ |
| $A_k = A_m$, $l_k = l_m$, $o_k = o_m$ | weakened waves: $F_k - F_m = T/4$, $A_k + A_m = 0$ |
| $\cup_k\rfloor$ and $\cup_m\rfloor$<br>$\cup_k\rceil$ and $\cup_m\rceil$, $o_k \approx o_m$ | beat: for $A_k = A_m$ with frequency $o_{k+m} = (o_{k+} o_m)/2$;<br>beat frequency $O = o_k - o_m$ |

2. The waves modified with **C**, undergoing **H**, turn into solitons and are stratified at the next stage - the influence of **C** on P consists of changing the transmitted $\pi$, or when the waves are mutually repaid, in eliminating the scenarios in which the bundle could go.
3. Remember that wave interactions are always nonlinear.

Non-binarity is secured with records of the condition stipulated by ***H*** and the progress of the present bundle, as well as the possibility for more than two waves to participate in the interaction. Irreversibility is caused by a change in the state of the waves under the influence of ***H.*** Thus, conditions of complementarity for superposition operation are executed.

## COINCIDENCE OPERATION (AXIOMATIC STATEMENTS 3.1, 3.6)

Coincidence differs from an accidental multitude of perpendicular waves and solitons due to its integrity which has a form of a definite concurrence of perpendicular components as for their length. Their coordination is reached as a result of the re-distribution of potentials of initial waves.

1. Suppose one coincidence interacts with {O*}, which is a set of objects from {O}. Then the conditions of coincidence operation are as follows.

   **Condition 2.2:** Suppose for bundles {P} which correspond to {O*} an ordered set of waves and soliton waves exists:

$$l_1 \perp l_2 \perp l_{2..} .. \perp l_n,$$

$$l_1 > l_2 > l_{2..} .. > l_{n,}$$

where $_n$ – the number of bundles in {P};

l– the length of ∪•.

Therefore, necessary and sufficient conditions of coincidence operation for single component ∪• will be execution:

$$l_n/l_1 = m_{11},\ l_n/l_2 = m_{12},\ \ldots,\ l_n/l_{n-1} = m_{1n-1}$$

$$l_{n-1}/l_1 = m_{21},\ \ldots,\ l_{n-1}/l_{n-2} = m_{2n-2}$$

...

$$l_2/l_1 = m_{n-11} \qquad (39)$$

$$\forall \Delta\pi_i = \sum_{j=1}^{n} \Delta\pi_j, j \neq j$$

{∪⟩}∨{∪⟋}

d<d*,

where *m, k, l* – the whole quantities, $\Delta\pi_i$, $\Delta\pi_j$ – changes of potentials.

As long as π of soliton waves are not changed, they are not considered in (39).

2. Special cases of coincidence operation are given in table 3.
3. In general, the multi-component wave doesn't create S- sets; its particular components – ∪ and • – may form such sets within their layers. Therewith coincidence operation is broken as concurrence is destroyed.
4. Non-binarity of coincidence is secured by the participation of several waves and records of their conditions; non-additivity – that it's impossible to add new components which would result in termination of concurrence.

*Table 3. Special cases of coincidence operation*

| $\{x(p_i)\}, \{x(p_j)\}$ | Description |
|---|---|
| $\pi_1=\pi_2,\ l_1=l_2,\ F_1–F_2=T/2$ | Coincidence operation of 2 waves (prototype – electromagnetic wave) |
| $\pi_1=\pi_2=\pi_3,\ l_1=l_2=l_3,\ F_1=F_2,\ F_1–F_3=T/8$ | Coincidence operation of three waves |
| $\pi_1=\pi_2,\ l_1=l_2=l\bullet, F_1=F\bullet,\ F_1–F_2=T/4$ | Coincidence operation of two waves and soliton wave |

## INTERFERENCE OPERATION (AXIOMATIC STATEMENTS 3.2., 3.7)

1. Within Interference operation only single or multi-component waves interact. They have properties that exclude superposition operation and coincidence operation. Conditions of Interference operation are as follows:

   **Condition 2.2:** Suppose $\cup_k \to P_i$, $\cup_m \to P_j$, $P_i$, $P_j \in \{P\}$, $l_k < l_m$, $\pi_k < \pi_m$. Therefore the necessary and sufficient conditions of Interference operation are as follows:

$\cup_k$, $\cup_m$||: (e>e*)∨(d>d*)∨((s>>1,~1)∨(p>>1,~1)) ∀i, j (40)

$\cup_k$, $\cup_m$⊥: (s=M)∨($\Delta\pi_\kappa \neq \Delta\pi_m$)∨($\cup_k$⟋∧$\cup_m$⟩), i=j (41)

$\cup_k$ and $\cup_m$: p=1, (42)

2. Special cases of Interference operation are given in table 4.

3. Interference operation influences the progress of bundle or convolution differently for different variants introduced in table 4.
4. Non-additivity and non-binarity of interference operation, as well as the absence of reverse operation, is secured by non-linear interaction of multi-component elements considering conditions and ***S.***

*Table 4. Special cases of interference operation*

| $\{x(p_i)\}, \{x(p_j)\}$ | Description |
|---|---|
| $\cup_k$ и $\cup_m$‖: | Nonlinear resonance, under $\cup_k$╯ and $\cup_m$╯ |
| $o_k=o_m$, | $\cup_k$) and $\cup_m$) |
| s=1 | $\cup_k$) and $\cup_m$) |
| s≈1, p≈1,F=const | Static fringe pattern, $Ai_j$, $\pi_{i,j}$ change |
| s>>1, $\cup_m$) | Consumption $\cup_m$; Ak, $\pi_k$ change and spread in a non-linear manner |
| s>>1, $\cup_k$╯ и $\cup_m$╯ | Reflection $\cup_m$; $Ai_j$, $\pi_{i,j}$ change and spread in a non-linear manner |
| $\cup_k$ и $\cup_m$⊥ | Properties remain the same |
| $\cup_k$ и $\cup_m$:<br>k=m,<br>π=1,<br>s=1,<br>e<e*,<br>d<d* | Coincidence operation of the consequent wave is retained; total π and $Ai_j$ are spread in a non-linear manner |

## DIFFRACTION OPERATION (AXIOMATIC STATEMENTS 3.4, 3.8, 3.9.)

1. Conditions to execute diffraction operation are operations for single and multi-component waves and soliton waves – are as follows:
   **Condition 2.4:** Suppose $\cup\bullet_k \to P_i$, $\cup\bullet_m \to P_j$, $P_i$, $P_j \in \{P\}$, $l_k < l_m$, $\pi_k < \pi_m$. Therefore, diffraction operation $\cup\bullet_k$ and $\cup\bullet_m$ occur in those cases only when the following is executed:

$\cup\bullet_k$, $\cup\bullet_m$‖: i≠j

$\cup\bullet_k$, $\cup\bullet_m$⊥: $(s \neq M) \vee (\Delta\pi_k \neq \Delta\pi_m) \vee (\cup\bullet_k$╯$\wedge \cup\bullet_m$)) $\forall i=j$ (43)

∪•k, ∪•m: for all pairwise parallel components $\cup\bullet_k$, $\cup\bullet_m$ (38) is executed.

2. It's correct that
   **Property 2.12:** Suppose diffraction operation occurs only for •╯, parallel or coincident ones for all components and each of them has a potential $\pi_i$ and movement velocity $v_i$. Then it is executed

$$\sum_{i=1}^{n} \pi_i v_i = \sum_{i=1}^{n} \pi_i v_i^* \quad (44)$$

where $v_i^*$ – movement velocity of soliton wave after diffraction operation.

4. Non-binarity and non-additivity of diffraction and absence of reverse operations are defined by the participation of several elements of different types considering conditions and superposition.
5. Special cases of diffraction operation are given in table 5.

*Table 5. Special cases of diffraction operation*

| $\{x(p_i)\}, \{x(p_j)\}$ | Description |
|---|---|
| $\bullet_k, \cup_m \parallel$: $s \sim 1$ | Waves bend round soliton waves and make diffraction operation pattern and change $\pi$, A |
| $s \ll 1$ | Soliton waves move through waves, $\pi_k$=const |
| $s \gg 1$: $\cup$) | Waves reflect; $\pi_m$, $A_m$ |
| $\cup$) | Disappearance of waves |
| $\bullet_k$), $\bullet_m$)$\parallel$ | Disappearance $\bullet_m$ |
| $\cup\bullet_k, \cup\bullet_m \perp$ | Properties don't change |
| $\cup\bullet_k, \cup\bullet_m$, $s=1$, $e<e^*$, $d<d^*$, $p=1$ | Coincidence operation of waves and soliton waves remains unchanged; $\pi_i, \pi_j, A_i, A_j$ are re-distributed in a non-linear manner |

## Classification of Self-Organization Scenarios for S-Space

In the above research, the main self-organization scenario of Sp was used, when space is created as a result of the interaction of one S with one O. For real systems, there is the interaction of one or several S with one or several O. Let's consider other variants of self-organization.

To see the whole picture let's present practically valuable cases in separate table 6.

The table shows that any combination of attributes is possible.

Let's briefly describe some of those scenarios.

*One S interacts with a set of {O}.* In this case, S-space is developed from one center in a different direction which number equals the number of objects. If the latter has similar differences from S, elements with identical properties emerge. As a result, interactions that are considered above are possible. As a consequence, the number of layer structures changes, their mutual membership, distribution of

*Table 6. Scenarios classification for bundle–convolution*

| Classification attribute | Scenarios | | | |
|---|---|---|---|---|
| quantity of {S} and {O} | *1S, 1O* | 1S, {O} | {S}, 1O | {S}, {O} |
| loss compensation $\pi$ | *absent* | incomplete | full | excessive |
| $\parallel$ or $\perp$ elements | *only* $\perp$ | *only* $\parallel$ | *both* $\parallel$, *and* $\perp$ | |

potential and properties related to it. “Topology” of S-space may be described as a tree-structured one with separately closed contours.

*Interaction of Set {S} with One O:* As long as S and O equally contribute to the bundle of Sp, this case is similar to the previous.

*Interaction of Set {S} with Set {O}:* Here S-space has several centers of development (by quantity S), with each having several different directions of development (according to the quantity O). Following the previous cases, interaction occurs and topological and metric properties change. In general, the “topology” of S-space is a plexiform one with separate tree-like structures.

*Partial Compensation of Potential Losses*: Within the development of one of the above scenarios quantity of elements in layers increases; a quantity of “exceeding” elements $\Delta n_i$ may be random; it is defined according to

$$\Delta n_i = \frac{\Delta\pi}{\pi_{in}}, \tag{45}$$

where $\Delta\pi$ – exceeding the part of potential; $\pi_{in}$ – the potential of an element of i- layer,and it is rounded off till the whole number.

*Full Compensation:* In this case, quantities of elements, as well as their potentials in the next layers, don’t change as compared to the current layer.

*Exceeding Compensation:* The quantity of elements in layers decreases; it may be a random one and it’s calculated in the same manner (45).

Consequently, compensation influences the number of elements in layers and evolution path of Sp – bundle occurs, or quite the opposite “convolution” does. It increases the total potential of Sp.

Let’s consider consequences of axiomatic statements in 1– 4 groups.

## Measuring Procedure

*Dimensionality r.* Link *r* with several perpendicular components, order and quantity of symmetry groups for Sp and for S- sets stipulated by conditions 2.2–2.4, properties, and its consequence. It is necessary word some special aspects of such understanding:

1. Single component wave (soliton wave) is considered one-dimensional; direction is measured along with amplitude. This implies:

   **Property 2.13:** In general, S- sets, and Sp are limited as for Universe. Their border is the surface of their structures.

2. It is necessary to distinguish *actual dimensionality* of Sp $r_a$, equal to the total amount of different perpendicular components which emerged within the bundle, and *potential* $r_n$, equal to the amount of different perpendicular components of potentially existing structures. There is:

   **Property 2.14:** $r_a$ Sp increases within bundle and decreases within convolution; $r_n$ decreases within bundle and increases within convolution, therewith for any stage of bundle or convolution it is executed

$$r_a + r_n = const \tag{46}$$

3. It's logical that dimensionality of emerged S- sets decreases within the bundle of Sp (consequence 1 in exposition 2.8.).
4. From (25, 27), and assumption of linear dependence *l, A* and $\pi$, it follows that

$$A_{i+1} = \frac{A_i \Phi}{n}, \tag{47}$$

where $n$ – index of $i+1$ layer.

Therefore, for a scenario when only perpendicular elements emerge, it's rational to define a length of coordinate segments proportional to amplitudes; for the first 10 numbers

$$\sum_{i=1}^{N} A_i \approx 1{+}0.\ 618{+}0.\ 382{+}0.\ 236{+}0.\ 146{+}0.09{+}0.\ 056{+}0.\ 035{+}0.\ 021{+}0.\ 013{=}2.\ 596$$

As long as, according to consequence 2 of exposition 2.3, there is a connection between relations $\Phi$ and the Fibonacci sequence with the quantity of elements $n$, and a sum of two terms of a series is bigger than the one of the following, so upper limit for $N=\infty$ as 1.2. 596+0. 034, i.e. dimensionality ≤1.2.630. So, the dimensionality of Sp may be considered as irrational (analogy with fractal object).

5. As within the bundle of Sp halves of waves of subjective and objective parts weaken, it's possible to interpret their dimensionality as the one having different attributes.

*Choice of a Reference System*: As noted above, a coordinate system (Cs) any set of soliton waves may be used which defines measurable space. The dimension of this space is less than the dimension Sp and is different from the dimension of the measured elements. This leads to non-representative measurements. To avoid it lets raise the following requirements to Cs:

- Suppose the measurements are made for elements of *k*- layer. Therefore Cs ={•)}, where • – soliton wave of the present layer.
- Suppose the measurements are made for elements of S- sets for one P. Therefore Cs ={$\bullet_n$)}, where $n$ – a number of stages which exist for the present P.
- Suppose the measurements are made for elements of all or part of S- sets {P}. Therefore Cs ={$\bullet_{in}$)}, where $i$ – numbers of all $P_i \in \{P\}$.

Such Cs will be *complete,* as long as they contain soliton waves of all layers (consequence 1 of properties 2.1); as the bundle is in progress they are refilled with new soliton waves, consequently, the number of their elements and dimensionality will increase. Other Cs will be called *incomplete.*

*Realization of Measuring*: Following axiomatic statements 4.1, 4.2, measuring represents diffraction operation. Therewith all special cases given in table 7 are realized. The fact that possible observe it means that solitons and waves which emerged as its result can be S- sets within following bundles

*Result:* Let's see how *{x (pi)}* is determined depending on Pc. There may be particular properties of S- elements: state, *l*, $\pi$; *A, F* and *o* of waves.

1. Perpendicularity of measured elements and soliton waves Cs are established upon the absence of Pc which is a consequence of permanence of Cs and elements after being measured.
2. Parallelism of measured soliton wave ) and soliton wave Cs are established by Pc and it represents as velocities changing of measurable element and soliton wave Cs. The potential of soliton wave is calculated with a formula which results from (44):

$$\pi_j = \pi_i \frac{v_i - v_{i*}}{v_{j*} - v_j}, \tag{48}$$

where * means changed the velocity of *i*– soliton Cs and j- of measured soliton wave.

3. The parallelism of wave and soliton Cs are established according to Pc and it means that condition or particular properties of the wave are changed. Special cases are given in table 7.

*Table 7. Measuring properties of waves parallel to soliton waves Cs*

| Property | Result | Assessment |
|---|---|---|
| Condition modality | Wave is retained<br>Wave disappears | )<br>( |
| Length of wave | Reflected from soliton wave Cs<br>Wave avoids soliton wave and diffraction operation picture emerges<br>Soliton wave moves through wave | $l\cup << l\bullet$<br>$l\cup \approx l\bullet$<br>$l\cup >> l\bullet$ |
| Potential of wave | Reflected from soliton wave Cs<br>Wave avoids soliton wave and diffraction operation picture emerges<br>Soliton wave moves through wave | $\pi\cup << \pi\bullet$<br>$\pi\cup \approx \pi\bullet$<br>$\pi\cup >> \pi\bullet$ |

4. Coincidence operation of measurable soliton waves and soliton waves Cs. In this case, Pc shows change $v_i$ for all components of measurable soliton. This operation takes place when Cs is incomplete.
5. Coincidence operation of measurable waves and solitons Cs is measured. Pc shows no changes $\{x(p_i)\}$ of waves which are measured. This operation takes place for complete and incomplete Cs.

## Comparison

Let's define analogies and differences between measurements in Euclidean space and S-space. As long as the measurements are decreased to the realization of particular groups of operations, let's compare the multitude {**DICS**} with the groups. In general, {**DIKS**} doesn't make groups, as it consists of non-binary operations, which result is affected by external factors via changes of condition modality of elements. Let's find out properties of similarity and difference as for associativity, the existence of single and reverse operations.

*Associativity*: The total of operations {**DIKS**} is not associative. For example, as a result of condition changes superposition operation of waves will give different results depending on their sequence. Only diffraction operation of pairs of solitons will be associative in all cases.

*Single Operation*: A single operation does not change the state of elements. A single operation takes place for the diffraction of pairs of solitons.

*The Reverse Operation:* which doesn't exist for {**DIKS**} in whole (according to the axiomatic statements 3.5- 3.8), it takes place for diffraction operation of pairs of solitons (axiomatic statement 3.9). Consequently, took place

**Property 2.15:** For the {**DIKS**} set of operations exists which makes a group – total of binary diffraction operations of soliton waves.

It doesn't make groups, {**DIKS**} has particular closeness:

- Operations Oi∈{**DIKS**} over ∪• gives the following result only ∪•.
- Operations $O_i$∈{**DIKS**} are commutated.

Let's mark "+" as presence and "- "absence of an attribute. The results of the comparison of properties for operations and measurements are in table 8.

Now let's compare the properties of the coordinate system and measurements. The results are given in Table 9.

*Table 8. Comparison of operation groups and {DIKS}*

| Attribute | Operations | {DIKS} | Diffraction Operation of Soliton Waves |
|---|---|---|---|
| Binarity | + | - | + |
| Closeness | + | + | + |
| Commutativity | + | + | + |
| Associativity | + | - | + |
| Existence:<br>Single operation<br>Reverse operation | +<br>+ | -<br>- | +<br>+ |

*Table 9. Comparison of coordinate system and measurements*

| Point of Comparison | Euclidean Space | S-space |
|---|---|---|
| Coordinate-system | A fixed multitude of elements of a particular type;<br>r= dimensionality of space;<br>Nothing changes within measuring;<br>an equal weight of coordinate axes | An evolving set of soliton waves;<br>r≤ dimensionality of space;<br>changes within measuring; the unequal weight of coordinate axes |
| Measuring | The procedure is similar;<br>Repeated results are similar | Two types of procedures;<br>Results of repeated measurements differ |
| Execution of metric axiomatic statements of symmetry, triangle, identity | Executed | Executed only in particular cases |

## INVARIANTS AND CLASSIFICATION OF SELF-ORGANIZATION SCENARIOS

**Calibration Invariants and Calibration:** Bundle-convolution of Sp may run according to similar or identical scenarios which are reflected in similarity or identity of properties. The following properties are often similar: membership and consequence of how elements form, distribution of $\pi$ and t, groups and symmetry patterns, etc. Let's call them calibration invariants similar to the notion in physics. Bundle having identical sets of invariants will differ as for absolute values and measuring units – calibration.

**Classification of Object Models According to Invariants:** As practice shows different tasks require models having varied sets of invariants, so, let's distinguish:

1. *Models:* Static models which retain:
   - Membership and consequence of the unification of elements i.e. bundle-convolution scenarios.
   - The number of elements and operations.
   - Dimensionality.
   - Groups and patterns of symmetry.
   - Distribution of relative values $\pi$, $t_0$, etc.

The form of such models is *C- graph* – a connected and unoriented graph, which tops correspond to potential and actual S- sets and S-elements, and its edges show the relation of their membership.

*PC- diagram* (Figure 3) – oriented consequence on C- graph which corresponds only to actual S-sets and S-elements, as well as relations of their membership. The diagram of the opposite orientation corresponds to convolution. In general, about graphs (Harari, 1973)

Using conventional symbols, they may be applied to reflect other properties.

*Figure 3. View of PC- diagram*

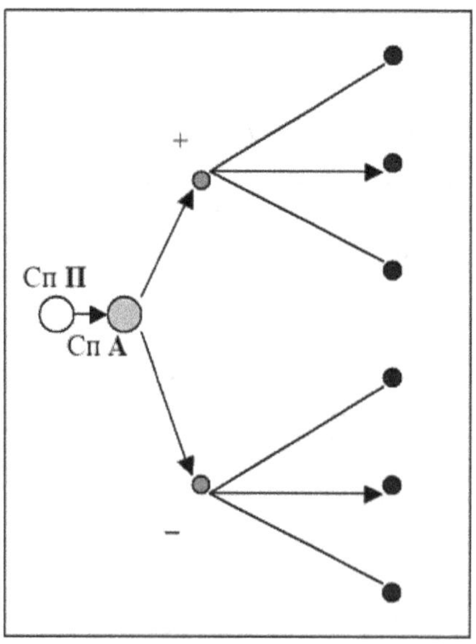

2. *Models 2:* Dynamic models which keep the following properties apart from invariants of model 1:

    ◦ Unambiguity of the adequacy of properties sets.
    ◦ Dynamics of structural changes for Sp.

This class includes varied analogous models which make it possible to reconstruct dynamics of wave processes due to their representation in form of waves with frequencies, amplitudes, etc. which are related to the initial ones in particular scales.

3. *Models 3*: Imitation models that partially or completely retain absolute values of system parameters.

These are models that reflect the values of properties in units that are typical for them. Models 3 as compared to previous ones contain qualitatively different components; they are the most similar to real systems.

## CONCLUSION

A comparison of a human organism structure and its interaction with an environment with properties of the complex system specified that human is a complex and open system and to make modeling if their mind and evolution respective methods are necessary.

It is shown, that existing modeling methods of complex and open systems have contradictions with properties of real systems in its axiomatic base. That's why it is necessary to develop new methods and models where these contradictions are eliminated.

The axiomatic wave model of S-space doesn't have such contradictions as it complies with all properties of complex and open systems.

This model represents an open and integral space, waves and solitons as elements of space, various modalities of elements, nonlinear operations of the interaction of elements, measurement procedures. The properties of space, elements, operations, and measurements are described by 20 axioms that make up 4 groups.

The self-organization scenarios of S-space are substantiated, which differ depending on the initial conditions. For the self-organization scenario (1S, 1O), calculations of the main parameters are given.

Three types of models of self-organization are introduced - expressing the sequence of a bundle of S-space; dynamic characteristics of bundle; imitation models.

## FUTURE RESEARCH DIRECTIONS

In the next chapter, the wave model of S - space and theory of self-organization of complex systems will be tested: how suitable they are for describing and explaining the features of human evolution given in the first chapter. It will also test its capabilities for modeling human interactions with the environment, including for clarifying the sequence, priorities, and number of interaction channels.

It is proper to use model 1 from three types of complex system models considered to simulate the human mind and human evolution.

Applying dynamic and imitation models may give more precise results but such work is stipulated for the future.

## REFERENCES

Abbott, L., & Sejnowski, T. J. (Eds.). (1999). *Neural Codes and Distributed Representations. Foundations of Neural Computation, Bradford Book*. Cambridge, MA: MIT Press.

Actistudy.info-vashhestvo. (n.d.). Retrieved from http://www.activestudy.info/pishhevye-veshhestva-i-ix-znachenie-dlya-organizma/

Aif-archive161698. (n.d.). Retrieved from http://www.aif.ru/archive/1671698

Anokhin, P. K. (1978). *Selected Works. Philosophic Aspects of Functional System Theory. Moscow, Russia:* Nauka.

Beauregard, M. (Ed.). (2003). *Consciousness, emotional self-regulation, and the brain*. Amsterdam, The Netherlands: Benjamin.

Capra, F. (1999). *The Tao of Physics*. Boston, MA: Shambhala.

Carter, R. (1998). *El nuevo mapa del cerebro*. Barcelona, Spain: RBA Ediciones.

Carter, R. (2014). *The Human Brain*. New York: DK Publishing.

Chalupa, L. M., & Werner, J. S. (Eds.). (2004). The Visual Neurosciences, 2 vol. Cambridge, MA: Bradford Book.

ChungS. J. (2014). Parallels between Confucian Philosophy and Quantum Physics.

Damasio, A. (1999). *The Feeling of What Happens: Body and Emotion in the Making of Consciousness*. Florida: Harcourt, Inc.

Edelman, G. M. (1989). The Remembered Present: A Biological Theory of Consciousness. New York: Basic Books.

Edelman, G. M. (2001). Building a Picture of the Brain. In G. M. Edelman, & J.-P. Changeux (Eds.), *The Brain*. London, UK: Transaction Publishers.

Fuster, J. M. (2003). *Cortex and Mind. Unifying Cognition*. Oxford, UK: Oxford University Press.

Gegenfurtner, K. R., & Sharpe, L. T. (Eds.). (1999). *Color Vision. From Genes to Perception*. Cambridge, UK: Cambridge University Press.

Georgieva, S. A., Belikina, N. V., Prokofjeva, L. I., Korshunov, G. V., Kirichuk, V. F., Golovchenko, V. M., & Tokaeva, L. K. (1981). Human physiology. Moscow, Russia: Meditsyna.

Gilbert, D. (1948). Foundations of Geometry. Moscow, Russia: OGIZ.

Gilbert, D., & Berneis, P. (1982). Foundations of Mathematics, vol. 1-2. Moscow, Russia: Nauka.

Gödel, K. (1931). Über formal unentscheidbare Sätze der Principia Mathematica und verwandter Systeme, I. *Monatshefte für Mathematik und Physik*, *38*(1), 173–198. doi:10.1007/BF01700692

Haken, G. (1985). Synergetics: a hierarchy of instabilities in self-organizing systems and devices. Moscow, Russia: Mir.

Hameroff, S. (2006). Consciousness, Neurobiology and Quantum Mechanics: The Case for a Connection. In J. Tuszynski (Ed.), *The Emerging Physics of Consciousness*. Berlin, Germany: Springer-Verlag. doi:10.1007/3-540-36723-3_6

Hameroff, S., & Penrose, R. (1996). Orquestrated Reduction of Quantum Coherence in Brain Microtubules: A model for Consciousness. *Journal of Consciousness Studies*, *3*, 36–53.

Hands, J. (2015). *Cosmosapiens: Human Evolution from Origin of the Universe*. New York, NY: Overlook Duckworth.

Harari, F. (1973). Graph Theory. Moscow, Russia: Mir.

Harris, L., & Jenkin, M. (Eds.). (2007). *Computational Vision in Neural and Machine Systems*. Cambridge, UK: Cambridge University Press.

Heyes, C., & Huber, L. (Eds.). (2000). *The Evolution of Cognition. Cambridge, MA: Bradford Book.*

Human Brain Project. (n.d.). Retrieved from https://www.humanbrainproject.eu/en/

Ivanitskiy, M. F. (2008). Human Anatomy (with foundations of dynamic and sporting morphology): Guide for Institute of Physical Culture. Moscow, Russia: Olimpiya.

Kauffman, S. (1993). *The Origins of Order: Self-Organization and Selection in Evolution*. Oxford, UK: Oxford University Press.

Kauffman, S. (1995). *At Home in the Universe: The Search for the Laws of Self-Organization and Complexity*. New York: Oxford University Press.

Kleene, S. (1957). Introduction to metamathematics. Moscow, Russia: Ed. John lit.

Kleene, S. (1973). Mathematical Logic. Moscow, Russia: Nauka.

Kleene, S., & Wesley, R. (1978). Foundations of Intuitionistic Mathematics. Moscow, Russia: Nauka.

Klein, F. (1956). A comparative review of the latest geometric research ("Erlangen program"). In *On the foundations of geometry. A collection of classical works on Lobachevsky geometry and the development of its ideas,* Moscow, Russia, pp. 399-434.

Korobko, V. I. (1997). Golden proportion and harmony issues of systems. Moscow, Russia: Publisher. Association of Building Universities of CIS states.

Kovalyov, Y. N. (1996). *Geometric modeling of human-machine systems: the development of apparatus. Kiev, Ukraine*: KMUGA.

Kovalyov, Y. N., & Grebenyuk, M. F. (2007). C-space description languages. Applied geometry and eng. Graphics, 77, pp. 29-35.

Kuzhekina, Y. I., Voronina, M. A., & Lukoyanova, A. S. (2016). The anthropic principle in cosmology. *Research and methodological online journal. Concept*, *2*, 531–535.

Libet, B. (2004). *Mind Time: The temporal Factor in Consciousness*. Cambridge, MA: Harvard University Press.

Lomov, B. F. (1979). Systematicity as a principle of mathematical modeling in psychology. *Matters of cybernetics,* 50, 3-18.

Lossky, N. O. (1992). *Intuitionism*. Moscow, Russia: Progress.

Luria, A. R. (1985). *Sensation and Perception*. Moscow, Russia: Publisher. MGU.

Mesarovic, M., Mano, D., & Takahara, Y. (1973). Theory of hierarchic multi-level systems. Moscow, Russia: Mir.

Mesarovic, M., & Takahara, Y. (1978). General system theory: mathematical foundation. Moscow, Russia: Mir.

Miller, E. K., & Cohen, J. D. (2001). An Integrative Theory of Prefrontal Cortex Function. *Annual Review of Neuroscience*, *24*(1), 167–202. doi:10.1146/annurev.neuro.24.1.167 PMID:11283309

Mkhitaryan, N. M. (2016). Man, and Dwelling. Gangemi Editore International Publishing.

Mkhitaryan, N. M., Badeyan, G. V., & Kovalyov, Y. N. (2004). *Ergonomic Aspects of Complex Systems. Kiev, Ukraine*: Naukova Dumka.

Monserrat, J. (2001). Engramas neuronales y teoría de la mente. *Pensamiento, 57*(2001), 176-211.

Monserrat, J. (2007), Neural Networks and Quantum Neurology: A Speculative Heuristic Towards the Architecture of Psyquism. In IWINAC 2007, 1, 1-20. Springer. doi:10.1007/978-3-540-73053-8_1

Olsen, S. (2006). The Golden Section, Nature's Greatest Secret", Wooden Books, Ltd.

Orlando, D. G. A. (2015). An introduction to universal, quantum-galactic mathematics. Retrieved from https://www.researchgate.net/profile/Giovanni_Orlando/publication/331928401_Orlando_Intro_Galactic_Math/links/5c935ae745851506d71ff7ba/Orlando-Intro-Galactic-Math.pdf

Pavlov, I. P. (1949). *Selected works. Moscow, Russia:* Academy of Sciences of USSR.

Penrose, R. (1994). *Shadows of the Mind: A Search for the Missing Science of Consciousness*. Oxford, UK: Oxford University Press.

Penrose, R., & Hameroff, S. (2011). Consciousness in the Universe: Neuroscience, Quantum Space-Time Geometry and Orch or Theory. In R. Penrose & ... (Eds.), *Consciousness and the Universe: Quantum Physics, Evolution, Brain, and Mind. Cambridge* , MA: Cosmology Science Publishers.

Polesn_inters_anatomy. (n.d.). Retrieved from http://www.polezen.ru/interes/anatomy.php

Radhakrishnan, S. (1956-1957). Indian Philosophy. Vol. 1-2. Moscow, Russia.

Schwartz, J. M., Stapp, H. P., & Beauregard, M. (2005). Quantum Physics in Neuroscience and Psychology: A Neurophysical Model of Mind/Brain Interaction. *Philosophical Transaction of the Royal Society of London, Series B*, *360*(1458), 1309–1327. doi:10.1098/rstb.2004.1598 PMID:16147524

Tarski, A. (1931). Sur les ensembles définissables de nombres réels I, *Fundamenta Mathematicae, 17,* 210-239 in Tarski (1983), 110-142.

Tarski, A. (1936). Grundlegung der wissenschaftlichen Semantik, Actes du Congrès international de philosophie scientifique, Sorbonne, Paris 1935, vol. III, Language et pseudo-problèmes, Paris, Hermann, 1936, pp. 1–8 in Tarski (1983), 401-408.

Todeschini, M. (1949). La Teoria delle Apparenze. Istituto Italiano D'Arti Grafiche.

Tononi, G., Boley, M., Marcello, M., & Koch, C. (2016). Integrated Information Theory: From Consciousness to the Physical Substrate. *Nature Reviews. Neuroscience*, *17*(7), 450–461. doi:10.1038/nrn.2016.44 PMID:27225071

Urmantsev, Y. A. (1974). The symmetry of Nature and Nature of Symmetry: Philosophic and scientific aspects. Moscow, Russia: Mysl.

Urmantsev, Y. A. (1988). Evolution study or General system development of nature, society, and mind. Puschino: Institute of phytophysiology Academy of Sciences of USSR.

von Bertalanffy, L. (1950). The Theory of Open Systems in Physics and Biology. *Science*, *111*(2872), 23–29. doi:10.1126cience.111.2872.23 PMID:15398815

Voronin, A. N., Ziatdinov, Y. K., Charchenko, A. V., & Ostashevkiy, V. V. (1995). *Complex technical and Human-machine systems: research methods. Germany:* Fakt.

Wikipedia. wiki. Complex. (n.d.). Retrieved from https://ru.wikipedia.org/wiki/Сложная_система

Wikipedia. wiki. Complex. (n.d.). Retrieved from https://ru.wikipedia.org/wiki/Системный_подход

Wolff, N., & Hameroff, S. (2001, November). A Quantum Approach to Visual Consciousness. *Trends in Cognitive Sciences*, *5*(11), 472–478. doi:10.1016/S1364-6613(00)01774-5 PMID:11684479

Zidermane, A. A. (1988). *Certain questions of chronobiology and chronomedicine.* Zinatne.

## KEY TERMS AND DEFINITIONS

**?,•:** S-elements: wave, soliton.

**?:** S-set.

**{DICS}:** All Sp operations.

**||:** S-elements parallelism.

**A, P Modalities of Existence:** Actual and potential of S-space, S-sets, and S-elements, rules of their mutual transition, depending on the effects of the external factors.

**C:** Coincidence operation.

**Co:** Convolution.

**Cs:** Coordinate system.
**D:** Diffraction operation.
**HME:** Human-machine-environment system.
**I:** Interference operation.
**IDS:** Information display system.
**IIDS:** The intuitive information display system.
**L:** Layering.
**ME:** "Man-environment" model.
**Mr:** Measurement result.
**PM:** Perception model.
**RDM:** Recognition and decision model.
**S:** Subject.
**S:** Superposition operation.
**Sp, °S-space:** Axiomatic wave model of a complex system. Includes 17 axioms that define the structure of space, state, and interaction of elements, the procedure for measuring.
**TF:** Target function.
**H Observation:** The influence of external factors (for perception and recognition tasks mostly).
**O:** Object.
**OM: Object Model:** Graph or layering-convolution LC-diagram, which show some invariants of S- space structure and evolution.
**⊥:** S-elements perpendicularity.
**π:** Potential - "captured" part of the universe; intuitive characteristic of the ability to interactions; is expressed in conventional units.
↗, ↘, ↓ **Modalities of Condition:** Expansion, weakening, disappearance.

## ENDNOTES

1 The similarity of ideas of ancient and modern scientists gives reason to think about the limits of human knowledge. As Ecclesiastes said: "Blowing toward the south, Then turning toward the north, The wind continues swirling along; And on its circular courses the wind returns." (Ecclesiastes 1:6 New American Standard Bible). Recall, that mental activity in natural philosophy was associated with the primary element of air.

2 According to general definition, (Wikipedia.wiki. Complex, n/d), complex system – is a system consisting of multiple interacting components (subsystems), thus complex system gains new properties which are absent on subsystem level and can't be named as features of subsystem level. Therefore, the system approach is necessary (in accordance with (Wikipedia.wiki. Complex, n/d), «system approach – a direction of the methodology of scientific cognition which is based on the object being considered as a system: an integral complex of interacting objects; a multitude of interacting objects; a total of beings and relations»).

3 For the first time psychical prototype of soliton wave was described by G.C. Russel, as a single wave, which behaves as a particle in many aspects (localization, motion without deformation, energy and impulse transfer, it retains structure within contact).

4 Other variants of interactions will be considered below, and summary classification is given in table 6.

5 The characteristic is also used, 1,618…, which equal to Φ+1, or $\Phi^{-1}$; the following relation exists between Φ and π: $\pi^2/4^2 \approx 0.617 \approx \Phi$.

# Chapter 3
# Self-Organization:
## Body and Mind

## ABSTRACT

*In the previous chapter, theoretically possible scenarios of self-organization of complex systems were described. Now, the one that is most suitable for modeling human consciousness are chosen. The validity of the choice is determined by the correspondence of the scenario to the facts related to the structure of the body, mental organization, internal self-regulation, and interactions with the environment. Verification shows that it is necessary to choose a scenario (1S, 1O). Based on it, the "human-environment" model is constructed, which is a 1-type model. Within the framework of this model, the characteristics of the levels and channels of human interaction with the environment, the mechanisms of the occurrence of exchanges and perceptions, the limits of self-regulation, and the definitions of important characteristics of the human personality are determined. The findings discuss some of the epistemological consequences of adopting the human-environment model.*

## BACKGROUND

First of all, it is a need to determine the facts that allow choosing the scenario of self-organization correctly. For that purpose, consider the data of several sciences:

- Anatomy related to the general plan of the body structure (Sinelnikov & Sinelnikov, 1996).
- Physiology related to the organs of perception and mechanisms of self-regulation (Georgieva et al., 1981; Ivanitskiy, 2008).
- Psychology related to the processes of perception and mental activity (Fechner, 1882; Arnheim, 1974).
- Physics related to the origin and distribution of light and sound, as well as conservation laws (Morz, 1949; Gorelik, 1959; Luria, 1985; Matveev, 1985; Gegenfurtner & Sharpe, (Eds.),1999).

DOI: 10.4018/978-1-7998-1706-2.ch003

When a self-organization scenario is defined and a model of the "human-environment" system is built, it will become possible to determine the characteristics of the levels and channels of human interaction with the environment, the mechanisms of occurrence of exchanges and perceptions, and the limits of self-regulation of the body. This is the main task of the chapter.

It should be remembered that a human is a holistic system, and all certain characteristics must comply with the conditions of unity.

Numerous studies are devoted to the determination of factors that contribute to the preservation of the unity of the system or are signs of unity. First of all, these are various symmetries (Weil, 1968; Urmantsev 1974, 1988; conservation laws related to them (E. Noether, 1918) harmonious relations, in particular, "the golden proportion" (Vasiutinskiy, 1990, Korobko, 1997; Bodnar, 2005), Fibonacci numbers and ratios derived from them (Soroko 1984; Olsen, 2006). Thus, verification of the model under the conditions of the unity of the system can be carried out.

The characteristics of the human mind are closely related to the main problem of epistemology - the question of the cognoscibility of a human and the world. This question has been raised in philosophy, psychology, theology ever since they arose (Aristotle, 1976-1983; Plato, 2006; Radhakrishnan, 1956-1957). And different schools defended opposing points of view - from the complete negation of the world's cognoscibility in view of its illusory nature, to a recognition of the fundamental possibility of knowing it (Hegel, 1913, 1916; Descartes, 1989-1994; Leibniz, 1982-1989; Locke, 1985-1988; Losskiy, 1992; Spinoza, 1999). The discussions of the Grand Unification, Anthropic Principle, and Artificial Intelligence, performed by modern scientists, are echoes of the controversies of ancient and medieval scientists and show that the issue is still far from being resolved, and, therefore, the insight resulting from the theory of self-organization is relevant.

## SUBSTANTIATION OF THE SCENARIO CHOICE

First, let's consider the general plan of the human body structure. It is simple and clear: the body consists of a symmetrical head and torso; each half of the body has a pair of limbs; each limb consists of three parts (shoulder, forearm, hand and hip, lower leg, and foot, respectively), connected by movable joints and ends with five fingers and toes consisting of three phalanges also connected by joints.

This structure is represented by the following graph (Figure 1):

*Figure 1. A general plan of the human body structure*

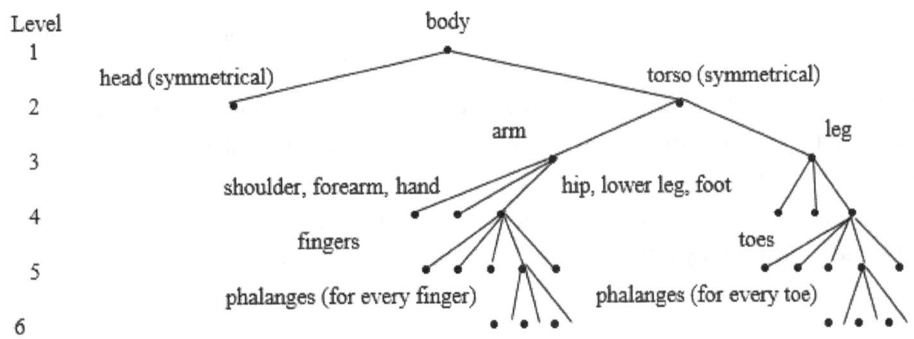

*Figure 2. Dependence of primary colors on light wavelengths (Symbiosis-online-publishing, n/d)*

| Color | Wavelength = nm (billionths of a meter) | |
|---|---|---|
| *Ultraviolet* | 280-380 | |
| Violet | 380-430 | |
| Indigo | 430-450 | |
| Blue | 450-480 | |
| Green | 480-530 | Human Visible Light Spectrum |
| Yellow | 530-580 | |
| Orange | 580-640 | |
| Red | 640-750 | |
| *Infrared* | 750-1000 | |

This graph is an implementation of the scenario (1 S, 1O), considered in the previous chapter. Now let's see if mental structures of a human group in a similar way, and thereby confirm or deny the assumption made in the previous chapter. First of all, let's consider the facts related to physics, physiology, and psychology of perception.

Sight and hearing are the basic perceptions of a human. Let's consider them in more detail.

Let's start with the *perception of colors* (Figure 2-3) (Ragain & James, 2015). There are seven of them. Let's see, if physical prerequisites are explaining why it is seven, or maybe it is because of the eye structure or physiology of perception?

1. The physical nature of color. Color depends on the wavelength. The wavelength of light is determined by the temperature of the radiating body. One should not forget about the corpuscular nature of light, which determines the discreteness of radiation, depending on the energy level of photons. Since there are many such levels, in reality, it is possible with some simplification to assume that a series of wavelengths in the part of the spectrum perceived by the eye from violet to red is continuous. Thus, in the series of waves, there are no peaks or dips that separate the ranges corresponding to the primary colors from each other.
2. Eye sensitivity (Georgieva et al., 1981). A human eye has about 110-130 million receptors - "rods" that ensure the perception of light in general, and only 5-7 million "cones" that ensure the perception of color.

The eye is not able to perceive a still image, therefore, even when somebody looks at one point, his eyes still move, making from 20 to 70 movements with a minimum amplitude per second.

In a perfectly transparent atmosphere, rods of the retina could react to the light of a candle located at a distance of more than 30 kilometers. The eye can distinguish between 130-250 pure color tones and 5-10 million mixed colors.

There is only one maximum in the field of yellow-green color, rather than seven, as one would expect. On the other hand, a human eye can distinguish between 130-250 pure color tones and 5-10 million mixed colors.

*Figure 3. The sensitivity of the eye to light waves of different lengths (different for day and twilight vision) (My-3-sotki, n/d)*

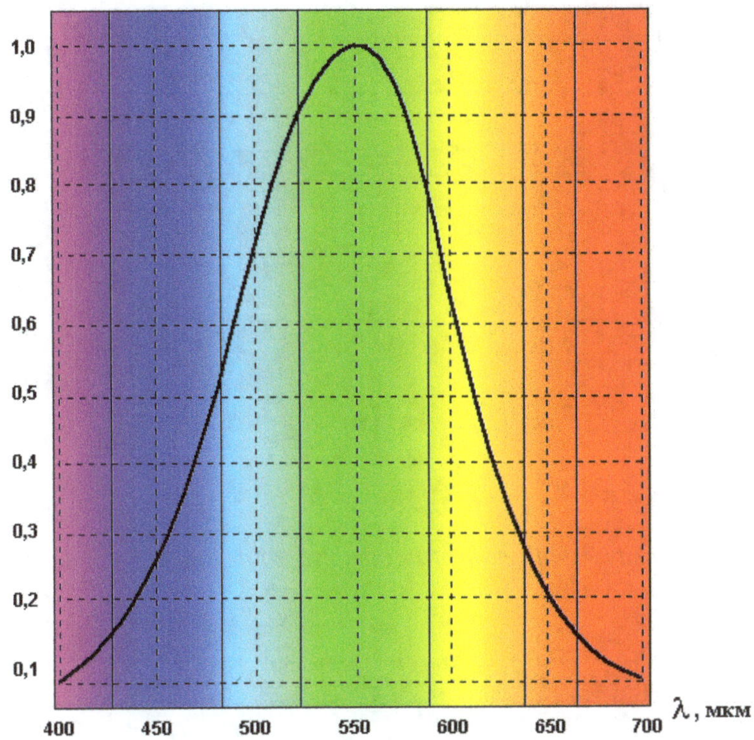


Thus, there are neither physical nor physiological prerequisites for selecting exactly seven colors. Adding that yellow-green color (chartreuse), corresponding to the best perception to the seven primaries, and thus it turns out that there are eight primary colors.

Interestingly, the following colors were considered primary in different historical eras:

- Two (dark and light or warm and cold).
- Three (preserved in many national flags, in the RGB system, etc.).
- Five (corresponded to elements or primary elements).

*The perception of sound* (Ramirez & Herbig, 2016). A human perceives sound nonlinearly; there are areas of relatively more and less high sensitivity compared to the sound pressure level measured with instruments (the first is measured in backgrounds, the second - in decibels, Figure4).

A human can distinguish between characters of the sound being perceived (Figure 5).

A human ear can capture sounds at a frequency of 10-20 Hz to 15-20 kHz and is most sensitive to the range of 2000-2300 Hz. All humans distinguish 3-4 thousand sounds of different heights.

Speech range 1-3 kHz. The best ear for music (the ability to distinguish height) is in the area of 80-600 Hz. Here an ear can distinguish, for example, two sounds at a frequency of 100 Hz and 100.1 Hz.

As with color perception, recognition of tones as high and low (2 gradations) or high, medium and low (3 gradations) is in no way related to the objective level of continuously variable sound pressure,

*Figure 4. Sound pressure level and sensitivity of organs of hearing (Biology-stack-exchange, n/d)*

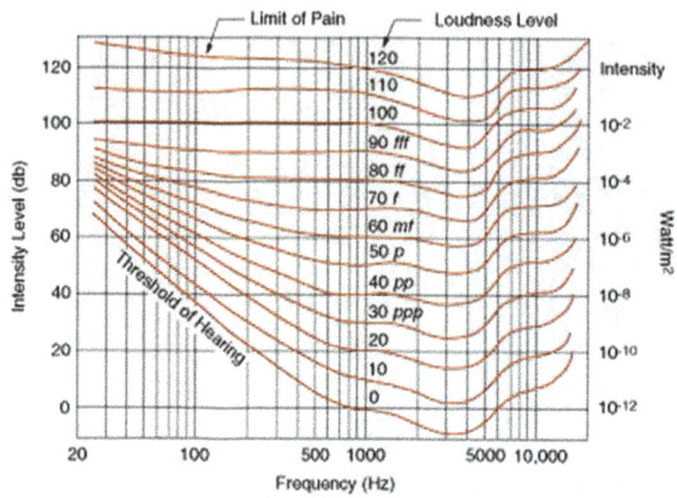


*Figure 5. The frequency-intensity visualization of the human audible range (Entanda-audiology-news, n/d)*

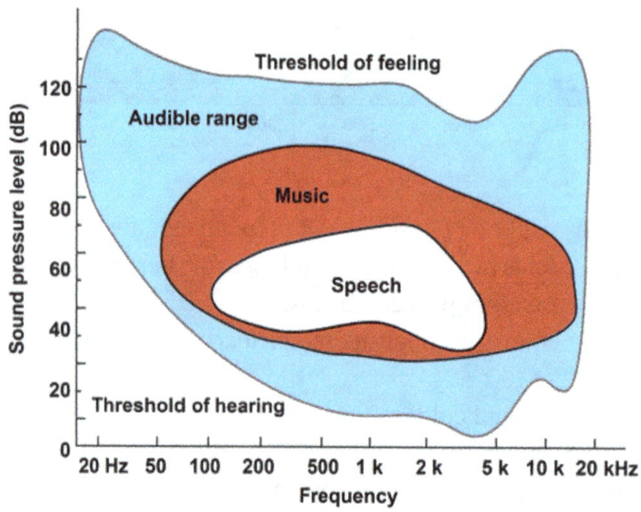


nor with the maxima and minima of the sensitivity of the organs of hearing. In music, only eight basic tones (octave) are distinguished, with continuous changes in pitch.

*Other sensory perceptions* are less studied. It is known, for example, that there are about 900 taste buds in a tongue, and their sensitivity depends on temperature.

The area of the olfactory zone of the nose is 5 square centimeters. There are about a million olfactory nerve endings. For an impulse to occur in the olfactory nerve fiber, about 8 molecules of an odorous substance shall come to its end, and for a sensation of smell - at least 40 nerve fibers should be excited.

A human also classifies these perceptions regardless of the physical effect and sensitivity of receptors. Different sources give a different number of tastes, aromas, etc., what is the evidence of a lack of clear perceptions. This is not surprising, given their generally minor nature.

*Sense organs*: Some receptors perceive hunger and thirst, as well as those in charge of balance, coordination, and orientation. They are not grouped into separate organs, but they send impulses to the human brain, which classifies them upon reaching a certain threshold of sensitivity. Different references treat their number and classification differently (from 21 to 33).

But there are only five external organs, and this is also inexplicable from physics or physiology. Why, for example, sight, hearing, touch - different feelings, if they all perceive external vibrations?

*Perception of space*: A human eye is a very sensitive instrument; there is evidence that it can perceive even single photons. Then, taking into account the difference in the size of such objects as (Figure 6):

- Human body – 1
- Body cell – $10^{-5}$
- Simple molecule – $10^{-9}$
- Atom – $10^{-10}$
- Atom nucleus – $10^{-14}$

It's should be expected that human does not see each other as shown in Figure 6, and probably as shown in Figure 7. Nevertheless, humans see each other exactly as shown in Figure 7, and it needs explanation.

The physicists do not give any explanation for the phenomenon of three-dimensional space. Moreover, if I. Newton considered space as three-dimensional, then A. Einstein's special theory of relativity is defined in the four-dimensional space-time continuum (G. Minkowski's space), and T. Kaluza operated

*Figure 6. Perception of volumetric objects as three-dimensional bodies in three-dimensional space (Wikipedia. Turinskiy avtoportret, n/d)*

*Figure 7. Perception of volumetric objects as discrete luminous points (Popmech-Ddesign-news, n/d)*

with five-dimensional space (adding the fifth dimension to the four-dimensional continuum allowed separating Maxwell's equations from other equations of general theory of relativity) (Wikipedia. Teoria Kalutsi, n/d). According to the theorem of E. Noether, there is a connection between symmetries of various types and conservation laws (Table 1) (Wikipedia. Teorema Neter, n/d). Modern physicists have calculated that the implementation of all known conservation laws requires ten-dimensional space (theory of strings) (Becker, Becker& Schwarz, 2007), and even twenty-six dimensions (theory of bosonic strings) are considered in some theories. Therefore, it is necessary to find an explanation for the perception of volumetric bodies as integral and three-dimensional in three-dimensional space.

The same is true for perception of time – human divide it into the past, present, and future. And just as space measurements are not completely equal (the third is less accessible due to gravity), so the present is less accessible to us in comparison with the past and future[1].

*Binary judgment* (good-bad, yes-no, black-white, etc.) is the basis of mathematical logic. However, its artificiality, relativity, and incompatibility in many cases of reality were clarified in ancient India (darshans of Nyaya and Vaisheshika), ancient China (Taoism) and ancient times (a series of Zeno's paradox). Also, modern image recognition specialists within the framework of mathematical logic can-

*Table 1. Correspondence of symmetries, invariances and conservation laws according to the theorem of Neter (Wikipedia. Teorema Neter, n/d)*

| Transformation | Invariance | Conservation Law for |
|---|---|---|
| Time broadcast | Uniformity of time | Energy |
| Space broadcast | Homogeneity of space | Impulse |
| Rotation of space | Isotropy of space | Moment of impulse |
| C, P, CP, and T symmetry | Isotropy of time | Parity |
| Lorenz group | Lorentz covariance | Mass center motion |
| Calibration conversion | Gauge invariance | Charge |

not solve the issue of "transforming a tadpole into a frog", but they are forced, as in many other cases, to use fuzzy logic, introduce categories of feature classes, etc. However, the binarity of valuation and conclusions is quite persistent and needs to be explained.

*Unity of truth*: A human is inclined to complete his/her judgments with an only true statement - God is one, the Universe is one, any decision is the only right one, etc. In most cases, such a reduction is not based on any objective facts and evincible judgments but responds to our intuitive feeling.

Knowing these ratios, one may determine the correspondence of perceptions with the following PC-graph (Figure 8):

This graph is also an implementation of a scenario (1S, 1O).

Comparing the topology of the graphs in fig.1 and 6, it is possible to claim that they are common at levels 1-5 - available 6$^{th}$ level and different numbers of elements at this level.

From here follow the following conclusions:

1. The formation of S-sets (body parts, levels) and a number of elements (channels - 1, 2, 3, 5) corresponds to the calculated one. There is uncertainty with the quantity of proprioceptive and interoceptive sensations, as they are understudied.
2. The sixth level of interaction with the environment is just being formed; the most advanced exteroceptive sensations of tones tend to number 8. This shows that the formation of the level has not finished yet - the number of elements is less compared to level 5, where interaction with the environment does not require any development.
3. A much greater similarity between a mental component of a human and structure of his/her body compared to the ratio of the mental component and structure of the body of animals (examples are discussed in the previous chapter), allows making the following conclusions:
    - The balance of mind and body (integrity) allows a more complex impact on the environment at the level of certain individuals[2].
    - This balance is an evolutionary advantage of a human that provides an increase in the probability of survival.

*Figure 8. Grouping of human perceptions*

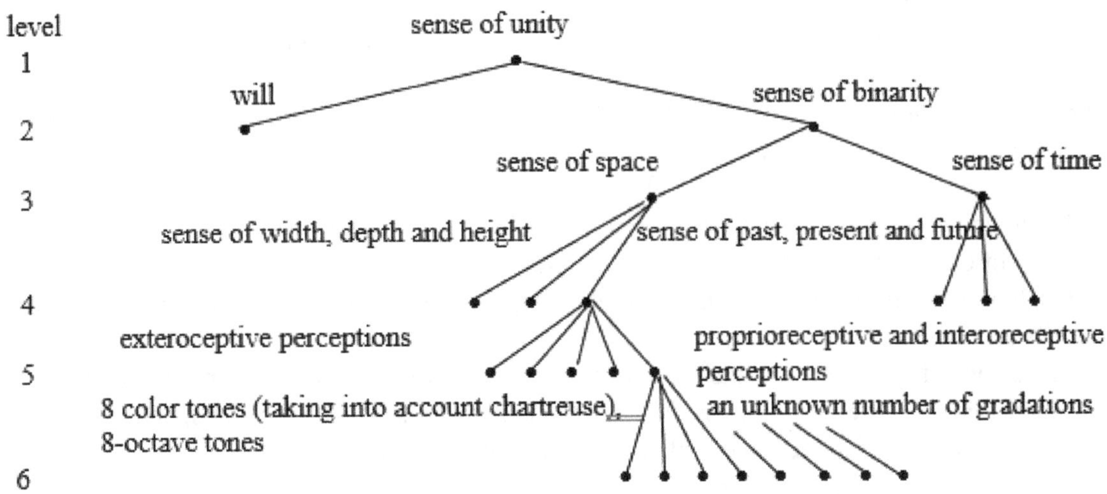

- An advantage of Homo sapiens versus other hominids probably refers to more perfect "settings" of channels and levels, which can be determined by their qualitative and quantitative correspondences to the signs of integrity - but this is still an assumption not supported by appropriate calculations.
- Striving for the most complete correlation of the body and mentality, expressed by the scenario (1S, 1O), as corresponding to the general laws of self-organization, is the most important factor of evolution, determining, along with the environmental conditions, the direction of natural selection.

## HUMAN-ENVIRONMENT MODEL

Now let's move to the construction of a human-environment model. But what is the scenario for the self-organization of the world around us? To answer, it is possible to use the anthropic principle: implementation of the scenario (1S, 1O) for a human would be impossible if this led to a serious inadequacy of the environment. Then the evolutionary advantages listed above would turn into disadvantages and natural selection would lead to the rapid extinction of people.

On the contrary, by generating genetic mutations and "pressure of natural factors", the environment "organizes" human evolution "in its image", that is, by implementing the scenario (1S, 1O).

When modeling relationships in the human-environment system, it is necessary to proceed from the following provisions:

1. The human-environment system is considered from the perspective of synergetics and the general theory of systems as an integral, complex, open system, in a state of dynamic equilibrium. The environment refers to the "close", cognoscible world, directly affecting processes of human life. The environment boundaries and system elements are not clearly defined.
2. The system is modeled as S-space; self-organization - as a set of bundle-convolution scenarios; components, relations, interactions - as S-elements, S-sets, relations, and operations displayed in the form of model 1.
3. The set of bundle-convolution scenarios is determined by elements of {S} and {O} - "remote" parts of the universe.
4. Invariants and details are determined by the specific features of the tasks.

The human-environment model (HEM) can be considered realistic provided that it can describe and predict such facts and phenomena:

- Heterogeneity, autonomy and, at the same time, the integrity of the system components (homeostasis of one's organism, self-sufficiency of nature).
- Qualitatively different interdependent levels of organization (physiological and mental processes, substance and flight).
- Prevalence of symmetry, constant Φ, and derivative relations, as factors of the system integrity.
- The qualitative difference in interactions of the components at different levels, both their coordination and globality of action (regulation systems and feedbacks).
- A permeable time scale of the processes for different levels - and their mutual consistency.

## Building HEM

The analysis of the actual data on an organization of a human and environment leads to the conclusion that the most common case of an organization takes place, that is, Sp is formed as a result of the interaction {S} and {O}, and integrity conditions take the form of:

$$B \to (\{C\}_o, Cn, \{O\}_p) \to CP \vee CP^* \vee CP^{**}, \quad (1)$$

whereas

$$CP \to n(\{C\}_{i=1,\dots,n}, \square_{xy}, \{O\}_{i=1,\dots,n}) \quad (2)$$

$$CP^* \to (m-n)(\{C\}_{i=n+1,\dots,m}, \square_{ixy}, \{O\}_{m-n}) \quad (3)$$

$$CP^{**} \to (o-m)(\{C\}_{o-m}, \square_{i=o-m,\dots,o, j=p-m,\dots,pxy}, \{O\}_{p-m}), \quad (4)$$

where CP with strokes represent types of scenarios; $n, m, o, p, o > p$ denote the amounts of S and O; $i, j$ are the sequence numbers of Sp bundles, which correspond to the interactions of various S and O; $x, y$ are the numbers of layers and corresponding S-elements and S-sets.

This means that within the framework of the general scheme (1), separation occurs in three ways: $n$ times according to the scenario (2), $m$-$n$ times according to the scenario (3), and $o$-$m$ according to (4). Let's consider them in more detail.

***The Bundle (2)*** is modeled as shown in Figure 9 graph, which displays general principles of the system, division into levels and elements, the number of independent characteristics to describe them. The dots show components, and the lines show their connections.

### Level 1

It corresponds to Fibonacci number 1 and the modality of Sp **II**. It means that Sp may exist as undifferentiated integrity.

### Level 2

The second Fibonacci number - 1 - and the modality of Sp **A** are the evidence of the dual nature of Sp: one holistic wave comprises two "skew-symmetric" halves. This creates the basis for differentiation of the system. Differentiation does not violate the unity, as the existence of $Sp^+$ and $Sp^-$ separately is impossible, and the symmetry is preserved in separation. System specifications comprise n sets of 1 parameter.

### Level 3

It corresponds to the number 2, and this means the actualization of differentiation of $Sp^+$ and $Sp^-$ as independent components of this layer. By the wave nature, both subject and object halves of S-space

*Figure 9. The general organization of the human-environment system*

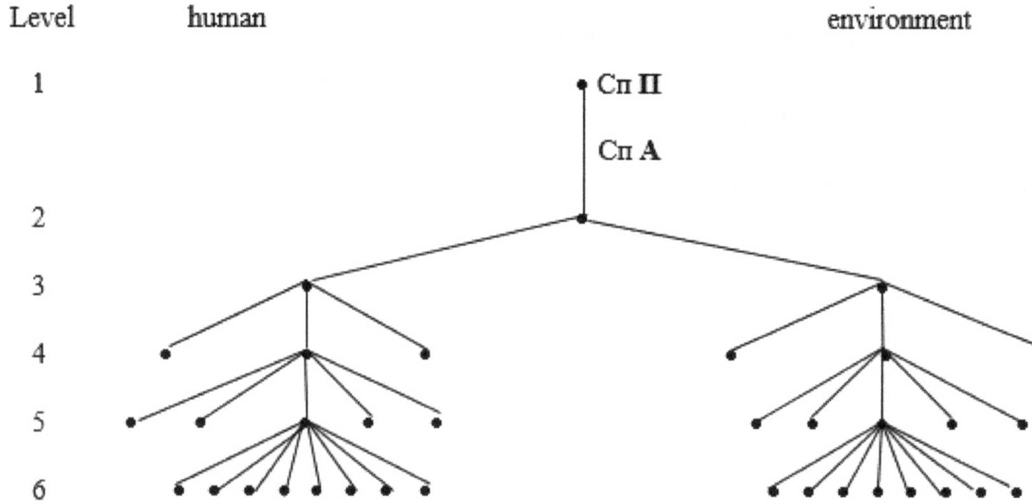


are integral. The order of symmetry is two. The state of the system is described by *n* sets of parameters, which include two independent characteristics. There is a correspondence between skew-symmetric characteristics Сп$^+$ and Сп$^-$.

## Level 4

The number of elements for Сп$^+$ and Сп$^-$ is equal to 3. The order of symmetry is 3. The system state is described by 2·n parameter sets of three characteristics that cannot be used to describe elements related to the previous layer. Irreducibility of different levels of the organization, keeping their unity within the same system, is also established.

## Level 5

The fifth layer corresponds to the set of 5 elements; the order of symmetry - 5; number of characteristics - 6·*n*·5.

These five levels characterize the general plan of the system organization. For a number of processes, the 6$^{th}$ and next levels, which correspond to the numbers 8, 13, 21... are traced.

The graph in fig. 8, as shown, characterizes the structure of a human body and organization of its feelings. In the Universe, there are single (for example, matter), binary (substance and field), ternary (three dimensions of space, energy, information, and entropy) and five-dimensional divisions (5 interactions, if separate electrical and weak).

***Bundle (3)*** is characterized by the presence of loops in the PC graph, which reflect interactions of S-elements and S-sets of different branches. The structure of a circulatory or lymphatic system of humans or natural river systems would be an example.

***Bundle (4)*** is modeled by a graph with network topology. This structure is typical, for example, for the nervous system, animal trails in the forest, etc.

## The Mechanism of Communication and Exchanges

Suppose, an external influence will result in a damage of either element, for definiteness, of *i*-th layer of $Sp^+$. Then:

1. As integrity is violated, the system would either cease to exist ($i = 1$, characteristic 3.3.), or restore integrity due to redistribution of potentials between layers of $Sp^+$. This redistribution, possible within certain limits, corresponds to the phenomena of regeneration, autoregulation, and fatigue, known to live organisms. For layers with numbers less than i, the impact of an external irritant will be less significant. Similar changes will occur For $Sp^-$, due to symmetry. The general potential Sp will reduce.
2. In the course of redistribution, modalities of existence and state will change, transformations $\bullet\rightarrow\cup$ and $\cup\rightarrow\bullet$, as well as interactions of S-elements, will occur. This will lead to the emission of waves and movement of solitons, i.e. exchange processes between the medium. Since qualitative characteristics of the layers are different, exchange and change processes of the layers with numbers i and i+1 $C\pi^+$ and $C\pi^-$ will also be qualitatively different.
3. Different exchanges and changes (preservation of dynamic equilibrium, restoration or destruction of systems and organs) can be described by {**DICS**} operations, proper characteristics of these processes. For example, normal, paradoxical and ultraparadoxical reactions of the nervous system to a replacement of the irritant values are interpreted as special cases **C**. Since the implementation of operations stipulated by restrictions (conditions 3.1-3.3) related to potentials of the elements, this leads to separation of interactions, respectively, separation of Sp. Conservation laws due to its inherent symmetry according to Noether's theorem will be fulfilled within the boundaries of either layer.
4. Interactions at resonant frequencies will be the most sensitive for the existing organization. Their compensation requires redistribution of the greatest part of the potential, and the impact will be significant for the next few layers. This effect explains the principle of operation of adverse effects of one type on different systems of the body.

Thus, the following provisions have a place:

**Provision 1**: *The reason for the existence: homeostasis and regeneration; direct and reverse links; qualitative differences in the exchange processes between components of the human-environment system in the context of unbalanced external influences is the need to preserve its integrity.*

**Provision 2**: *Those environmental influences that contribute to a restoration of the integrity of $Cn^+$ should be considered favorable; and influences that violate the integrity of $Cn^+$ - negative.*

**The limits of self-regulation of the system** are determined based on the need to keep a dynamic equilibrium of the system, which is formalized as prevention of any further stratification (or convolution). This corresponds to the change of the potential for every element of every layer within the boundaries of $\pi\div\pi\cdot\Phi$. Thus, for indicators derived from the potential, the following provision takes place:

**Provision 3**: *The boundaries of changes of indicators belonging to the same level should not exceed the ratio 1:$\Phi$ for scenario (2) or its derivatives for scenarios (3), (4).*

Now let's consider the question of the role of the external environment in the formation of the structure of the human psyche.

Let's use the graph in fig. 8 to correlate levels and channels with the structures and "abilities" of the human psyche in charge of perception of the environment. Let's call them a set of *subjective human space* - to avoid misunderstandings associated with the concepts of "psyche", "soul", "spirit". So:

1. Subjective space is a gestalt caused by the general laws of self-organization of complex systems, structuring interactions with the environment;
2. This gestalt includes a set of irreducible to each other levels, structures and channels, either thereof has its object of perception;
3. Namely:
    - intuitively perceived relationship of unity with the environment (1 channel);
    - by sensing one's "ego" - the inner unity of the subject (1 channel);
    - will and reason affect the object and estimate the feedback (2 channels);
    - the nous operates with space-time categories (2*3 channels - three dimensions of space, past, present, and future of time, grouping sensations as well as intero- proprio- and exteroceptors);
    - organs of perception determine specific reactions to direct effects (1*5 + 1*(?) + 1*(7) - there are five external sense organs; it is not known exactly how many internal (hunger, thirst, etc.) sensations are there and how many sensations refers to boundary (for example, a sense of balance). Once again, note that internal and boundary sensations are not represented by specific organs;
    - then there is a change in the principle of organization, perceptions become homogeneous and separate organs of perception are no longer formed (although their beginnings can be seen, for example, in different responses to the light of rods and cones of a human eye). Nevertheless, one can, already with some approximation, trace correspondence of sensations to the graph in fig.6. Thus, here the structuring ends;
4. Thus, the relationship between the channels is described by a graph or diagram of the self-organization scenario (1S, 1O), and the number of channels by a number series of Fibonacci - 1-1-2-3-5-8. That is, for example, reason operates with binary categories according to the principle of "good/bad", spatial-temporal relations of the nous are ternary, the number of external sense organs is five, etc. Note that the scope of a particular channel covers all subsequent levels, and channels related to subsequent levels inherit some of its features since they are formed on its basis. A more detailed diagram is shown in table 2;
5. All these channels are present and active all the time, but only one of them, which determines the corresponding "state of mind", is dominant at some time. In its turn, interaction with a specific object determines the content of the state of mind - in the form of an idea, action, word, emotion, including those related to the past (memory) or future (prediction) of time.
6. Psychologists do not have a precise understanding of what is meant, for example, by "intuition" and whether it is a sensation, feeling or ability, if any physical structure corresponds to it and where such a structure is located, and how it relates to "understanding" and "knowledge". Therefore, fundamentally different definitions are proposed - by pointing to the object of perception, the level of an organization according to, and also characteristics of perception. Thus, each level or channel (S-set or S-element) corresponds to one or another stage of S-space separation and is determined according to the table. 2.

*Table 2. Levels and channels of subjective human space*

| Levels of Perception | Objects of Perception | Characteristic of Perception | Channels, Structure and Their Number |
|---|---|---|---|
| 1 | Unity | Human and environment are not separate, but the possibility of separation potentially exists. | Intuition 1 |
| 2 | Separation | Awareness of a person as a self-sufficient whole | + Ego (mind, homeostasis) 1 |
| 3 | Impact and reaction | Impact on "not yourself", the response of the environment in response | + Will (action-reaction) and reason (good-bad) 2 |
| 4 | Space and time | Organizing impacts on the environment and its reactions by space and time categories | + Nous and emotions (tracing forms and changes) 2*3 |
| 5 | Sensory impacts | Organizing the effects of the environment, internal and borderline sensations by color, sound, etc. | + Feelings and emotions 2*3*5(?) |
| 6-7 | Tones of color, sound, etc. | Distinguishing between shades, tones, etc. | + Feelings and emotions 2*3*5*8(?) |

*Figure 10. Weighing coefficients of perception channels*

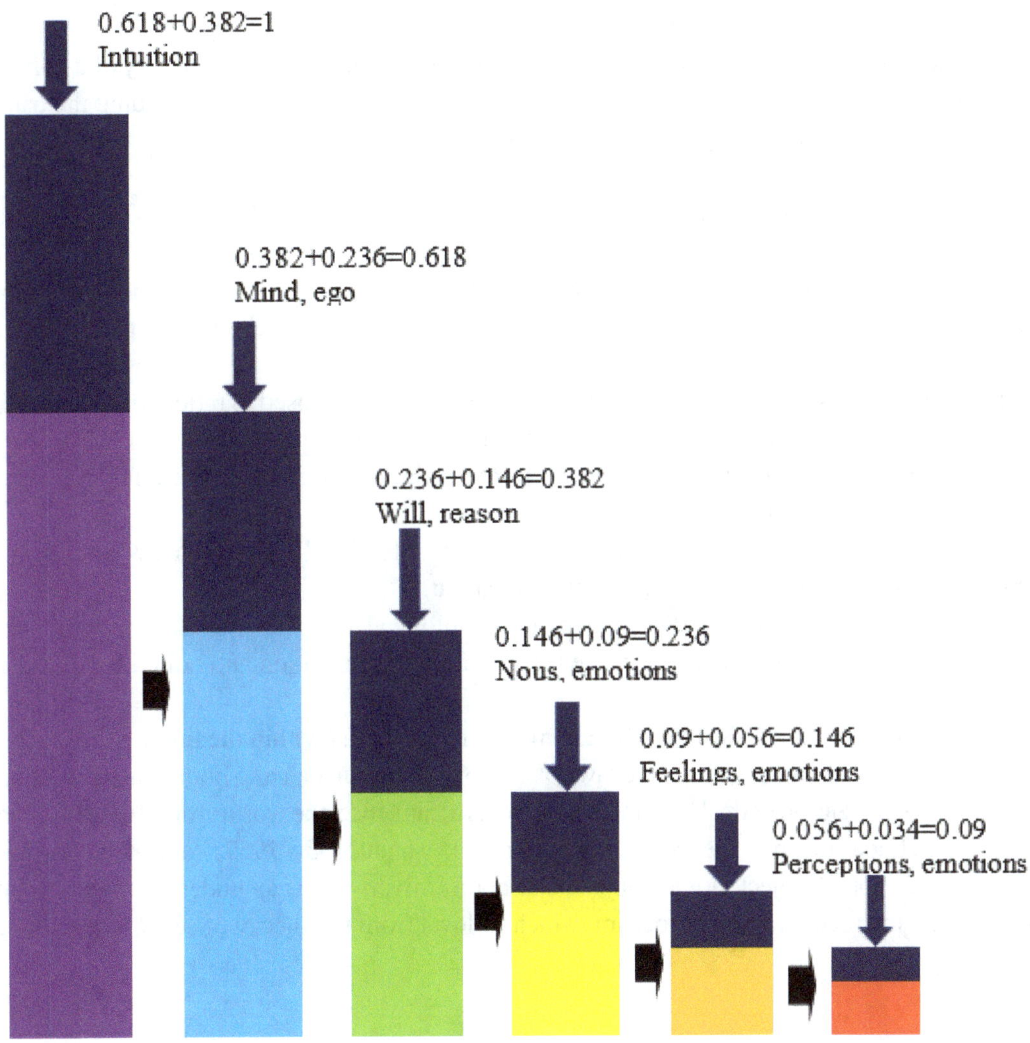

7. To start the process of updating the structure, external influence is required (axioms of the first group), i.e. some metabolic process. At the same time, keeping integrity requires a ratio of the potential levels determined based on the proportion of the golden section (chapter 2, *Self-organization of complex systems*). The corresponding weighting factors of every level are proportional to the series 1, 0.618, 0.382, 0.236, 0.146, 0.09. They also determine the contribution of either perception channels to the creation of a complete image (Figure 10). At the same time, since the potential of the previous level is greater than all the elements of the next one, there is a qualitative difference in the characteristics of both levels (non-additivity), which makes it possible to perceive heterogeneous and multi-quality objects. Thus, the inheritance of the characteristics of the previous level is caused by the transfer of its "qualitatively colored" potential, and the appearance of new features - external compensations.

## CONCLUSION

So, the substantiation of the structure and priorities of the human subjective space obtained based on a human-environment model is the main result of this chapter (Table 2, Figure 9).

Based on these results, let`s offer our answer to the question about the cognoscibility of a human and the surrounding world, specifying it as follows: which human abilities provide for finding the truth and what are the limits of understanding the world in general?

Let's present, in brief, the main existing views.

There are, on the contrary, rationalistic views, dating back to Socrates (469-399 BC), and presented by such thinkers as Aristotle, B. Spinoza, G. Leibniz, R. Descartes, G. Hegel, (Hegel, 1913; 1916; Aristotle, 1976-1983; Plato, 2006) and others who recognize Reason is the basis of cognition and behavior.

There are sensual views that go back to the well-known statement of the philosopher John Locke (Locke, 1985-1988): Nihil est in intellectu, quod non prius fuerit in sensu, i.e. "There is nothing in the reason that would not exist in the feelings" - which is understood as opposed to rationalism, and affirmation of feelings and perceptions as the main form of authentic cognition.

Intuitivism Bergsson (2001) and Losskiy (1992): Recognize intuition as the most reliable means of knowledge and rejects formalization of the act of cognition.

At the same time, there is no clear definition of "reason", "feelings", "intuition", which also makes them practically inapplicable for creating an optimal habitat.

There is a similar discord among 10-12 existing psychological schools.

The option of an answer, following from the theory of self-organization and built a HEM, the next.

1) The world around us is cognoscible for a human only insofar as within the limits of S-space, both for the Universe and human, the same laws of self-organization, causing both structures of human psyche and organization of the Universe, take place. In this sense, the world around us can be likened to what the Hindus mean by the word "Maya" (Radhakrishnan, 1956-1957) - not an illusion, but not a true world - not a Subject, not an Object, and not a Universe. It is an understanding of relations between the microcosm and macrocosm, which follows from the theory of self-organization.

2) A human perceives the world around him/her due to the symmetry of his/her mental structures and structure of the Universe; spontaneous symmetry breaking leads to a metabolic inherent perception process.
3) The process of cognition develops as a process of self-organization; first, intuition is used, which determines unity with the object of knowledge, then ego, which establishes the subject-object relationship, the object is affected and its reactions are evaluated (will and reason are involved), reactions and evaluations are related to the categories of space and time (nous), and finally, complemented by sensual impressions. Thus, the process of cognition corresponds to the scenario of stratification Sp. The first levels set the context within which the data of channels located on subsequent levels are interpreted.
4) There is also a reverse process corresponding to the convolution scenario - when sensory impressions are ordered by the nous and evaluated, based on previous experience, by reason operating with binary categories. The rules of such operations relate to formal logic, so mathematics based on such rules stands behind the phenomena of the surrounding world;
5) Different people with significantly different experiences recognizes the world more or less equally, since the processes of restoring internal integrity, understood as recognition, are fundamentally the same, although they differ quantitatively for different people.

## FUTURE RESEARCH DIRECTIONS

Will people living in different historical epochs perceive the surrounding world stereotypically, and will this stereotype change from epoch to epoch? Will their behavior and, in particular, their creativity also be limited to certain frameworksб specific to each era? Yes, if each epoch will correspond to a certain state of mind - the "spirit of the age" - then the evolution of mind occurs, and not if such stereotypes do not exist, and our mind, having individual differences, remains unchanged for humanity as a whole since our ancestors became Homo sapiens.

These questions will be considered theoretically in the next chapter, with the help of the apparatus of the theory of self-organization already involved. It is necessary to find markers that allow judging the evolution of mind or lack thereof.

Finally, it is necessary to explain the features of the evolutionary process noted in the first chapter that was not considered here.

So, if the "static" aspect of self-organization is involved in this chapter, its "dynamics" nature will be highlighted in the next chapter.

## REFERENCES

Arnheim, R. (1974). *Art and Visual Perception. Moscow, Russia*: Progress.

Becker, K., Becker, M., & Schwarz, J. (2007). String Theory and M-theory: A Modern Introduction. Cambridge University Press. Biology-stack-exchange. (n.d.). Retrieved from https://biology.stackexchange.com/questions/53062/why-is-the-human-ear-most-sensitive-to-4000-hz-tones

Bergson, A. (2001). Consciousness and Life. *Moscow, Russia*: ROSSPEN. Entanda-audiology-news. (n.d.). Retrieved from https://www.entandaudiologynews.com/features/ent-features/post/optimising-hearing-aid-processing-for-music-appreciation

Bodnar, O. Ya. (2005). Golden section and non-euclidian geometry in science and art. Ukrainian technologies. Lviv.

*Descartes, R. (1989-1994).* Works in 2 (Vol. M). Thought.

Fechner, G. T. (1882). *Revision der Hauptpuncte der Psychophysik, Leipzig Aristotle. (1976-1983). Works in 4 Volumes.* Moscow, Russia: Misl.

Gegenfurtner, K. R., & Sharpe, L. T. (Eds.). (1999). *Color Vision. From Genes to Perception.* Cambridge, UK: Cambridge University Press.

Georgieva, S. A., Belikina, N. V., Prokofyeva, L. I., Korshunov, G. V., Kirichuk, V. F., Golovchenko, V. M., & Tokayeva, L. K. (1981). Human Physiology. Moscow, Russia: Medicine.

Gorelik, G. S. (1959). *Oscillations and Waves. Introduction to Acoustics, Radiophysics, and Optics.* Moscow, Russia: Fizmatlit.

Hegel. (1913). Phenomenology of the Spirit. Saint Petersburg.

Hegel. (1916). The science of Logic. Saint Petersburg, 1916.

Ivanitskiy, M. F. (2008). *Human Anatomy (with foundations of dynamic and sporting morphology): Guide for Institute of Physical Culture. Moscow, Russia*: Olimpiya.

Kant, I. (2007). *Criticism of Pure Reason. Moscow, Russia*: Eksmo. doi:10.1007/978-1-137-10016-0

Korobko, V. I. (1997). The Golden Proportion and Problems of Harmony Systems. Moscow, Russia: Ed. Association of Universities of Civil Engineering of the CIS countries.

Leibniz, G. V. (1982-1989). Works in 4 Volumes. *Moscow, Russia*: Misl.

Locke, D. (1985-1988). Works in 3 Volumes. *Moscow, Russia*: Misl.

Losskiy, N. O. (1992). *Intuitivism. Moscow, Russia*: Progress.

Matveev, A. N. (1985). Optics: *Moscow, Russia*: Higher School. My-3-sotki. (n.d.). Retrieved from http://my-3sotki.ru/?p=2042

Morz, F. (1949). Fluctuations and Sound. M.L.: GITTL.

Olsen, S. (2006). *The Golden Section, Nature's Greatest Secret.* Wooden Books, Ltd., 2006.

Plato. (2006). Works in 4 Volumes - Saint Petersburg: Publishing House of the University of Saint Petersburg. Publishing House of Oleg Abyshko, Popmech-Ddesign-news. (n.d.). Retrieved from https://www.popmech.ru/design/news-359292-potryasayushche-krasivye-snimki-zvezdnogo-neba-iz-kabiny-pilota/

Radhakrishnan, S. (1956-1957). Indian Philosophy. V. 1-2. *Moscow, Russia*: Nauka.

Ragain, C. J. (2015) A Review of Color Science in Dentistry: The Process of Color Vision. Retrieved from https://symbiosisonlinepublishing.com/dentistry-oraldisorders-therapy/dentistry-oraldisorders-therapy34.php

Ramirez, T., & Herbig, R. (2016) Optimising hearing aid processing for music appreciation. Retrieved from https://www.entandaudiologynews.com/features/ent-features/post/optimising-hearing-aid-processing-for-music-appreciation

Sinelnikov, R. D., & Sinelnikov, Ya. R. (1996). *Atlas of Human Anatomy in 4 Volumes. Moscow, Russia*: Medicine.

Soroko, E. M. (1984). *Structural Harmony of Systems. Moscow, Russia*: Nauka.

Spinoza, B. (1999). Works in 2 Volumes. Saint Petersburg, Russia: Science. Symbiosis-online-publishing. (n.d.). Retrieved from https://symbiosisonlinepublishing.com/dentistry-oraldisorders-therapy/dentistry-oraldisorders-therapy34.php

Urmantsev, Y. A. (1974). *The symmetry of Nature and Nature of Symmetry Philosophical and Natural Scientific Aspects. Moscow, Russia*: Misl.

Urmantsev, Y. A. (1988). *Evolutionary or General Theory of Development of Systems of Nature, Society, and Thinking*. Pushchino, Russia: Institute of Plant Physiology, USSR Academy of Sciences.

Vasyutinskiy, N. A. (1990). *Golden Proportion*. Moscow, Russia: Molodaya Gvardiya.

Weyl, G. (1968). *Symmetry*. Moscow, Russia: Nauka.

Wikipedia. Teoria Kalutsi. (n.d.). Retrieved from https://ru.wikipedia.org/wiki/Теория_Калуцы_Клейна.

Wikipedia. Teorema Neter. (n.d.). Retrieved from https://ru.wikipedia.org/wiki/Теорема_Нетер

Wikipedia. Turinskiy avtoportret. (n.d.). Retrieved from https://ru.wikipedia.org/wiki/Туринский_автопортрет#/media/File:Leonardo_self.jpg

## ENDNOTES

1. The memory serves to navigate in the past, anticipation – to navigate in the future, whereas there are no motions in the present – it moves along with our consciousness, from the past to the future.
2. Ants solve the problem of complex impact on account of strict specialization: queen mother, workers, warriors, etc. have their specific limited functions.

# Chapter 4
# Self-Organization:
## Evolution and Behavior


## ABSTRACT

*The paradoxes of evolution are considered from the perspective of self-organization theory. The S-elements having several modalities, and this circumstance will help to understand why evolution is happening exactly this way. The evolution does not stop at the formation of human of a modern type. Slight changes in the human body and huge changes in his mind have occurred; the evolution markers should be defined. Behavior is another aspect of evolution. A simple criterion for determining psychological types may be the relative development of channels of human-environment interaction. The way of interaction determines preferential needs, motivations, and patterns of behavior, what, in turn, determines social structure and place of a specific human in it. Here a reservation is necessary - of course, the ideology, origin, historical and other circumstances are of great importance, after all, a particular human may have two or more dominant channels, but his inclinations will still determine his social behavior.*


## BACKGROUND

The references containing certain facts related to the evolutionary process are analyzed in chapter 1, together with the features of evolution. It makes no sense to repeat this analysis, so we'll discuss the reference where human behavior is considered in more detail.

John Watson (1913), and his followers William Hunter and Karl Spencer Leshley were the founders of behaviorism - a school that considered mental processes solely through a prism of behavioral patterns. The main message of this work - behavior is a reaction to a stimulus - is confirmed in many cases; however, there are quite a few examples when this is clearly not the case: self-sacrifice in a war, ascetic way of living of a certain number of people, accumulation of money and property to transfer to close people after one's death, etc.

In general, behaviorism has been justly criticized: for not paying attention to the inner world of human, bringing human down to the level of an animal (at the same time Pavlov (1949) developed a theory of conditioned reflexes, similar to behaviorism in terms of methodological background, based on experi-

DOI: 10.4018/978-1-7998-1706-2.ch004

ments with dogs), disregarding one's innate abilities, lack of interest to study motivations and needs, as well as manifestations of creativity in science and art, etc.

Skinner (1965, 2002), recognizing this criticism, suggested that personal events, in particular thoughts and feelings, are controlled by the same factors as human behavior (radical behaviorism), and that became the basis for experimental behavioral analysis.

Then, a theory of language and thinking was created. These approaches are used for scientific, as well as applied purposes, such as the elimination of problems related to development and autism, clinical psychology, speech behavior, management of organizational behavior, etc.

However, in view of the complexity of human behavior, other approaches to the study of his behavior are being developed, compensating for any inherent shortcomings of behaviorism.

In particular, Maslow (1954) investigated human motivation and needs in detail.

Gestalt psychology (Vygotsky, Köhler & Koffka, 1988; Koffka, 2014) pays great attention to the innate abilities of human perception of the environment, as well as the act of intuitive perception, having justified a number of principles and laws. It is also used in a number of therapeutic practices.

A. Augustinavičiūtė (Augustinavičiūtė, 2016) was the founder of socionics, having a based classification of psychological types on the binary principle. This makes it possible to predict the professional and social behavior of a human. In general, there are many classifications of psychological types, and the oldest one – in terms of temperament - was developed by Hippocratic and Galen in Antiquity.

Many psychologists pay great attention to age-related behavior and cognitive process associated with it (Piaget, 1951; Hollis, 2013).

The different psychological schools in the world, one way or another paying attention to human behavior. Note social psychology (Sherif, 1954; Asch, 1955; Festinger & Carlsmith, 1959; Homans, 1961; Gergen, 1973; Higgins & Wilson, 1975; Tajfel & Turner, 1986; Triandis, 1994; Heaven, Furnham, 1998; Andrews, 2001; Baron & Kerr, 2003; Parygin, 2010; Fromm, 2012; Waude, 2017); the socio-psychological consequences of the self-organization of complex systems theory will be discussed in the conclusion and continued in subsequent chapters.

It should be recognized that either school has both experimental and theoretical achievements in the fields of study that have chosen. However, they lack a systematic approach, which allows considering various aspects of mental activity and behavior from the same methodological positions.

The theory of self-organization of systems gives hope that this drawback will be eliminated.

## PARADOXES OF EVOLUTION AND THEIR INTERPRETATION

Remember (chapter 1, *Human evolution: facts and problems*) that upright bipedalism, fabrication of the tools; binocular vision, the marked predominance of vision over other senses, changes in the larynx structure, loss of hair are the most important changes. And complication of the brain structure, increase in its volume and energy consumption replaced by their degradation from the time of the Upper Paleolithic are even more important changes.

Correlations between anatomic changes and further external manifestations of reasonable activity are traced.

The paradox of the changes taking place is that the positive effect manifests in some generations and is useless initially or even harmful to the survival of the species.

Thus, the first Homo sapiens did not show any superiority of their more perfect brain over Neanderthal men at least for 130-150 thousand years, and only after their painting and stone processing went far ahead.

And just at this time - and this is another paradox - the human brain began to degrade, simplifying and decreasing from the times of the Upper Paleolithic to our days by 400-450 $cm^3$.

But from the point of view of the theory of self-organization, these and similar paradoxes are regularities, rather than paradoxes.

Remind that human, his systems and organs within the framework of this theory are waves and solitons with a certain potential, turning into each other under the influence of non-systemic factors that exist in a boundary environment for a limited time and are characterized by the modalities of existence and state (chapter 3, *Self-organization: body and mind*).

Thus, the potential possibility of upright bipedalism and hands release manifested quickly resulted from a genetic mutation (wave actualization and modality of its excitation) and became fixed (the wave transformed into a soliton). At the same time, the potential use of hands to make tools appeared - and after a long time, after another external impulse that changed the potential of the system - it became relevant, and the potential was sufficient for making the Olduvai tools. Similar external impulses preceded other changes, such as anatomical (binocularity of vision, its predominance over other senses, changes in the larynx structure, loss of hair) and technological (transition to Asheulean, Mousterian and Upper Paleolithic stone processing techniques).

The same is true of the improvement, and the degradation of a human brain - an external impulse resulted in genetic mutation and stayed in force thereafter, and as a result of the brain gradually increased and became more complex, from generation to generation, and the body adapted to "this parasite", and only in a long while, the Cro-Magnon brain could show its usefulness. And then the impulse exhausted, and it was time for degradation.

This whole process can be compared to the development of a human personality: in childhood, the organism grows for no apparent reason, and this happens without conscious effort, "itself". It is also easy to study, one has enough resources to go in for sports and so on. It is time for maturity, and a human has to spend his resources to keep his intellectual and physical form. "Throes of art" is the witness. And finally, old age comes - again, for no apparent reason, everything starts becoming useless - neither training nor medicine help. The body is degrading, creative activity is fading away, and nothing can prevent the inevitable death.

The transition of a wave in the stage of excitation (childhood) to the stage of a stable soliton (maturity) and a wave in the stage of extinction (old age) and disappearance (death) are evident.

At the same time, all changes are coordinated by the scenario of self-organization.

The authors are aware that such reasoning is only a hypothesis. They would be more convincing with temporary estimates based on models of the 2nd type. Unfortunately, such estimates are still premature: it is difficult to be confident in the chronology of both biological evolution and assessment of the duration of cultural eras on the basis of episodic findings. Were there any populations that were the ancestors of modern humans found? When exactly did mitochondrial Eve and Y-chromosomal Adam live? When did the Paleolithic, and when did, for example, Neolithic begins for the modern people? The estimates in chapter 1 differ by tens of thousands of years. And even genetic research cannot always help. And this eliminates the calibration of models of the 2nd type.

It can only be argued that human evolution is accelerating, and this confirms the prediction of the theory of self-organization in a qualitative way - alas, not in quantity yet.

## EVOLUTION OF MIND AND ITS MARKERS

So, the mind of modern human had time to go through several periods.

A magical childhood, when everything grew and tuned up as if by itself, but about which there were no material traces.

Maturity, which lasted from the time of occurrence of anatomically modern humans with a larger brain than now (230-140 thousand years ago according to genetic data and 200-100 thousand years ago according to archaeological materials), and up to the times of the Upper Paleolithic. What did a human do with his so perfect brain? Archaeological findings leave us at a loss - the same Mousterian technique, cave paintings, almost indistinguishable from the drawings of Neanderthals.

And only the legends of many nations seem to have retained some shadows of this past. And, unlike modern theories, they show our ancestors as primitive savages, dealing with hard struggling for existence (Didenko, 1999) is the best illustrative example), they talk about Paradise, Golden Age, Sattva Yuga, when gods and spirits communicated with humans, lived among them were and taught them on the grounds of civilization, and humans were careless and happy. These traditions allow different interpretations (and will be discussed in more detail in the next chapter). However, let's ask: under what conditions are such communication (if any, of course) possible? Obviously, that one may perceive matters that are outside the system, having a very developed intuition (when it suppresses even the mind of ego). And - again hypothetically - it was a highly developed intuition that needed such a perfect brain.

And then the human mind went down to the third level (learning good and evil - binary assessments speak of the development of will and understanding on account of intuition), and communication with the gods became impossible. And it was reflected in the myths as the expulsion of human from Paradise.

Old age began, and the human-made even greater efforts trying to get the Paradise back and avoid death, well, at least to maintain heavenly comfort, and this, hypothetically, is the cause of both technical and artistic culture. But nothing can be changed, and the brain is degrading irreversibly, and the legends tell about common people, rather than gods and heroes. This hypothesis and its rationale will be considered in more detail in the following chapters.

Thus, on the basis of material characteristics, one may study only the last, old-aged, period of a human mind, which covers approximately 40-30 thousand years of his evolution. This period will be studied with the use of the theory of self-organization, which gives the following prediction of the activation of channels of interaction with the environment: intuition ego will and reason nous senses. In this case, we recall that a person has all the channels, and activation means only the domination of one of them; at the same time there are obvious differences between the state of the channels of an individual and some kind of averaged "spirit of the age".

But what are the markers that would test theoretical predictions? They have the following requirements:

1. Have an expressed correlation with the state of mind.
2. Cover the period from the Upper Paleolithic to the present day.
3. Spread throughout the world.

To be represented in most cultural activities (this condition is not imperative, but preferable).

There are not so many phenomena that meet at least the first three conditions in the history of mankind.

Painting, or, more precisely, characteristics and way of depicting the space is one of them. Firstly, the artist always depicts his subjective space, secondly, painting has been with mankind throughout its entire history, thirdly, one can meet pictorial images even in the most remote corners of the earth.

The myth of the World Tree is another similar phenomenon. Reflecting ideas about the structure of the Universe (condition 1) and originating in the Paleolithic era (condition 2), it is common in many cultures around the globe (condition 3) and is expressed in different forms of culture (religion, literature, painting, oral art, popular culture, etc.).

Both these markers complement each other, and they will be used as markers of the mental evolution in the future.

## PSYCHOTYPES AND BEHAVIOR PATTERNS

The systematic approach to analyzing behavior is based on correlations between channels of interaction of human with the environment (defined according to the theory of self-organization of complex systems in chapter 3, *Self-organization: body and mind*), psychological types, needs (including comfort conditions) and motivations, position in the public hierarchy, and behavior patterns that change with change of age and from one epoch to another.

Channels of Interaction and Psychological Types

If we base the definition of the psychological type on the principle of the dominance of either channel, we will get the following six psychological types:

1. "Balanced personality": Here, potential distribution across the levels is perfect, and all channels are harmoniously developed (fig. 8 in chapter 3, Self-organization: body and mind), such a person does not experience any internal conflicts and health problems, he is equally successful in all activities. His successfulness is determined by the overall potential of the individual. In reality, such harmonious personalities are extremely rare.
2. "Contemplator": Here, as for the following types, the balance between the levels is already broken. Since the total value of the potential for a person is unchanged, the strengthening of some qualities - in this case, intuition - will occur at the expense of others - above all, ego and will. Therefore, contemplators tend to be calm, contemplative, not active, lifestyle, preferably in a quiet and beautiful, if possible, deserted place.
3. "Egoist": Strengthening of the ego, will, and reason comes through intuition and, to a lesser extent, reason, and feelings. This is fraught with health problems - mental and physical. The egoist is active, he is always in the center of events and tries to rebuild the world to fit his needs.
4. A "Fighter" is characterized by an increased obsession with self-realization, conflict, and aggressiveness (will prevails over reason), which cannot but lead to both physical and mental disorders, since intuition, ego, and reason are oppressed. This is a "Martian" type, the motto thereof, like that of Portos, is "I'm fighting because I'm fighting". Vigorous and diverse activities place increased demands on both stimulations of the right mood and relaxation.
5. An "Adventurous researcher" is not inclined to escapes, he loves predictability, orderliness, stability – "money likes quietness". But within these limits, it can be quite energetic, active and rational (reason prevails over the will) both in business and science. "The nous is added" at the expense

of senses, and some people are able, consciously or not, to sublimate sexual energy for "creative purposes," which was noticed by Freud.

6. An "Artistic personality" is prone to some exaltation, cultivation of his emotional states and sensory experiences. All this can get along with extreme mercantile spirit. Sometimes such people have good intuition, although they are more often inclined to take different superstitions for granted. From this, it follows that amplification of sensory sensations occurs due to the reason, will and analytical part of the nous. The maturity of channels depends on the overall potential of the individual. In turn, only a high degree of maturity of the channel leads to its domination - for a particular psychological type, individual, or historical era.

Different extent of the channel maturity leads to the need to identify different subtypes within a single psychological type. With some extent of conditionality, we will distinguish four such degrees: excellent, high, medium and low. The maturity of the channel can be expressed by a series of words describing its weakening as the potential decreases. We'll show some tables indicating the relationship between the word-characteristics, the extent of maturity of the channel, and subtypes for either psychological type.

*Table 1. The perfection of the channels and subtypes of a balanced personality*

| Characteristics of the Channels (Higher Potential is at the Top, Lower – at the Bottom) | Degree | Subtypes |
|---|---|---|
| Ability to commit fundamental discoveries | Excellent | Genius |
| Ability to create original works | High | Creator |
| Ability to fulfill any comprehensive task | Medium | Jack-of-all-trades |
| Ability to perform any simple work | Low | Multipurpose worker |

*Table 2. The perfection of the intuition and subtypes of contemplators*

| Characteristics of the Channels (Higher Potential is at The top, Lower – at the Bottom) | Degree | Subtypes |
|---|---|---|
| Communication with God | Excellent | Mystics |
| Adoration of God | | |
| Contemplation | High | United with nature |
| Anticipation | | |
| Associativity | | |
| Insight | Medium | United with humans |
| Love of neighbor | | |
| Sociability | | |
| Dullness | Low | Incoherent |

*Table 3. The perfection of the ego and subtypes of egoists.*

| Characteristics of the Channels (Higher Potential is at the Top, Lower – at the bottom) | Degree | Subtypes |
|---|---|---|
| Autism | Excellent | Living in own world |
| Pride | High | Rulers |
| Self-approval | Medium | Able to achieve success |
| Self-confidence | | |
| Self-esteem | | |
| Lack of faith in one's own strength | Low | Able to meet the failures |
| Self-deprecation | | |

*Table 4. The perfection of the will/reason and subtypes of the fighters*

| Characteristics of the Channels (Higher Potential is at the Top, Lower – at the Bottom) | | Degree | Subtypes |
|---|---|---|---|
| Will | Reason | | |
| Aggression | Wisdom | Excellent | Warrior |
| Expansion | Assertiveness | High | Invader |
| Adventurism | Appraisement | Medium | Battler |
| Perseverance | | | |
| Stubbornness | Indetermination | Low | Troublemaker |
| Weakness | Non-distinction | | |

*Table 5. The perfection of the nous and subtypes of the adventurous researcher*

| Characteristics of the Channels (Higher Potential is at the Top, Lower – at the Bottom) | Degree | Subtypes |
|---|---|---|
| Faith in cognoscibility of the world | Excellent (plus the excellent extent of the reason) | Philosopher |
| A feeling of the harmony of the world | | |
| Able to establish regularities | High (plus the high extent of the reason) | Scientist |
| Ability to abstracting | | |
| The ability for analysis and synthesis | | |
| Ability to determine relations | Medium (plus the high or medium extent of the will) | Businessman |
| | Medium (plus the low extent of the will) | Bureaucrat |
| Confused perception | Low | Fool |
| Stupidity | | |

*Table 6. The perfection of the perception and feelings; subtypes of artistic personalities*

| Characteristics of the Channels (Higher Potential is at the Top, Lower – at the Bottom) | | Degree | Subtypes |
|---|---|---|---|
| Perceptions | Feelings | | |
| Extrasensory | Rage | Excellent | Crazy |
| Ability to differentiate more than 7-9 shades and tones | Exaltation | High | Actor |
| Ability to differentiate 7-9 tones | Sensitivity | Medium | Philistine |
| Loss of sight, hearing loss, etc. | Coldness | Low | Disabled |

## Hierarchy of Needs and Motivations

Hierarchy of needs and motivations. According to the conclusions of A. Maslow (Maslow, 2002), there are groups of needs (feeling of lack of something) and motivations (impulse for an action to meet the needs), forming a pyramid of five levels:

Physiological needs (hunger, thirst, intention to avoid any pain and discomfort, including cold, noise, etc.);

1. The need for security. Maslow treats it like this: "Once physiological needs are satisfied, their place in the motivational life of an individual is taken by the needs of another level, which in the most general form can be combined into the category of security (need for security; stability; being dependent; protection; freedom from fear, anxiety and chaos; the need for structure, order, law, restrictions, and other needs). Similar to physiological needs, these desires can also dominate. They can usurp the right to organize one's behavior, subordinating all the possibilities of the organism to their will and aiming them to achieve security. For such a person there is no more pressing need than the need for security".
2. The need for love. Another quote: "Once the needs of the physiological level and needs of the security level are satisfied, the need for love, affection, belonging is actualized, and the motivational spiral starts a new round. A person as never before starts feeling the lack of friends, absence of a loved one, wife or children. He is longing for warm, friendly relations, he needs a social group that would provide him with such relations, a family that would accept him as an integral part. It is this goal that becomes the most significant and most important for a person; he may no longer remember that once when he endured need and was constantly hungry, the very concept of "love" did not cause him anything but a contemptuous grin. Now, he is tormented with a feeling of loneliness, painfully experiencing his rejection, looking for his roots, soul mate, or friend".
3. The need for recognition. The following quote: "Every person constantly needs recognition, a stable and, as a rule, high appreciation of his own merits, each of us needs the respect of the people around and ability to respect ourselves. The needs of this level are divided into two classes. The first includes desires and aspirations associated with the concept of "achievement". A person needs a sense of his own power, adequacy, and competence; he needs a sense of confidence, independence, and freedom. We refer to the need for reputation or prestige, need for gaining status, attention, recognition, and fame to the second class of needs. Satisfying the need for appreciation,

respect gives the individual a feeling of self-confidence, self-worth, strength, adequacy, a feeling that he is useful and valuable in this world. Any unmet need, on the contrary, gives him a feeling of humiliation, weakness, helplessness, which, in turn, serve as a ground for despondency, trigger compensatory and neurotic mechanisms".

4. The need for self-actualization (self-expression). And the last characteristic: "Even if all the above human needs are met, we can rightly expect that he will soon feel dissatisfied again because he is not doing what he is predisposed to. A human must be who he can be. A human feels that he must conform to his own nature. Obviously, this need is expressed differently for different people. One person wants to become a perfect parent, the second one seeks to achieve sporting heights, the third one is trying to create or invent. Normally one starts to feel the need for self-actualization only after he has satisfied the needs of the lower levels".

Correlation Between the Needs and Motivations and Levels of the World's Perception:

- Physiological needs correlate with sensory perceptions. Thus, physiological comfort, as well as appropriate motivation is entirely related to level 5.
- The need for security, as Maslow describes it, correlates with a sense of disorder, irregularity, anxiety, and chaos. And this is the level of intellect, both in its rational (left-hemispheric) and sensual (right-hemispheric) aspects. When a motivated action leads to the opposite feelings, the need disappears, and the state becomes comfortable. Therefore, such motivations and needs related to level 4.
- The needs and motivation for self-realization, as an activity in the outside world, correlate with the will and reason. Here, preventing any self-realization, whether in the form of conflicts with people, lack of resources, etc., is uncomfortable. The need refers to level 3.
- The need for recognition and motivation to its satisfaction, taken either as the attainment of power or "seizure" of fame, respect, high marks, with a sense of self-importance and power, clearly relates to the sphere of ego and level 2.
- The need for love and corresponding motivations are not uniform. On the one hand, they clearly relate to biological, sensory needs, and instinct of procreation. On the other hand, the desire for self-denial and self-sacrifice, manifested by a small part of humanity throughout its history, shall be correlated with the highest manifestations of love for the Motherland, humanity, and God. And between them, there is a need for friendship, recognition, social adequacy, i.e. unity with the world at the "everyday" level. But they all correspond to the gradations of intuition, i.e. level 1.

Obviously, the main motivations are related to the needs. However, the inner world of a human is not limited to them. The attitude to the environment, due to motivation, is not neutral: certain feelings are accompanied by an understanding of both external signals and their context, and their internal state.

So, the hierarchy of needs can be integrated into the general structure of human interactions with the environment. This enables establishing a scientific approach to the issue of a comfortable habitat, in particular, comfortable dwelling, which will be done in chapter 13, *Practice as a criterion of truth: the choice of a comfortable dwelling.*

Now let's check our classification - are the distinguished types fundamental and multi-purpose? To do this, we turn to a fundamentally different source - the history of human communities. The development of mankind shows that were a social organization occurred naturally, rather than as defined by an "obsession", a class, and then caste structure was always formed. Moreover, the social separation occurred depending on the motivations determined by the prevailing channels of communication with the environment.

**Correlations Between the Channels and Place in the Social Hierarchy**

Let's consider India as an example, where caste structure was valid almost to the present day. So:

- Brahman priests are those who are inclined to serve the gods, self-denial, and contemplation (intuition).
- Rajah and Kshatriya warriors are those who crave for power and able to subjugate others in a tough fight (ego, will, and reason).
- Vaishya: free traders, farmers, and artisans are those who are adventurously seeking material wealth (nous). Note that all of them are able to make important decisions on their own.

These are behind the boundary:

- **Shudra:** unfree servants and slaves, whose main goal is to get sensual pleasures, as well as an outcast - Chandalas.

If we look at feudal Europe, Russia, Muslim countries, we can observe a similar picture there. Thus, the fundamental nature and universality of the identified psychological types are confirmed by the historical experience.

Let's summarize all the obtained correlations in table 7.

*Table 7. Correlations of psychological types, channels of perception, needs and motivations, estates, types of activity and basic conditions of comfort.*

| Psychological Type | Prevailing Channel | Basic Need | Basic Motivation | Estate | Form of Activity | The Basic Condition of Comfort |
|---|---|---|---|---|---|---|
| balanced personality | channels are harmonious | harmony | creative work | "perfect" | any | harmony with the world |
| contemplator | intuition | love | self-denial and contemplation | Brahman, monk, and mystic | spiritual | peace, openness |
| egoist | ego | recognition | lust for power | Rajah, king, and ruler | managerial | prestige, self-sufficiency |
| fighter | will, reason | self-realization | proof of one's competence | Kshatriya, noble, and warrior | military | transformation |
| adventurous researcher | nous | security | lust for knowledge and wealth | Vaishya, businessmen, scientist, tradesman, artificer, farmer | entrepreneurial, scientific, and production | orderliness, proportionality |
| artistic person | sensations and feelings | physiological comfort | sensual pleasure | Shudra, artistic, servant, and outcast | imitation of creativity, service | sensory comfort |

*Table 8. Age-related specific features of the needs and motivations (according to beliefs of Hinduism)*

| Period of Life | Needs | Motivations |
|---|---|---|
| Birth, baby | Physiological, for security and love | Be with one's mother |
| Child | For love | Learn the world |
| Householder | For self-actualization and recognition | Continue one's race, getting familiar with the world |
| Gone to the forest | For breaking ties with the world | Leave the family |
| Forswear the world | For renunciation of the world | Renounce the worldly life |
| Death, corpse | For ritual burning | Get a better birth or merge with Brahman |

## Age Adjustment

There are very detailed studies of the characteristics of all periods of life (Piaget, 1951; Hollis, 2013). They all fit into the wave nature of human. Leaving all extra details aside, let's give a correlation of the needs and motivations for the traditional age periods for India (Table 8). Remember, that for Hindus, death goes into birth and the cycle is repeated.

Such adjustment allows determining the parts of premises of one's house or apartment, as well as functions and types of buildings in the urban environment.

## The Patterns of Behavior

The patterns of behavior directly follow from table 7. Let's talk, taking into account the limited volume of the book, about the most vivid, illustrative manifestations corresponding to the excellent and high levels of development of the channels.

- Balanced personalities behave like geniuses: they replace the existing paradigms in science, art, society, and religion with their own, not worrying about any material wealth and continuation of the race. Examples: Imhotep, Pythagoras, Leonardo da Vinci, Nikola Tesla.
- Contemplators demonstrate self-denial, humility and asceticism, lust for a solitary life, and only occasionally - if such is God's will - return to the world, promoting new teaching, sometimes with fanaticism. Examples: Buddha, Lao Tzu.
- Egoists are eager to remake the whole world for themselves, for that they need supreme power. For that purpose, they make personal sacrifices, and, of course, do not count with the victims of other people, are rightly considered tyrants. Examples Alexander the Great, Julius Caesar, Napoleon.
- Fighters make a state of war a path to self-development and value dynamics and transformation most of all. They behave proudly and arrogantly, always emphasizing their superiority. In brief moments of rest, they approve themselves as refined natures and create works of painting, poetry, and music. Examples: samurai, nobles.
- Adventurous researchers do not make any revolutionary discoveries, but always win, acting within the framework of a paradigm, whether they create a business empire or work in the field of science. Examples: Mayer Amschel Rothschild, Bill Gates, medieval scholastics, in particular, Thomas Aquinas.

- Artistic personalities prefer theaters, art salons, and journalistic activities, enveloping them with a veil of continuous scandal. It is no need to give examples; everyone can see them anyway.

## Temperament Adjustment

Temperament is one of the most popular characteristics of a human. In antiquity, there were 12 temperaments, and now only 4: sanguine, choleric, phlegmatic, and melancholic. They can also be associated with the magnitude of the potential - in this case, vital energy - of a human. The influence of temperament on human behavior is as follows:

- The sanguine person introduces vigor, activity, optimism into any activity.
- Choleric acts calmly, and them emotional uncontrollable outbreak and rabid activity come, and after it - emptiness and inaction.
- Phlegmatic is calm, patient, predictable and conservative.
- Melancholic paints his life in tones of sadness, pessimism, and passivity.

Finally, let`s note that during abrupt changes in life - in revolutions, wars, natural disasters, under stress (Selye, 1982), - human behavior would change dramatically. However, discussion of these features is beyond the scope of this work.

# CONCLUSION

Thus, within the framework of the theory of self-organization, it is possible to describe and explain some paradoxes of human evolution. Of course, this is a result of independent value. But also, this proves that the theory can be used to study the next period - the evolution of the human mind from the Upper Paleolithic to the present day. Markers that could confirm or refute the conclusions made on its basis were also proposed.

Another important result - the implementation of the self-organization scenario (1C, 1O) can be considered as a kind of "goal" of evolution. How does evolution fit into the general pattern of self-organization of complex systems? Perhaps there is a slightly higher frequency of birth of beings whose structure is closer to the graph expressing this scenario or their better survival? This can be verified experimentally. The examples in Chapter 1 confirm the feasibility of such studies.

Now let's look at the use of a systematic approach and the theory of self-organization to study human behavior, both social and personal.

It turned out that the results of different quality obtained by different methods (the concept of socionics of the psychological type, Maslow's motivation and needs, adaptation of temperament and age, taking into account and explaining the social structure) can be integrated into the framework of one model. The scenario of self-organization, which underlies the study of subjective space, is extra-personal and non-social, and, as a "great gestalt", determines a person's ability to interact with the environment.

All this made it possible to find a simple approach to the definition of social groups - each estate corresponds to a predominant channel of interaction with the environment. The patterns of group behavior of estates are well known in history and repeatedly described in the literature - these are the patterns of the prophet, king, nobleman, etc.

Testing and compiling a psychological portrait of a particular person gives the key to understanding his personal social identification and personal behavior, which has many practical applications: professional selection, therapeutic practices, choosing a comfortable home, etc. Examples are given in chapter 13.

## FUTURE RESEARCH DIRECTIONS

Firstly, further research will be aimed at checking the assumptions of the theory of self-organization about the evolution of mind using historical data.

Secondly, this theory will be used to study behavior and determine changes in human behavior patterns in different historical eras.

And the third goal, beyond the scope of this book, is to conduct joint research with representatives of the behaviorism and social psychology schools. To clarify the results of behavioral experiments, a systematic approach is needed, which gives many predictions regarding the dependence of the behavior of human communities and specific people on the psychological type, age, temperament, needs and motivation; on the other hand, experiments are necessary to eliminate the overly theoretical nature of the systems approach. Such cooperation seems promising.

## REFERENCES

Andrews, P. W. (2001). "The psychology of social chess and the evolution of attribution mechanisms: Explaining the fundamental attribution error" (PDF). *Evolution and Human Behavior*, *22*(1), 11–29. doi:10.1016/S1090-5138(00)00059-3 PMID:11182572

Asch, S. E. (1955). Opinions and social pressure. *Scientific American*, *193*(5), 31–35. doi:10.1038cien tificamerican1155-31

Augustinavichute, A. (2016). Socionics. Moscow, Russia: Black squirrel, 455- 464.

Baron, R. S., & Kerr, N. L. (2003). *Group Process, Group Decision, Group Action*. New York: McGraw-Hill.

Didenko, B. (1999). *The civilization of cannibals.* Humanity as is.

Festinger, L., & Carlsmith, J. M. (1959). Cognitive consequences of forced compliance. *Journal of Abnormal and Social Psychology*, *58*(2), 203–211. doi:10.1037/h0041593 PMID:13640824

Fromm, E. (2012). *Anatomy of human destructiveness*. Big Press.

Gergen, K. J. (1973). Social Psychology as History. *Journal of Personality and Social Psychology*, *26*(2), 309–320. doi:10.1037/h0034436

Heaven, P., & Furnham, A. (1998). *Personality and Social Behavior*. Routledge.

Higgins, A. C., & Wilson, E. O. (1975). Sociobiology: The New Synthesis: Harvard University Press.

Hollis, J. (2013). Pass in the middle of the road. How to overcome the midlife crisis and find a new meaning of life: Kogito-Center.

Homans, G. C. (1961). *Social behavior: its elementary forms*. New York: Harcourt, Brace and World, Inc.

Koffka, K. (2014). Principles of Gestalt Psychology: Mimesis International.

Maslow, A. (2003). Motivation and Personality: Peter.

Parygin, B. D. (2010). Social Psychology. Origins and prospects. Saint Petersburg, Russia: SPbGUP, 2010.

Piaget, J. (1951). *The Psychology of Intelligence*. London, UK: Routledge and Kegan Paul.

Selye, H. (1982). *Stress without Distress*. Moscow, Russia: Progress.

Sherif, M. (1954). Experiments in group conflict. *Scientific American, 195*(5), 54–58. doi:10.1038cie ntificamerican1156-54

Skinner, B. F. (1965). *Science and Human Behavior*. Free Press.

Skinner, B. F. (2002). *Beyond Freedom and Dignity*. Hackett Publishing.

Tajfel, H., & Turner, J. C. (1986). The social identity theory of intergroup behavior. In S. Worchel, & W. G. Austin (Eds.), *Psychology of Intergroup Relations*. Chicago, IL: Nelson-Hall.

Thorndike, E., Watson, D., & Karelin, A. (1988). Behaviorism.

Triandis, H. C. (1994). Culture and Social Behavior: McGraw-Hill College.

Vygotsky, L., Köhler, F., & Koffka, K. (1988). Gestalt psychology: ozon.ru.

Watson, J. (1913). Psychology from the Behavioral Perspective.

Waude, A. (2017). *Social Influence: Why We Conform in Groups*. Retrieved from www.psychologist-world.com

# Chapter 5
# Paleolithic:
## Unity


## ABSTRACT

*The examples of Paleolithic painting, their dating and cultural context are given. Basically, these are the works found in the caves of Chauvet, Altamira, Lascaux. The features of space depicting and some other features of the images been analyzed. Their difference from the contemporary patterns of space depicting been discussed. The connection of such signs with the levels of subjective space is established, which allows us to judge the development of its channels. The origin of the World Tree myth dates back to the Paleolithic epoch, there are very few images to judge upon its plot. Nevertheless, it is possible to reconstruct the Paleolithic version of the myth, based upon indirect signs. It also gives the chance to judge upon the state of mind of the Paleolithic humans. The results of the reconstruction of the mind and behavior of the Paleolithic human are presented in the form of a generalized psychological portrait and description of the behavior pattern.*


## BACKGROUND

The sources for Paleolithic painting can be divided into four groups.

First, those are the reproductions of works presented both in albums Pomerantseva,(2006) and in electronic form.

Second, these are the related historical and archaeological publications, that allow us to specify the periodization of Paleolithic epoch and the dating of the painted works (Leroi-Gourhan, 1967, 1978, 1982; Joiner, 1985; Blednova et al., 1998; Coulson & Campbell, 2001; Maugh & Thomas, 2009), as well as to view the images in the cultural context of this era (Taylor, 1980; Golan, 1993). Surely, there might be some discrepancy in the dating and hypothetical nature of culture reconstructions.

The next group of sources belongs to the attempt to understand why a primitive human began to paint at all, and why he did it in such a way, but not otherwise.

A range of hypotheses are evident here - he used to paint in such a way because a primitive human did not know how to do differently (being based upon some similarities with the children's drawings

DOI: 10.4018/978-1-7998-1706-2.ch005

(Osorina, 2016), he also used his drawings as the elements of cult and magic rituals (Abramova, 1971; Stolyar, 1985; Levi-Strauss, 1994; Gutrie, 2005), as well as he painted in an altered state of mind, which caused a significant difference from the contemporary painting (McKenna, 1995).

The arguments pro et contra of such hypotheses are also widely present in the related publications.

Some publications are devoted to the technique of depicting space – without any reference to the human mind (Rauschenbach, 1986, 2001) and with this reference (Kovalyov & Nitsyn, 2012).

The third group of sources relates to the research of the myth of the World Tree. Both the images of the World Tree and the hypothetical reconstructions of the myth plot have been presented (Golan, 1993).

And the fourth group of publications is the reconstructions of the human mind and behavior (Heaven & Furnham, 1998; Boehm, 1999; Kelly, 2000; Lewis-William, 2002; Eshleman, 2003).

## SPACE IN PALEOLITHIC PAINTING

Samples of a primitive painting are known all over the continents except Antarctica. The oldest of them date from 60 thousand years ago, and the new ones are being created by our contemporaries who live in the distant "corners" of our planet. According to its geographic scope and time evidence, the primitive painting has been the unprecedented human creative works. At the same time, it remains amazingly homogeneous, recognizable, and makes the viewer feel strong impressions.

The phenomenon of primitive painting has been studied for more than a century. It should be noted that any work is unconditionally individual - and nevertheless, it possesses a certain "spirit of the epoch", evident in the canonized artistic methods, social tastes, typed of mentality. The similarity of these features determines the homogeneity and recognition of primitive painting and makes it possible to evaluate the mind of a primitive human.

To perform such a reconstruction, it is necessary to solve the following tasks beforehand:

- Consider the localization and periodization of the works of primitive arts.
- Present a visual series of the most outstanding and illustrative works.
- Identify the features of space painting, including the comparison of the paintings with the other manifestations of material culture.
- To give the interpretation of a primitive human's state of mind.

### Localization and Periodization

In 1879, the amateur archaeologist M. S. de Sautuola discovered the Cave of Altamira (Spain, Figure 1), there were many paintings of the ancient people. One year later, Sautuola, along with J. Villanova-e-Pierre, published the results of their research, where they referred to the images to the Paleolithic epoch. Many scientists considered this message to be incredible.

Some similar caves with the paintings exist in different parts of Europe and all over the world, but most of them are located in the Pyrenees on both sides of the Spanish-French border.

Apart from the images found in the caves, there are many rock paintings. They look close to the bone carving and relief patterns, that have been traced on a soft stone surface, such as sandstone.

The traditional periodization of the Stone Age includes the Paleolithic, Mesolithic and Neolithic epochs, which have been defined primarily by stone processing techniques. Therefore, each of those epochs

*Figure 1. Cave painting from the Cave of Altamira. Español: Replica of the ceiling at the Altamira Cave, exposed in the gardens of the National Archaeological Museum of Spain.*
*Author: José-Manuel Benito (José-Manuel Benito, 2006)*

*Figure 2. Tools from the Cave of La Mouthe (the south of France): 1-7 - Périgord; 8-14 - Aurignacian; 15 - Solutré; 16-22 – Magdalenian (Dons-maps-lamoute, n/d)*

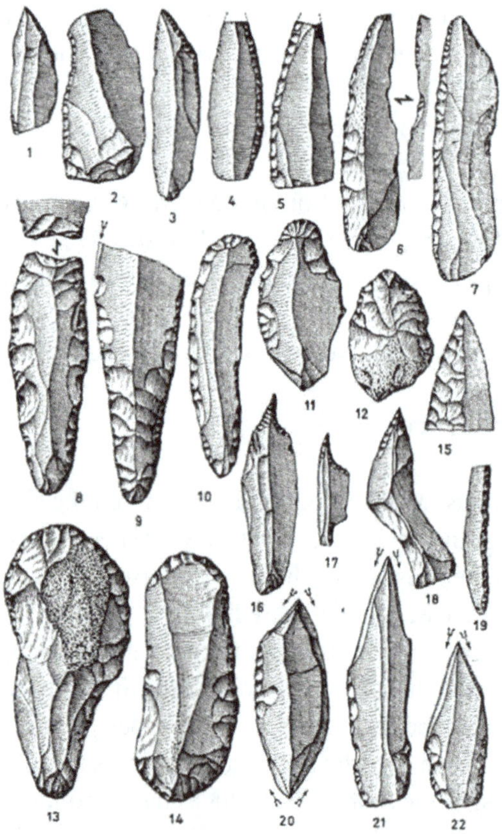

has been divided into several periods. For the Upper Paleolithic (50-10 thousand years ago, Figure 2), the following periods have been defined:

*Aurignacian*: 32-20 thousand years ago; typical tools made of large plates, scrapers, etc., edge retouching was applied more often along with the round contouring of the tools; ancient encampments: Geissenklösterle, Riparo Mochi, La Ferrassie, La Comba del Bonito, Laugerie-Haute, Krems.

*Périgord*: Mixed concept, sometimes being implied as to the earliest Aurignacian.

*Gravette*: 29-22 thousand years ago; typical bone darts with a split base, arrowheads with a wire edge, knives with a side notch, chisel, scrapers, small knives, bone arrowheads, awls, spatulas, and decorative pieces; ancient encampments of La Gravette, Le Flageolet, La Von Robert and Riparo Mochi.

*Proto – Magdalenian*: 23-20 thousand years ago; typically elongated tips, scrapers, drills, small chisels, sometimes with small teeth; despite the name, it is not the early phase of Magdalenian; ancient encampments of Laugerie-Haute and Abri Pataud.

*Solutré*: 22-16 thousand years ago; typical engravings, "Upper Palaeolithic Venus", leaf-shaped tips, pressure with bone or horn squeezers arrowheads with "laurel-leaved" and "willow-leaved" tips, tips with a handle and with a side notch, processing on both sides with a pressure flaking technique; ancient encampments of Laugerie-Haute, Combe-Sagnier.

*Magdalenian*: 18-10 thousand years ago; typical polychrome painting, tools with the shape of thin blades, lots of tools made of mixed materials, bone and horn, microlithization of tools, no tips, no pressure flaking technique, silica chisels prevail, a lot of piercing tools and scrapers; ancient encampments: Laugerie-Haute, Parpallo, Cueva Neria.

*Final Paleolithic*: 15-10 thousand years ago; typical microlithization of tools, tips with a notch, piercing tools, tools made of Buckhorn with silica inserts, one-sided notched teeth, Thich harpoons, tips with a stalk, with hoe-like liners made of buckhorn, thick harpoons, bows; ancient encampments of Ahrenshöft, Stellmoor and Meiendorf, Alt Duvenstedt.

The primitive painting reached its full flowering, during this epoch not only the technique of stone processing improved, but also the first real works of arts appeared, in particular, the feminine figurines called "Paleolithic Venus" (Maugh, Thomas, 2009), Figure 3.

*Figure 3. Paleolithic Venus (also called the figurines of the Great Mother) (Traditionalist-lvejournal, n/d)*

Nearly all of those figurines have similar features - wide hips, abdomen, chest, lack of feet. Their faces, arms, and legs are slightly designed. Sometimes the whole head looks like a hairstyle, however, the features of reproductive function have been depicted in rather an exaggerated and even grotesque way. A huge rear part, thighs, a pregnant abdomen, pendulous breasts... Probably, Venus was the image of Earth Mother, being pregnant with the dead ones, who should be born again to their eternal life. They were the objects of worship: the ancient people buried their dead ones into the ground, performed funeral rituals, and believed in their resurrection, which means they worshiped the Raw-Earth-Mother who gives us food, life, and rebirth.

Male sculptures are extremely rare. There are also some animal figurines. Almost all the sculptures were made of stone or bone, sometimes of fired clay.

It can be noted that both stone tools and sculptures, being the voluminous bodies, prove that their authors were well aware of the three-dimensional pattern of the surrounding world.

## Visual Series

As the most representative and studied images, the images of the paintings of the previously mentioned Cave of Altamira and the no less famous Cave of Lascaux use. In general, the most famous caves include Rouffignac (paintings were discovered in 1956), Del Romito (1961), Kapova Cave in the Southern Urals (1959) and Hoit-Tsenker Agui (1972) in Western Mongolia. Nowadays, there are more than one hundred and fifty caves, that had been painted by primitive artists in Western Europe only.

Altamira is a limestone cave located in the Cantabrian mountains of Spain, 30 km far from the city of Santander (Figure 4).

The cave is 270 m long (according to other sources its length is 400 m) and it consists of corridors and halls, the main hall is 18 m long and from 2 to 6 m high. The arches of the halls and corridors are painted with the drawings made with coal, hematite, ochre and some other natural dyes.

The drawings in Altamira depict bison, wild boars, horses and fallow-deer mainly; as well as the fingerprints and palmprints; some vague images depicting humans and persons with animal heads were discovered. Various signs were depicted also (Figure 5, they are interpreted by some researchers as symbols, while some others specify them as the so-called synoptric symbols).

*Figure 4. Plan of the Cave in Altamira*
*Photo: E. Pfeiffer. The Creative Explosion (Dons-maps-lamoute. n/d)*

*Figure 5. Signs (Quesabesde-notitias-nomada, n/d)*

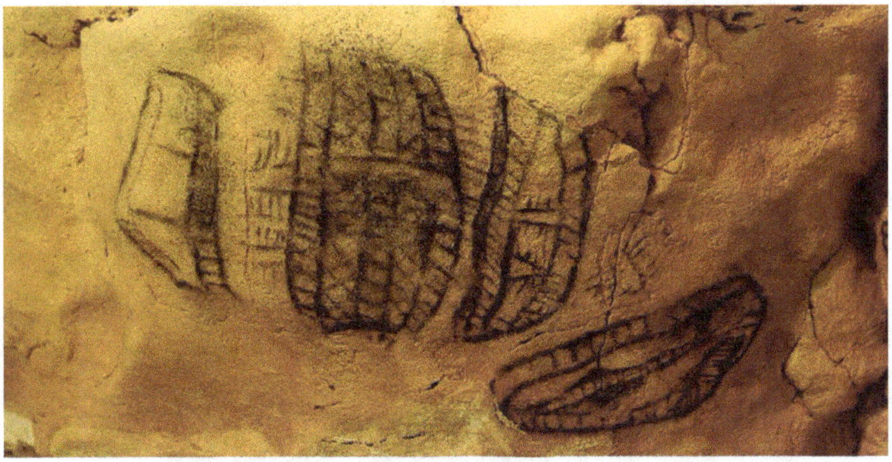

The dating of the Altamira rock paintings has been changed a number of times; according to the recent data, the oldest image is that one of a "key-like" symbol (35.6 thousand years ago - an early **Aurignacian,** or even being older), and much of the images belong to the Solutré epoch (since 22 thousand years ago).

In the first hall, there is a Greater Plafond decorated with 31 images of bison, deers, horses, boars (Figure 6).

Figures 7-9 represent some other images.

It's evident that the condition of the pictures leaves much to be desired, the color rendering looks different in various photos, therefore, let's try to apply trace drawing technique on the spots where the figures would be vague, as, for example, when looking at the image of a man with a bird's head (Figure 10).

The space depth is not evident within the pictures, it seems that space itself is very fragile, modified, wave-like moving, there are no fixed points of view and directions of the glance. There is neither a horizon or any landmarks. The images themselves are flat as well; there is no attempt to convey the volume by the play of light and shadow. The handprints don't mean the authorship of the artist, but

*Figure 6. Greater Plafond (trace drawings)*
*Photo: 1911 Encyclopaedia Brittanica Permission: Public Domain (Dons-maps-lamoute. n/d).*

*Figure 7. Handprints were performed by blowing ochre mixed with water over the hands, resulting in the effect shown here (Quesabesde-notitias-nomada, n/d)*

*Figure 8. Bison in Altamira*
*Photo: Original, Leroi-Gourhan (1992) (Dons-maps-lamoute. n/d*

rather an ownership sign for the cave. As for the shaman polymorphic images, perhaps, they had some personal characteristics.

The images often overlap each another, they are embedded into the relief, the orientation related to the floor, arches and mutual dimension is free, the proportions and scale are not maintained (Figures 11-14).

There seem to the real masterpieces; there might even be the animal "portraits" (Figure 15-16)

Those images help us trace the stylistic features of Paleolithic painting in the best way. The pictures are made with "stable" and "long" lines (some images of bison are 2.5 m high). Some sophisticated

*Figure 9. Horse*
*Photo: Guinea (2011). (Dons-maps-lamoute. n/d)*

*Figure 10. Bird-man of Altamira*
*This drawing was made originally by H. Breuil (Litos-perigord, n/d).*

colors have been applied, yet considering the limited palette of natural dyes; accurate details have been depicted, animals' images look recognizable, the mood and even the expressions of their "faces" become recognizable quite well. However, the natural proportions have been distorted and the images of the figures seem to "grow" out of the walls, taking into account relief and cracks over there. The compositions possess certain dynamic features, neither symmetry nor centering is evident. Besides, there are no reference points that could "fix" the sky or the earth, as well as the time motion or a point of view within the eternal continuum of pictures. However, the depth illusion is created - due to the relief, texture and color shades (as well as related to photo-editing in collections of illustrations, enhancing the brightness and contrast of images, which should be adjusted).

This style can't even be called the "naïve" realism, it looks rather like impressionism, being the message of the artist's living impression. Nevertheless, there is an evident contrast between the course and grotesque-looking Paleolithic Venus and the synchronous, elegant and confident cave paintings.

*Figure 11. Image overlay Photo: Guinea (2011) (Dosoavtor-2018-11, 2011)*

*Figure 12. A free orientation of images (Dons-maps-lamoute. n/d)*

*Figure 13. The image embedded into the relief (Kaplya-sveta-posle, n/d)*

The images in the Cave of Lascaux have been depicted in a similar technique.

The cave is situated in France, 40 km far from the town of Perigueux. It was discovered in 1940. Its galleries are about 250 m long, their average height is 30 m. The painted part of the cave is located mainly in its upper part (Figure 17).

Over a thousand images have been identified in the cave, there were 597 different figures among them. The archaeological findings in the cave have been related to **Solutré** and Magdalenian cultures,

*Figure 14. Disproportions and scale distortions (Quesabesde-notitias-nomada, n/d)*

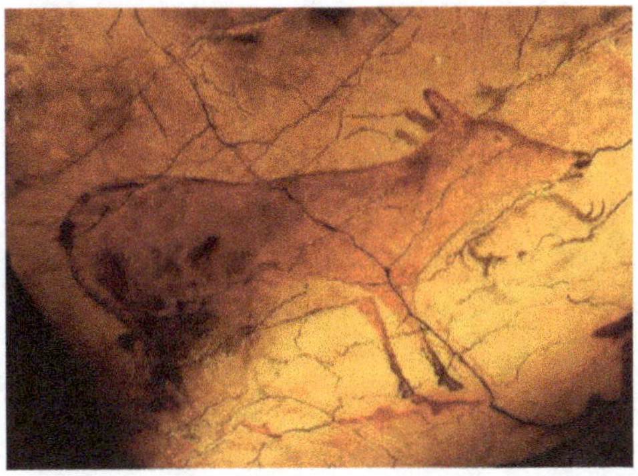

*Figure 15. Bison*
*Photo: Guinea (2011). (Dons-maps-lamoute. n/d)*

*Figure 16. Head of a bison (Cueva-museo-Altamira, n/d)*

according to the latest data, they date back to 20-21 thousand years ago, however, they correspond to **Solutré** period.

The presence of everyday objects, such as dart points, scrapers, chisels and lamps, dyes, drilled shells, and food remains, prove that those people stayed in the cave for quite a long time, although they hardly lived permanently there.

*Figure 17. Cave of Lascaux (Lascaux-culture, n/d)*

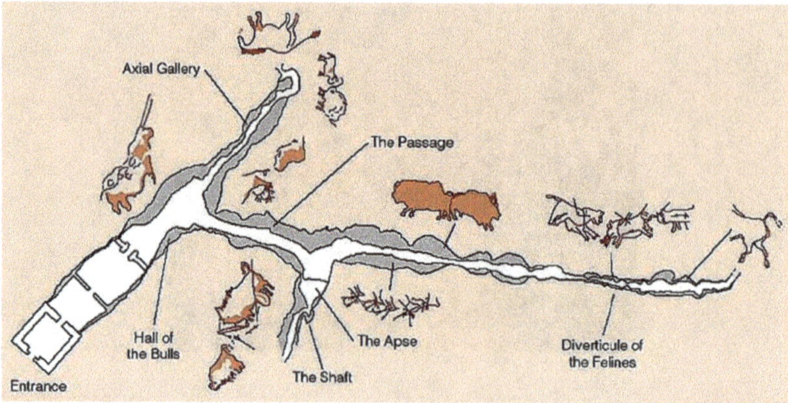


*Figure 18. Signs in the Cave of Lascaux, as classified by A. Leroy-Gourhan (Leroy- Gourhan, 1978)*

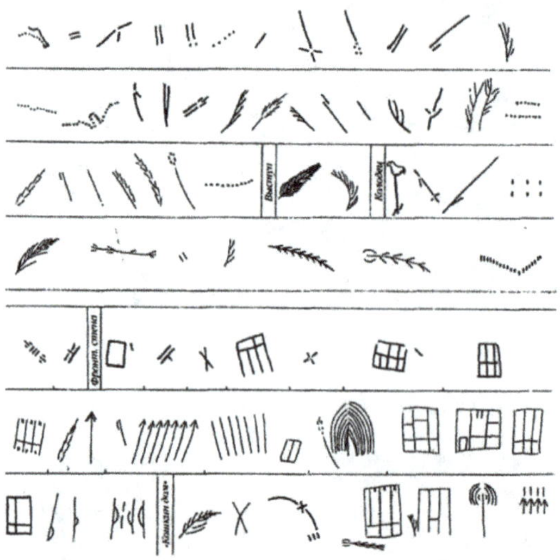

Similar to those found in the Cave of Altamira, the images of animals predominate over there: those of bulls, turs, deer, horses, goats, bears; the pictures of birds, persons, polymorphic images are fewer. There is also a series of signs that have been classified by the researcher A. Leroi-Gourhan (Figure 18).

It is quite probable that the combination of signs and animals, being mixed with the natural relief, is specific for the “hunting” magic, as well as the ideas about the food – the Mother Earth, who is giving birth to all new generations of people and animals. Some other interpretations might be possible.

Lets’ give a general idea of the composition of images and their sizes (Figure 19).

Similar to the images in Figures 6-16, space looks very changeable, however, there is a lack of depth and point of view (Figure 20)

It may ascertain the overlay of various images, disproportions and scale distortion, the free orientation of images, the use of the wall texture (Figures 21-23).

*Figure 19. Composition and scale of images (Pikabu-Story-sentyabr, n/d)*

*Figure 20. Lack of depth and point of view (Religion-peace-pervobytnaya, n/d)*

Similar to the paintings found in Altamira, it is necessary to notice the "portrait" characters of animals (Figures 24-26).

The general level of skills is rather low. Nevertheless, the same vivid character of impression, some dynamism, the presence of expressive images and certain details, a similar manner of performance should be underlined. Neither attempt to depict the dimension of figures is evident.

By way of conclusion, let's give an example of a polymorphic image (Figure 27).

Let's try to interpret this important scene (it's about 3 m long). A man wounded a bison (it can be seen the falling guts and a broken spear), and then a bison killed a man. And probably a rhino leaving to them should be to blame for everything?

However, let's have a more precise look at it (Figure 28). Why does a man have a bird's head (similar to the later gods of Egypt)? Why is there still a bird on a vertical pole? If a man takes a dead position, why is his phallus hard, is it agony? And what do those six points near the outgoing rhino mean?

*Figure 21. Image overlay. Puncturing technique along the contoured shallow lines (evident in Altamira) (Kaplya-sveta-posle, n/d)*

*Figure 22. Disproportions and scale distortion (Ngorod-doistoricheskie-pamyatniki, n/d)*

*Figure 23. Free location of images and use of the wall relief of the cave (Yandex-collection-cart, n/d)*

*Figure 24. Head of a bull (Kaplya-sveta-posle, n/d)*

Based on similar drawings, André A. Leroi-Gourhan Leroi-Gourhan, (1967, 1978, 1982) believed that the Cave of Lascaux was the sanctuary for the people of those times, and the image probably depicts a certain magic ritual or a rite with a complex plot, thus, it can only guess about its meaning.

However, there is such data that follow below (Blednova et al., 1998):

*Figure 25. Head of a deer*
*Author: HTO (2009) (Wikimedia. Lascaux. Megaloceros, 2009)*

*Figure 26. Scene (Vektork-naskalnaya-zhivopis, n/d)*

*Comparison of the composition of the fauna species with the animals' images immediately causes one of the cardinal questions. Due to the fact that there are a few or nearly zero animals' images depicted on the walls and ceilings, which would be a large proportion in the hunting species, it completely contradicts the concept that the rock paleolithic arts served as a magic means of hunting or fertility in the generally accepted sense found in the literature issues. It happens even though there are some images of the animals being stabbed with spears or arrows in the cave. It should be mentioned that 77.8% of reindeer bones found among all the animal bones of various species were discovered in the layer. However, there are no images of a reindeer on the walls, except for one. Thus, 60% of images of horses are completely*

*Figure 27. Bird-man (a man with a bird's head) (Yandex-collection-cart, n/d)*

*Figure 28. The right part of Figure 27 (zoomed image) (Yandex-collection-cart, n/d)*

*inconsistent with the actual remnants of those animals: the ratio of their bones was 0.8% only. A red deer: its images on the walls are depicted by 11%, while the images found in the layer are 1.5%; bulls; bones, including bison's ones: 6% are represented by the images, while 0 pieces were found in the layer; images of mountain goats are depicted by 2.8%, while the images found in the layer are 0 pieces. There are no images of a mammoth on the walls or found in the layer.*

Probably, maybe better interpret the pole with a bird being a totem (the generally accepted opinion - totemism arose much later)? Then the image is quite easy to read: the ancestor (a hard phallus) of the Birds' family (it is not clear which one he belonged to) was dead while hunting, having broken his spear. Thus, his spirit was embodied in a totem and continues to help the family. And here comes his advice - make the spears stronger!

The Neolithic era ends with the "Dark Ages" of the Mesolithic. The epoch of the Middle Stone Age (Mesolithic, 14-10 thousand years ago), due to its short duration, doesn't have any periodization; however, some cultures have a certain difference, in particular, the Azilian culture is implied (13-9 thousand years ago).

There are two opposing opinions concerning the Mesolithic arts. According to the first opinion, after the end of the Paleolithic in the development of arts, a certain pause took place that covered the Mesolithic epoch. The opinion is based upon a small number of monuments, that usually have a controversial dating. The least doubtful is the Petroglyphs of Stone Grave (Ukraine), Kobystan (Azerbaijan), Zarautsay (Uzbekistan), Bhimbetka (India).

According to another point of view, the evolution of the arts still takes place. A few images of people, animals, plants have been found. It is possible to note, that, the compositions of a great number of figures and scenes dominate within the cave paintings of the Mesolithic, which depict various moments from the routine lives of the hunters — hunting, violence, magical rituals. After that, the big works have been replaced by the smaller ones, with a detailed drawing technique and a large number of characters. Images are very conditional and dynamic, sometimes polymorphic, their meaning refers to certain symbols.

*Figure 29. A sculpture found in Lepensky Vir (Serbia), Late Mesolithic, possibly early Neolithic (Serbia-Lepensky-Vir, n/d)*

The sculpture works, in contrast to the Paleolithic Venus, become more complex, images of fantastic creatures appear (for example, a figure found in Lepensky Vir, being interpreted as a "fish-man" (?), Figure 29).

During this epoch, the climate changes, along with many animals and composition of their species alter, the hunting and fishing traditions are spreading all over. New tools are created, a bow and arrows appear, dogs are being tamed. Stone processing techniques are being improved.

For example, some silica spear points with a blunt bowed edge and a basement notch, as well as harpoons made of buckhorn, being shorter and harder than their previous model of the Magdalenian epoch are typical of the Azilian culture. The colored or engraved pebbles are particular elements of the time. The drawings are mostly painted in red color and look like oval spots, transverse zigzags, grids, stars, and they sometimes look like the stylized figures of the persons or animals. Those pebbles might have been used for magical purposes and could be compared with the Australian churinga, as the aborigines believed that they embodied the souls of their relatives. The beliefs of the Azilian people could also be judged by the discovery in the Cave of Grouse of Nette, where the burial of 33 skulls thickly covered with red paint was found. Those skulls were kept with the decorations of drilled shells and deer teeth. There is a presumption that the skulls' burial was related to some special culture or the trend for skulls' hunting. Maybe the Mesolithic people believed that they could steal the vital power of a murdered person.

It is necessary to note that by the time when the Mesolithic epoch ended in Europe, the first agricultural communities had already appeared in the Middle East.

Let's make the summary. Paleolithic Painting:

1) Has been characterized by spontaneity, bright colors, vivid features of plots and space itself, being dynamic, polymorphic and chaotic in its indefinite pattern.
2) Space is represented by two-dimensional and saturated polymorphic patterns. There are no elements of perspective images (point of view, horizon lines, vanishing points, principal ray). However, the depth illusion is created - due to the relief, texture and color shades.
3) Compositions possess dynamics, there is no symmetry or centering.

4) There are no reference points that would help to fix a skyline or a ground level, a timely motion or a point of view. Let's suggest some way – the top and bottom lines could be quite distinguishable in some images.
5) Natural scales and proportions have been distorted, as if the figures "grew" out of the walls, considering their relief and cracks.
6) Demonstrates the understanding of the very "soul" of animals, unity with them. Though totemism and the taming of dogs and other animals has been traditionally dated back to an earlier period, it can be supposed that this unity was the necessary condition for "convincing" the animal or feeling it as an ancestor of the family.
7) Polymorphic images are present.
8) There are magical or cult scenes.
9) There are abstract images.

It should be noted that the Paleolithic drawings make a very strong impression: the spectator feels the hunting excitement, he becomes involved in the magic ritual, he feels sympathy for the wounded animal. It is surprising a sensation of fresh spirit and significance of the event has been preserved since such ancient and primordial times.

They say that Pablo Picasso, having visited the Cave of Altamira, once exclaimed: "After the works found in Altamira all artistic works began to decline!".

There are questions related to the meaning of the primitive painting, the reason why a human used to become an artist, what kind of message he wanted to convey to the audience, why those cave paintings affect a modern person in this way and influence him so hard?

Let's leave this issue for a while and interpret the Myth of the World Tree.

## THE MYTH OF THE WORLD TREE: INCIPIENCE

There is a very small number of images similar to those depicted in Figure 30 (Golan, 1993), which could be interpreted as a reference to this myth.

The literature (Taylor, 1989; Golan, 1993; Propp, 2000; Eshleman, 2003) presents various points of view on beliefs of the Paleolithic era in general and related to the World Tree, in particular.

*Figure 30. The first images that might be interpreted as the World Tree (12-17 thousand years ago) (Golan, 1993)*

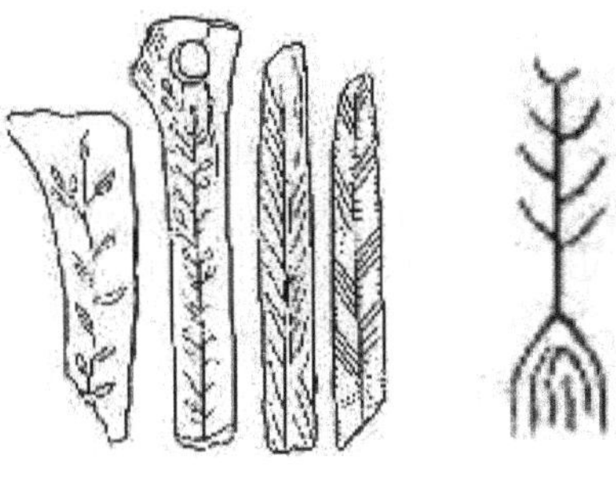

Naturally, no written evidence could convey the specific plot of the myth. Therefore, it can be reconstructed based exclusively upon the indirect data from later times. They include:

- the almost universal prevalence of the myth – as evidenced both in space (typical for many people's traditions) and in time (common for the practice of contemporary shamans), that proves both the ancient plot of the myth and its archetypal character;
- The Tree itself makes the space-and-time events to be structured and mean a symbol of their unity; the crown, the trunk, and the roots mean the heavenly, earthly and underground worlds, the identification of the Tree with the Axis of the world is quite common. The cardinal directions world and the seasons are being grouped around the Tree;
- relation to the cult of ancestors; their attitude towards the living humans is well-disposed;
- relative availability to travel along the Tree from one world to another, searching for eternal life in the world of heavenly spirits or reincarnation on the Earth.

A human was always been concerned about the thoughts of death and his desire to live forever ...

It can be supposed that at the very beginning of times the World Tree was not treated as a myth, but served rather a visual illustration – just like a tree drops its leaves every autumn, so a human body dies; just like the stem and roots keep themselves, so a human soul keeps on existing; as in the spring the leaves are in blossom, so the soul finds a new incarnation. The changes pass by, while the tree remains to unite and eternal.

This concept was an intuitive comprehension and everyone did not even require any complicated notions or cults. Besides, such explanations described a life cycle for any tree. So, the people were never afraid of death, because they knew that only their body was mortal.

As the intuition became weaker, the accessibility for the average person disappeared, and only shamans reserved the ability to travel through their "thoughts (squirrel) along the tree". The significant detail about this was that those shamans used to travel along an inverted tree, that is, they entered the underground world at once, then they ritually died and gained access to the other parts of the Tree. Thus, the Tree used to become a personal "shaman tree", and then it turned into the unique world concept. So, the people who had a weaker intuition and lost their access to the practice could only believe in shamans and worship the Tree. Thus, appeared a Myth itself.

Within the framework of this paradigm, a different interpretation of the scenes depicted in Fig. 27-28 could take place. The shaman dies ritually, turns into a bird, and the bird is free to fly, reaching the top of the Tree (the heavenly world) and can come back (the tree trunk is a symbol of the visible world, i.e. the world of the living creatures).

## MARKERS ANALYSIS

Before giving our explanation to the peculiarities of Paleolithic painting and the above reconstruction of the myth of the World Tree, let's consider the way how they have been interpreted by contemporary researchers. It can be defining the following points of view.

## Depicting of Space

*The primitive hypothesis* is focused upon the most obvious feature - the inadequate depiction of space - it explains this by the fact that the primitive people (similar to contemporary kids) did not know how to do this without being aware of the properties of space and perspective techniques or axonometric approach.

It would be reasonable to have a look at the pictures of the kids who have not yet been taught to draw and compare them. With the purpose of proper understanding, let`s apply (Osorina, 2016) some features of child psychology, it will help us determine the chronological boundaries of the sampling.

The first stage (up to 1-year-old) of children's drawing means scribbles – they are presented by the points, spots, and lines.

The next stage (1-2 years old): a child purposefully leaves visible traces of his presence in this world; he possessed the space of a paper sheet and identifies himself with those lines and points.

The next stage (2-3 years old): a child discovers the boundaries of a paper sheet. On approaching the limits, the lines lay along the boundaries of a sheet, go around its corners, trying to return inside the sheet. The child is aware of the boundaries of the situation with his actions. Further, the child opens the sign function of a drawing – he realizes that the lines, spots, and points mean something. They become the elements of graphic language through which the child begins to create the images of persons, animals, objects, and abstractions.

The next stage (3.5 - 5 years old): a child begins to represent the space of a sheet as the space of the world, which should be arranged to settle the characters. The main structure-forming principle in this "world" is the vertical separation of a paper sheet into the top, middle and bottom zones. This is the oldest principle of the symbolic organization of space, being embodied in various cultures in the image of the World Tree. This is the earliest dimensional scheme through which a child attempts to build a model of the world.

The next stage (5-6 years old, before going to school): a child's efforts in making the model of the world are combined with the abilities specific for the kids' subculture. Children appreciate the help of adults, but when they begin to learn drawing and use standard ways of building perspectives, the precious kids' drawings (being the mirrors of a child's soul) turn into inconspicuous images. So, the most valuable presentation is the drawings of the kids from 2 to 5 years old (Figures 31-33).

The looking at these figures, show the evident influence of adults.

A comparison of the children's drawings with the primitive images shows that they contain all 9 features which are specific for Paleolithic painting. It means that this hypothesis can be certainly justified.

Such points of view were followed by many of the first researchers; traditionally, they were skeptical about the artistic value of drawings. Here is a typical opinion concerning the motives and objectives of the primitive painting (Blednova et al., 1998):

*One of the main backgrounds for appearance and the initial rise of the figurative activity can be considered the leisure and some way of settled life, which could have appeared not earlier than certain progress was achieved in the struggle for existence. To spend some time on bone carving, carving the bas-reliefs, creating multi-colored compositions, an ancient human should have possessed a feeling of security and be provided with the most necessary things. All those activities appeared due to the formation of more or less permanent human groups and the emergence of new hunting methods which helped them keep the food reserves. The relation between the development of arts created in the caves and hunting for big animals was noted by the prominent specialist Henri Breuil. He underlined that this relation was not*

*Figure 31. Drawings of the kids (2-3 y.o.)*
*Author: Yegor. (Ufa-mama-forum, n/d)*

*Figure 32. Drawing of a girl (3.5 y.o.)*
*Author: Alice. (Baby-lessons_zima, n/d)*

*Figure 33. Drawings of the kids (5 y.o.) (Isweet. Risunki detej, n/d)*

*accidental and explained it by the fact that hunting for big animals developed the human imagination and skills, enriched the memory with lively, deep and keen impressions necessary for creating beautiful images of figurative art.*

In the Cave of Montespan located in France, the archaeologists found a statue of a clay bear with the signs of spear stubs. The primitive people might have associated the animals with their images, and by way of "killing" them, they could provide their success in hunting. Such findings prove the existing connection between hunting rituals and artistic activities".

Let's suppose that the ancient human "had nothing to do" or "trying to gain his success in hunting" used to come into a dark and cold cave, actually risking to meet a predator, get lost or fall into a rock crevice and spend days or weeks there, somehow getting up to the high ceiling, in order just to draw? Probably, he should have had some stronger, or just the otherwise, a vital motivation.

One should not exaggerate the ignorance of ancient people (as well as the rationality of contemporary humans). "The representatives of a few civilizations could be compared with the Australian aborigines in their taste for knowledge and speculative thinking, which are so close to amazing intellectual dandyism, if we come to remember that we mean the people who were at a rudimentary level of their material life," – C. Levi-Strauss concluded (Levi-Strauss, 1994). However, the Australian aborigines developed up to the rear lines of the world progress; while the tribes of Altamira, Lascaux or Levant achieved its highest peak – notwithstanding their similar material scarcity, their creative abilities were not worse than those of the aborigines.

It is impossible to survive in the two-dimensional space, and the three-dimensional Paleolithic sculptures confirm that the ancient people realized this basic feature, and our contemporaries do the same. It should be underlined, that the ancient arts developed over dozens of thousands of years, and if the artists felt the need to depict the three-dimensional space, they would find a way how to do it.

And, finally, the primitive crafts could not have had such a specific and powerful effect on a contemporary human. Therefore, such a hypothesis should be denied.

However, which concept could explain the similarity of children's drawings to the Paleolithic painting? The similarity of the states of mind: their intuition is still strong (spontaneous impressions and bright colors, lack of a point of view), but their ego (images of hands on the walls of caves and the possession of a paper sheet space, as well as their "self-portraits" and the family images in children's drawings) clearly manifested the will and the reason (while drawing, the author tries to make the world better, makes it structured in his way). No one should ever doubt that the Paleolithic painting has been created by the adults, and it's reasonable to compare it with the works of mature artists who have their standard methods of building perspectives.

Subsequently, three main approaches for comprehension of the primitive painting have been developed:

- Study and analysis of images, their classification, chronological and typological correlation with each other, comparison with the materials obtained from archaeological excavations.
- Understanding the images as the evidence of the spiritual world of our ancestors, the search for ethnographic and folklore relations to the drawings.
- Analysis of images as a specific cultural phenomenon, combining the features of archaeological monuments and works of art. Some interesting facts have been gradually collected.

In the Paleolithic art of France, the images of fish, reptiles, insects, and representatives of flora are extremely rare. Scandinavian, Siberian, and Ural rupestrian drawings and petroglyphs (Abramova, 1971) that have been found among taiga forests never contained the images of trees. The images of the rock arts rarely reflect the landscape elements (Coulson, Campbell, 2001). There is neither landscape, no background, the images are not merged with the natural environment and contain the main figures only.

In many cases, it can be saying that they are not real animals, but some mythical characters and fantastic creatures. Thus, in the Cave of Lascaux, there is an image of a fantastic beast with long straight horns and spotted skin. The mythological characters are presented by the dragon swallowing the Sun on the rupestrian drawings of Shishkino, or a similar monster from the Stone Island of Angara. Fantastic predators are engraved on the rocks in the Minusinsk Basin. These images excited the ancient artist and made a part of his view of life.

In this respect, the figures of animals with strange marks look quite interesting. Some Paleolithic images of Europe are covered with lines and dots, it can be seeing the images of swans with semicircles inserted into the contour on the petroglyphs of Onega. The hips of some deer in Siberia are decorated with spiral curls. All those elements have no relation to reality. However, it can find such characters in the plots of fairy tales and epics.

The African images of bulls and sheep with solar signs on their horns are particularly interesting in this respect. When a hunter was hunting after them, having seen a similar sign on their bodies, he realized that it was not an ordinary beast, but "the master of animals".

A significant part of the cave paintings is composed of various types of signs, ideograms, and symbols. There are the geometric figures next to the images of animals, birds, persons: they are rhombs, spirals, lines, zigzags, grids, etc. There are many such signs in French and Spanish cave paintings belonging to the ancient stone age. The works of the subsequent epochs contained even more of them. For example, the non-figurative trends predominate over the images of persons and even animals in the rupestrian drawings of Ural. Such a combination of realistic drawings with signs-and-symbols makes a definite distinction for the primitive art from the arts of the era of civilization.

Such statistics confirm that such drawings were the magical instruments for achieving a specific goal and gives the background for the hypothesis of magic and cults.

According to this concept, the primitive paintings are not the works of art in the modern sense, but the remains of witchcraft or a magical rite that has survived until our times. This view was held, in particular, by Leroi-Gourhan, who believed that the Cave of Lascaux has been an ancient sanctuary. The meaning of some images can be defined through the analogy from the ancient mysteries and rites, while the meaning of the other ones has been lost forever. The researchers Fraser and Taylor (Taylor, 1980) compared the cave paintings with the ancient myths collected by them (being the contemporary researcher, Golan performed such comparative analysis based on the material evidence from the Caucasian monuments (Golan, 1993).

There is a grain of truth in this hypothesis, though it lacks some definite features and fails to depict either the state of mind and spirit of the ancient artist or the mechanisms of influence on a contemporary human. The problem of the reality of the myth also remains unclear - is it a primitive explanation of the natural phenomena or a reflection of some non-systemic reality? This hypothesis also does not present any explanation of the specific transfer in space.

McKenna (1995) underlined that the importance of the state of mind of the ancient artist, thus this hypothesis became rather popularized by the notorious author G. Hancock. Based upon the research data about the influence of psychotropic substances on the state of mind (seeing synoptric signs and unusual

*Figure 34. Pictures by Pablo Amaringo (Amaringo, n/d)*

images, as well as comparing it with a peculiar technique of cave paintings), T. McKenna suggested that primitive painting has been created in a state of trance.

It is possible to agree that there is a rational aspect, in order to recognize the facts that shamans are known to use drugs, long dances, rhythmic music, and other techniques to enter the state of trance, besides, they are capable of receiving various types of visions, predictions, etc. in this state.

However, a comparison of the paintings made by the artist Pablo Amaringo (Figure 34) under the influence of a psychotropic drink called "ayahuasca" shows not only the similar features (distortion of space and image polymorphism) but also defined the difference (three-dimensionality of space); synoptric signs look similar to the images found in caves or churingas.

Let`s continue to find the comparisons based upon the analysis of the drawings of people who suffer from schizophrenia (Figures 35-36) and the difference will be evident again.

Here the difference is evident again – it's a sick and fantastic character, and the absence of truth is not typical of the Paleolithic artists. The reason is blocking a human intuition due to the overwhelming development of the ego.

In such a way, the mentioned hypothesis explains just the part of the facts.

Let's try to rationalize our point of view concerning the primitive painting.

*Figure 35. Drawings of a girl who suffers from schizophrenia*
*Author: Kate. (Trinixy. Risunki devushki, n/d)*

*Figure 36. Drawings of a man who suffers from schizophrenia*
*Author: A. Wolfley. (Pulson. Isrusstvo-risunki, n/d)*

Let us refer to the above-said idea that the artist always depicts his subjective personal space with all its levels and channels. However, he emphasizes that very part which means much to him due to some reasons, where his attention and mind remain focused.

What is this part exactly?

The space of Paleolithic painting is lively, changeable, emotional, rich in colors and power; it seems to create the images within the everlasting inexhaustible well-being, and a person, being united with him, "is begging" him to create the desired thing and to add this desire into his everyday existence.

Namely, this synergy of a man and space (features 1-6) indicates an intuitive connection and the first level. It means the predominant channel.

The sensations of one's abilities and power (despite the lack of a point and direction of vision, the author underlined his presence in the very fact of creating a picture) correlate with the second level (i.e. the ego). This channel has not yet been completely developed.

The features 7-8 mean the volitional impulses in an attempt to change the surrounding world for the better, but not for the worse (i.e. mind). Those attempts are strong enough to make us survive, but not sufficient to create the civilization – just like a contemporary man could not have done it (so, he could not have survived out in the wildlife). Accordingly, those channels are still stronger than ours.

The presence of abstractions and symbols (feature 9) means the development of the mind. So far, only "skill of knowledge" is being formed from those signs, it makes the subject synchronized with the perception of a particular object (the ternary relationship, chapters 2, 3). It means that the mind has not yet acquired its full power.

Finally, sense impressions appear here. Bright colors, dynamics, emotional reflections demonstrate their powerful character. Besides, a human needed a higher level of sensations in comparison to a contemporary human, to survive out in the wildlife.

What kind of motives and incentives could the artist possess?

By the moment when the phenomenon of the primitive painting appeared, the human brain underwent some degradation (chapter 1) – just due to its internal wave-like nature. The relation with the non-systemic factors, that offer energy and life, grew weaker. The most sensitive representatives of the human race had its sensation in such a way if they stayed at the transition stage after youth, when they could easily achieve everything, towards maturity, when they required effort - and they began to search for gods, eternal life and the lost paradise. This became their burning passion then.

But where did the ancient people look for these "non-system factors"? They exist neither on earth nor in the world of living creatures.

Then they used to depict the death scenes, those images appeared in deep and dangerous caves, in those places which were inaccessible for the living creatures and were natural for the dead ones, as if inside the womb of the Mother Earth the spirits of the ancestors live, who helped their family member in every way... probably, the dead were supposed to be the "grandfathers", "house spirits", "patron ancestors" (they could also be the animals), who had already found their way to eternity (the dead ones never change, but their continues and timeless features are a sign of eternity).

And then, the source of life itself is depicted in bright colors, being the hope of rebirth after the death of the body, while the grotesque images of Venus are created on purpose, as the symbols of the fertility of the Mother Earth, the burial mounds or hills mean the symbols of the pregnant womb.

But how did all those practices help the artist achieve his goal?

Since ancient times, two magical rules have been established - "both above and below," and "that is what the gods did before, the people are doing now," that means that when becoming similar to something, you can also possess its properties (isn't it the intuitive understanding of self-similarity and fractality of the worldwide systems?).

Now that the artist becomes similar either to the source of life in the depths of the Mother Earth, or to the dead one inside her womb, he is begging this "non-systemic" well-being and finally acquires an external impulse – i.e. the "inspiration" - and materializes it in his pictures, implementing "translation" from the language of the dead into the language of the living ones.

For a better understanding, it is possible to note that many techniques of augmented cognition, both the ancient and the modern ones, are based upon modeling the state of the dead. For example, let us describe the stages of achieving the augmented cognition (Samadhi), based on the classical description by Patanjali (Yoga Sutras by Patanjali (2000).

The first step is the isolation of one's senses from external stimuli (pratyahara). The dark, deserted and quiet cave is ideal media for this.

Next comes the cease of internal dialogue (nirvitarka, nirvichara), staying in the womb of the Earth and the dead world correlates with the silent and solemn mood.

Then comes the stage of purification of mind from our previous impressions and habits (wasana and samskara), which leads to the direct perception of concepts (samprajnjata). According to the modern experiments of putting humans into hydrospace media, a person begins to experience some color and auditory hallucinations after a while (purification from wasans). A long-lasting staying in the cave helps achieve a similar effect.

The next step is an immersion into a human ego and the feeling of some kind of enlightenment (adhya atma prasadanah) when the ego becomes dissolved. The experience of staying in hydrospace media, as well as the monks' practices of all times, prove this effect. Psychotropic substances might have been used for the same purposes:

"*We drank soma, we became immortal,*
*We went out to the light; we found the gods*!"

one of the hymns of the "Rigveda" describes this stage in such a way. It is important to note that the drawings of the people who suffer from schizophrenia or made by the artists who stayed under the influence of "ayahuasca" drink belong to the level of previous, unrefined and unenlightened sensations.

After the ego has ceased, an overwhelming feeling of unity, love, grace, life, “inspiration” (nirbija samadhi) arises, and it is the goal. All those achievements do not depend on a person only but appear due to the synergy with the non-system factors. The whole process is followed by the external signs of death - cessation of breathing (kevala kumbhaka) and pulse, immobility of the body, etc.

If the ancient artists managed to reach the final stage, their drawings looked like some kind of iconography, that depicted the non-system factors and seemed familiar and ordinary images, without any doubt about their heavenly nature, along with the signs - yantras (tools) that help structure one’s mind and make it feel some kind of “inspiration”. Eventually, those picture galleries could have been used as the sanctuaries, where they used to perform the initiations and so on.

However, this is not a categorical statement., because the above-mentioned interpretation refers to a significant part of the cave paintings, but it does not cover all the concepts.

Now let’s try to interpret the impact of Paleolithic painting upon a contemporary human being.

When we say “eternal” and “imperishable” beauty, we recognize its timeless nature. And this timeless nature itself makes an impact upon the most delicate human heartstrings and meets the most intimate desire of a human being – i.e. to live an eternal life. That’s why the primitive painting itself, being “swept away” from its timeless essence and being “charged” over with eternal life, makes its strong and emerging impact upon us and stays “symmetrical” to the channels of the first, second and third levels, being invulnerable to the devastating course of time ...

## World Tree

Does the reconstruction of the myth confirm the above-mentioned concept? Yes, it does. The same feeling of unity (the intuition), the same thirst for the highest power and eternal life (ego), the same expansion into all of the three worlds (the will) and the same question of choosing good or evil (the reason).

Any other facts known about the Paleolithic period haven’t been contrary to this point of view.

The drawings of children under the age of five years old who have developed intuition, but not developed their ego, will, mind and personal characteristics, confirm the concept.

It is time to make a summary of the above said and present the view of the Paleolithic mind.

## PALEOLITHIC MIND

It would be convenient to imagine the state of mind in comparison with a contemporary human being. But what kind of psycho type should be chosen? The “balanced personality” seems to be the best choice - a person should look like “callos kagatos” – a symbol of unity, harmony, and measure for the ancient Greeks, the ideal symbol for yogins - “balance means yoga”, as Krishna used to teach Arjuna in “Bhagavad Gita”, this concept is also fundamental for the contemporary self-perfection schools. Such an ideal ratio is illustrated in Fig. 8 of chapter 2, *Self-organization of complex systems*. It will be used as a background. The pale blue color will illustrate the development of channels during the Paleolithic era (Figure 37).

It is necessary to underline the previously mentioned term once again: the relative development degree can be valued qualitatively; the time for quantitative estimation has not come yet. It is some kind of a common portrait, whereas the mind of a particular person may be different.

*Figure 37. Paleolithic mind*

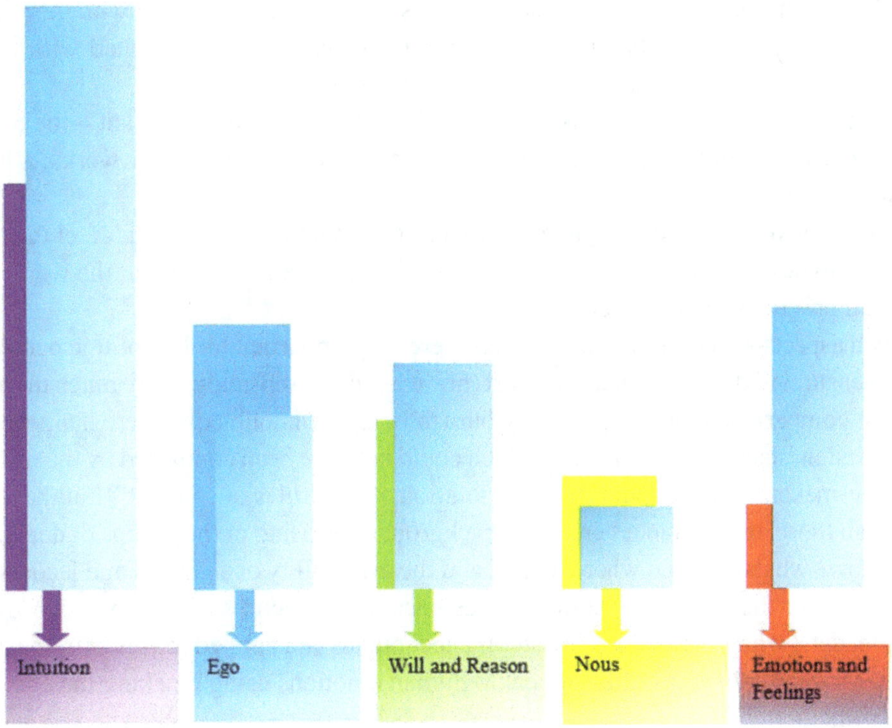


## BEHAVIOR PATTERNS

Let's use Table 7 of chapter 4, *Self-organization: evolution and behavior*, which correlates the perception channels with the social groups and their behavior patterns.

The dominant intuition correlated the Paleolithic human as a person who belongs closer to the contemplative type of character. Thus, such a relationship as love, mutual help, readiness to sacrifice themselves for the good of the family (at least for the sake of the members of one family) prevails even after death. This interpretation has been confirmed by the numerous findings of the remnants of people with severe lifetime injuries, of very ancient persons' findings, funeral rituals, patron - ancestors. Those people who possessed the perfect intuition were the mostly worshiped persons in those times because they were able to predict whether the hunting would be successful, warn about the natural disasters, avoid conflict with neighbors, tame an animal, speak to the ancestors. There were the persons who gradually became the shamans. Women, who possessed good intuition, also played an important part. Probably, did such relations predetermine this era that would ever be mentioned as the Golden Age?

The ego has not yet become a dominant factor, but the "trend" developed in this respect, some persons managed to surpass the "norm" and this event contributed much to the emerging leaders. However, their power was not unconditional yet; the common opinion of the family was much more important. Accordingly, the leaders played their significant role mainly while hunting, being the fellow countrymen.

With time, being in hunting and suffering from the hardships of hunger and cold, in transition periods and battles, the human will become stronger. The scope of aggression was not aimed at the family members; the people identified themselves mainly as related to "us" and "them" because their ego has not been developed yet. Cro-Magnon people devastated or exterminated the Neanderthals, the remnants of people with pierced skulls, fractures, stuck arrow tips have been found, along with the recorded cases of cannibalism, it confirmed the fact that the aggressive behavior was associated with a strong will. However, the times of wars of all against all has not yet begun.

The desire for self-perfection and making the world better was also significant – the dangerous environment of the caves could not prevent the ancient artists from creating their works, a human being never gave up in all life situations.

The mind has not yet acquired its dynamics; therefore, the human being didn't feel the need to have the whole things to be duly arranged. The beliefs, speech and general ideas about the world were primitive, precise and pragmatic, respectively.

Finally, such aspects as sensations and feelings were very important for Paleolithic people, as well as their sensual pleasures and physiological comfort (however, those sensations were much more significant and resistant in comparison to a contemporary human being). Undoubtedly, there happened numerous outbreaks of passion, scandals, and disputes which could not have been controlled by the mind yet. However, the trouble-makers were not punished for violation of the "rules of conduct", but they were exiled.

In general, all those circumstances created a background for rising of the groups of dependent people who did not know "what to do and where to go", and the availability of shamans and leaders was a mere survival condition for them, the people themselves became the victims of manipulation by the part of shamans and leaders, who were quick to master the techniques and applied for the extreme emotionality of the human. Modern politicians often appeal to human emotions using blocking the ego, reason, and mind of a person.

There is also another aspect related to the discussion of matriarchy tradition in the Paleolithic era. Highly developed intuition, as well as extra sensations and feelings - all those features, apart from sexism, correlate with the female type. Intuition helps us find a solution in an unusual situation, relief a conflict, preserve the knowledge and convey it, which is rather important for the survival of the community.

Besides, the men are much more predisposed to aggression and binary logic, took their leadership role in military operations, hunting and in some other situations when the decisive power actions were required (Boehm, 1999).

If also take into account a considerable number of figurines of the Great Goddess and the absence of the images of "male gods", then the matriarchy (not the absolute, but the situational one) took place.

But if the comparison of the Upper Paleolithic with the Golden Age and Paradise is true, then the end of this era was the expulsion from paradise. Population growth and deteriorating climatic conditions (almost like in our time) provoked migratory waves and numerous conflicts over hunting grounds and places convenient for living. Brutal wars were the result of this crisis (Kelly, 2000). The social consequences did not slow down either - the importance of leaders increased; the influence of women decreased. The behavior has also changed, but this will be discussed in the next chapter.

## CONCLUSION

The features of Paleolithic painting and, in particular, the space images have been determined. It managed to make the explanation for those features within the framework of the approach suggested in chapter 4, *Self-organization: evolution and behavior,* as a consequence of the certain status of the channels of human interaction with the environment. While analyzing the contemporary interpretations of Paleolithic painting, it managed to apply a range of conclusions, to reconstruct the state of mind of the Paleolithic people, while some of them had to be declined. It was possible to verify and confirm the reliability of the selected markers of the evolution of the human mind.

The additional material for the reconstruction of the state of mind has been obtained during the analysis of the myth of the World Tree - the efficiency of this marker has been also confirmed in such a way. The diagram illustrated in Figure 37 demonstrates the state of mind of a Paleolithic human that can be assessed in qualitative, but not in quantitative terms. It confirms the forecast mentioned in chapter 4, *Self-organization: evolution and behavior*. Despite the absence of a quantitative component, the scheme will help perform the comparative analysis of the state of mind of human beings from another epoch, as soon as any similar schemes would have been developed for them. This factor will be significant for determining the evolution of the mind as a whole.

Having reached the peak of its development by the beginning of the Paleolithic epoch, the human brain was structured and "optimally designed" for the intuitive perception of non-systemic factors. That is the reason for its extreme complexity as compared to the survival demands. The subjective space was mostly intuitive at this time. Subsequently, due to the wave nature of the human-environment system, the human brain underwent a certain degradation and his mind passed to the descending levels. As we're already noted, this fact is not surprising - the development of structures at any level and the increase in their potential is followed by the decrease in the potential of the previous (parent) level and, probably, by the decrease of the subsequent level either, so, this effect could be partially compensated by means of exchange with the environment. This process took place within the subsequent periods. The sacred painting is a means to prevent the separation and stimulate the reverse process (convolution). The primitive painting possessed this feature to a greater extent.

## FUTURE RESEARCH DIRECTIONS

Since the efficiency of the selected markers has been confirmed, they should be further applied, to obtain the actual data about the state of the human mind within other epochs. Therefore, those data will be applied for verification of the evolution forecast mentioned in chapter 4, *Self-organization: evolution and behavior*.

Another goal of the research is an attempt to establish the quantitative indicators of the status of certain channels - however, there are no required data for this, and probably they will never appear.

## REFERENCES

Abramova, Z. A. (1971). *Primitive Art*. Novosibirsk, Russia: NSU.

Amaringo. (n.d.). Retrieved from http://cannabismoon.net/ru/gallery/psycho/amaringo.html

Baby-lessons_zima. (n.d.). Retrieved from http://www.babylessons.ru/zima-detskij-risunok-alisa-35-goda/#more-1171

BenitoJ.-M. (2006). Retrieved from https://commons.wikimedia.org/wiki/File:Techo_de_Altamira_(replica)-Museo_Arqueológico_Nacional.jpg?uselang=ru

Blednova, N. S., & … . (1998). *Primitive Art. Problem of Origin*. Kemerovo, Russia: KPI.

Boehm, C. (1999). *Hierarchy in the Forest: The Evolution of Egalitarian Behavior*. Harvard University Press.

Clayton, E. (2003). *Juniper Fuse: Upper Paleolithic Imagination & the Construction of the Underworld Paperback*. Wesleyan University Press.

Coulson, D., & Campbell, A. (2001). *African Rock Art*. Harry N. Abrams, Inc.

Cueva-museo-Altamira. (n.d.). Retrieved from http://www.cuevamuseoaltamira.com/

Dale Gutrie, R. (2005). *The Nature of Paleolithic art*. Chicago, IL: University of Chicago Press.

Dons-maps-lamoute. (n.d.). Retrieved from http://donsmaps.com/lamouthe.html

Dosoavtor-2018-11. (2011). Retrieved from http://dosoaftor.ru/2018/11/28/naskalnaya-zhivopis-altamiryi/

Golan, A. (1993). *Myth and Symbol. Moscow, Russia:* Russlit.

Heaven, P., & Furnham, A. (1998). *Personality and Social Behavior*. Routledge.

Isweet. Risunki detej. (n.d.). Retrieved from http://isweet.com.ru/risunki-detej-5-let-foto.html

Joiner, A. D. (1985). The origin of art / A. D. Joiner. Moscow, Russia: Art, 1985.

Kaplya-sveta-posle. (n.d.). Retrieved from http://kaplyasveta.ru/posle-altamiry/altamira-impressionizm-15-000-let-do-n-e-chast-2-2.html

Kelly, R. C. (2000). *Warless societies and the origin of war*. Ann Arbor: University of Michigan Press. doi:10.3998/mpub.11589

Kovalyov, Yu. N., & Nitsyn, A. Yu. (2012). Depicting the Subjective Space in Painting. *Technical Aesthetics and Design, 8*. 49–56.

Lascaux-culture. (n.d.). Retrieved from http://www.lascaux.culture.fr/#/fr/02_00.xml

Leroi-Gourhan, A. (1967). *Treasures of Prehistoric*. New York: Harry N. Abrams.

Leroi-Gourhan, A. (1982). *The Dawn of European Art: An Introduction to Palaeolithic Cave Painting*. Cambridge, UK: Cambridge University Press.

Leroi-Gourhan, A. (1978). Signs in Prehistoric Art: Maseno.

Levi-Strauss, C. (1994). Primitive Thinking. Moscow, Russia: Republic.

Levi-Strauss, C. (2001). Structural Anthropology. Moscow, Russia.

Lewis-William, D. (2002). *The Mind in the Cave*. London, UK: Thames and Hudson.

Litos-perigord. (n.d.). Retrieved from http://lithos-perigord.org/spip.php?rubrique25

Maugh, I. I., & Thomas, H. (2009). Venus figurine sheds light on origins of art by early humans. *Los Angeles Times*. Retrieved May 14, 2009.

McKenna, T. (1995). Food of Gods. Moscow, Russia: Publishers of Transpersonal Institute.

Ngorod-doistoricheskie-pamyatniki. (n.d.). Retrieved from http://ngorod.net/doistoricheskie-pamyatniki/peshhera-lasko/photo-4/

Osorina, M. (2016). Secret Children's World of in the Space of the Adult's World: Peter.

Pikabu-Story-Sentyabr. (n.d.). Retrieved from https://pikabu.ru/story/12_sentyabrya_1940_goda_byila_naydena_peshchera_lasko__quotsikstinskaya_kapella_naskalnoy_zhivopisi_5335111

Pomerantseva, N. A. (2006). Primitive Art. From Ancient Cultures to Early Civilizations: [album]. Moscow, Russia: White City.

Propp, V. Ya. (2000). Historical Backgrounds of the Fairy Tale. Moscow, Russia: Labyrinth.

Pulson. Isrusstvo-risunki. (n.d.). Retrieved from https://pulson.ru/iskusstvo/risunki-shizofrenika-12-foto.html

Quesabesde-notitias-nomada (n.d.). Retrieved from http://www.quesabesde.com/noticias/nomada-altamira-museo,1_5259

Rauschenbach, B. V. (1986). Systems of Perspective in Visual Arts. General Theory of Perspective. Moscow, Russia: Science.

Rauschenbach, B. V. (2001). The geometry of Picture and Visual Perception. Saint Petersburg, Russia: Azbuka-Classic.

Religion-peace-pervobytnaya. (n.d.). Retrieved from https://religionpeace.ru/pervobytnoe/pervobytnye-pecshery.html

Serbia-Lepensky-Vir. (n.d.). Retrieved from http://44serbia.ru/lepenski-vir-nahodki-i-zagadki/

Stolyar, A. (1985). The origin of the visual arts. Moscow, Russia: Art.

Taylor, E. B. (1989). Primitive culture. Moscow, Russia: Science.

Traditionalist-livejournal. (n.d.). Retrieved from https://ltraditionalist.livejournal.com/617658.html

Trinixy. Risunki devushki. (n.d.). Retrieved from https://trinixy.ru/143795-risunki-devushki-stradayuschey-shizofreniey-14-foto.html

Ufa-mama-forum. (n.d.). Retrieved from https://www.ufamama.ru/forum/messages.aspx?topicid=9392

Vektork-naskalnaya-zhivopis. (n.d.). Retrieved from http://vectork.org/naskalnaya-zhivopis-peshher-lasko/

Wikimedia. Lascaux. Megaloceros. (2009). Retrieved from https://commons.wikimedia.org/wiki/File:Lascaux,_Megaloceros.jpg?uselang=ru

Yandex-collection-cart. (n.d.). Retrieved from https://yandex.ru/collections/card/5a07c8eb0c1ed2002fed3a3e/

Yoga Sutras by Patanjali. (2000). Moscow, Russia: Janus books.

# Chapter 6
# Neolithic:
## Separation and War

## ABSTRACT

*The examples of Neolithic paintings created on different continents are given. Particular attention is paid to clarifying the dating and identification of paintings. Other types of artistic and production activities of people are considered, as well as data on their social, religious, and cultural achievements. Along with the well-known galleries of Neolithic painting in Tassili (Algeria) and the Spanish Levant, the lesser-known ones, in particular, the Stone Grave (Ukraine), are considered. Changes to the World Tree myth in the Neolithic are closely related to the cult of the dead. The correlations of markers and the state of levels and channels of subjective space are determined. A generalized psychological portrait of a man of the Neolithic epoch is compiled, and his behavioral patterns are described.*

## BACKGROUND

Sources of Neolithic pictorial art are presented in three groups as it was for Palaeolithic age.

These are reproductions of works presented in digital form.

Further, a historical and archaeological publications which makes it possible to specify chronology of Neolithic age, dating and identification of pictorial artworks as Neolithic ones (Lhote, 1962; Abramova, 1971; Mykhailov, 2005), and consider examples within other achievements of the epoch (World history, 1955; Blednova et al., 1998).

Let's pinpoint multiple publications about Kamyana Mohyla (geological and archaeological dating, description of petroglyphs, their interpretation, and data on other findings) (Veselovskiy, 1893; Rudynskyi, 1961; Danylenko, 1986; Mikhailov, 2005).

The next group of sources is reconstructions of techniques of drawings, the ways of space relations depicting are expressed as well as interpretation of their motivation of the authors (Rauschenbach, 2001; Kovalyov & Nitsyn, 2009).

The last group of the sources is dedicated to the myth of the World Tree and cult of the dead. Note that many of them describe later representations, but based on them, it is possible to model Neolithic version (Golan, 1993; Propp, 2000; Popol-Vuh, 1993; The Gilgamesh epic, 2015).

DOI: 10.4018/978-1-7998-1706-2.ch006

## NEOLITHIC PAINTING: EXPANSION AND CULTURAL CONTEXT

**Dispersion.** The numerous samples of Neolithic pictorial art are famous all around the world. Imagine the samples and their dating, trying to cover a wide space both in a geographical and in a chronological sense.

Europe (Figure 1).

American continent (Figures 2-3).

*Figure 1. Painting in Magura cave (Bulgaria). More than 700 paintings were discovered, dating from 8000 to 4000 years ago. Plots – hunters, dancing people (shamans?), multiple animals (Wikiway. Bolgaria-peshera, n/d).*

Asia (Figure 4).

*Figure 2. "Cave of hands" – Cueva de las Manos (Argentina). 13000 to 9500 years ago. Plots – left hands, hunting scenes and pictures of animals. (Proistoria.Assets-galeries,n/d).*

*Figure 3. Caves of Serra da Capivara (Brazil). 25 000 years ago. Plots – rituals, hunt, trees, animals. (World-rock-art, n/d).*

Africa (Figures 5-6).

*Figure 4. Caves of Bhimbetka (India). More than 600 paintings. The oldest painting is 12 000 years old. Sometimes there are ranked as Mesolithic ones. Plots - people, bison, tigers, lions, and crocodiles. Author: Aleksey Verbov (Lsvsx-livejornal, n/d).*

Australia (Figure 7).

*Figure 5. Cave drawings in Laas Geel (Somalia). Dating – from 11 000 to 5000 years. Plots – cows, people in ceremonial dress, domesticated dogs, giraffes (World-rock-art, n/d)*

*Figure 6. Cave drawings in Tadrart Akakus (Libya). The most ancient ones – approx. 14 000 years ago. The paintings reflect the changing climate of Sahara. About 9 000 years ago Sahara was full of green and lakes, forests and wild animals, in particular, giraffes, elephants, and ostriches. (World-rock-art, n/d).*

Besides the caves there are many rock paintings; they are related to bone carving and reliefs pressed

*Figure 7. Cave painting of Kakadu (Australia). Dating – up to 20 000 years. Plots – rituals, animals. (Megalitica. Naskalnaya zhivopis, n/d)*

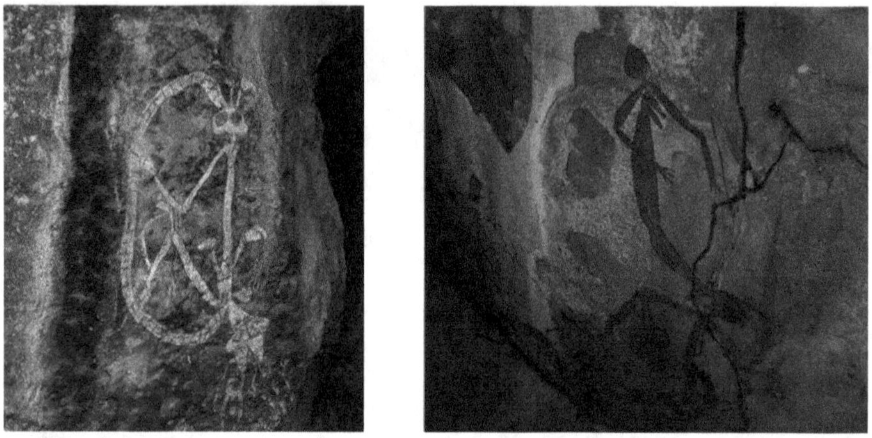

in soft stone, for example, sandstone. Such a technique is known on the territory of Ukraine as well, in particular, in Kamyana Mohyla near Melitopol city.

Maybe that some of the presented drawings are considerably older than the nominal line of Neolithic age - usually dating is 10-8 thousand years.

## Culture Context: Material Culture

As compared to Palaeolithic age, human achievements in Neolithic age are presented in the following directions:

1) **Stone Processing**: Techniques of two side chipping off, nagging, drilling, polishing is accomplished. For example, polished axes (Figure 8) look like jewelry. By the end of this age, the first megaliths emerge. Stone is used to building structures of different purpose.
2) Selection of plants containing carbohydrates and proteins (North Africa, Middle East, Northern Iran – wheat and barley, and inter stream area of the Yangtze River and Huang He and Indus Valley - legumes and rice, Mesoamerica - maize and batata (corn and potato).
3) Domestication of animals in which meat and milk contain proteins necessary for a human.

## Culture Context: Sculpture and Ceramics

Bizarre Palaeolithic Venus is not the only topic of sculptures within that age; more realistic figurines appear, perhaps, even toys (Figure 9).

Molding in clay and pressing out were used to make ceramic storage jars covered with diversified ornaments and sometimes painted with mineral pigments (Figure 10). Particular ornaments may be considered as cult ones.

*Figure 8. Stone axes. Great Langdale. Ambleside, Great Britain. Author: Michael Greenhalgh (Wikimedia. Neolithique-001, n/d)*

*Figure 9. Bear. Samuskyi grave field, 4000 BC. (Civilka. Samurskiy medved, n/d)*

## Culture Context: Magic and Cults

Without going into an analysis of the plots of myths and the content of cults and magical practices, it is possible to note their development: the complication of the plot, no longer hypothetical, but the explicit appearance of shrines, the allocation of shamans and priests as a separate estate.

## Culture Context: Social Relations

It is necessary to highlight two aspects.

*Figure 10. Storage jar. Culture of pit-comb ceramics. Russia (Civilka. Sosud-neolit, n/d)*

First of all, an increased craving for enlargement and aggressiveness. Neolithic pictorial art is rich in battle scenes, there are also pictures of execution (Figure 11), and entombments contain numerous skeletons with penetrating skull injuries and heavy intravital injuries.

The research results of modern primitive tribes indirectly witness the scale of violence (Warspot. Pervobytnaya-voyna, n/d):

"According to (Lawrence, H. K. (1996; Gat, 2006), about 90–95% of studied societies related to 37 traditional cultures take part in military activities against their neighbors.

This general tendency is typical for the most different tribes: aboriginals of Australia, Indo-China, and New Guinea, Indians of Amazon region, grain growers of African savanna and hunter-gatherers of tropical forests.

Conflicts with neighbors arise permanently in all monitored groups and it is the main cause of death for 24-35% of men between 15 and 49 years old. Among Yanomami of Ecuadorean Amazonia, 15% of the adult population (24% male and 7% female) die a violent death within several generations in sequence from the moment anthropologists started to study them. On New Guinea in Dani tribes 28,5%

*Figure 11. A scene of execution. Neolithic age (?). Rock pictorial art of the Levante in Spain (according to Nash 2005:Figure 3) (Anropogenez, n/d)*

men and 2,4% women die a violent death, Euga – 34,8% men. In the tribe of Goilala 29 people from 150 have become victims of tribal conflicts within 35 years of observation. Although the female death rate is considerably lower – from 4 up to 7%, they encounter high risks of intra-tribal violence. It is also typical for men of the tribe. And in this case, inter-communal riots may cause more deaths than even conflicts with other tribes".

Consequently, war chiefs became a separate social category; all adult and healthy men were the fighters.

Then, large settlements emerge. Perhaps, Çatalhöyük in Turkey (Schmidt, (2015) is the example worth attention (Wikipedia. Chatal-Khyuyuk, n/d).

The settlement had existed for 1400 years (7100-5700 BC) and it was located on two hills); its area is estimated as 13-20 ha, and only its minor part is excavated.

The general view of the settlement is shown in Figure 12.

There are no passes between the houses, also there were no doorways. And they entered the house from the roof where stairs lead inside. The entrance was also a ventilation hole and it was meant to extract smoke from open fireplaces.

The houses have only one floor, area – 11-48 м$^2$, there are main premises with the benches, platforms, containers for food, kitchens, and warehouses. Walls were built of sun-dried earth brick; bearing construction of the roof - wooden beams. Presumably, 5-10 people lived there.

The houses were also used for burial. The departed ones were buried under the floor of the houses, mainly, under the fireplaces and other internal high points, with gifts: precious and semi-precious stones, weapons, clothes, and wooden jars.

An about the third part of the houses had paintings and sculptures of bullheads. Some of the frescos expressed geometrical design, other – scenes of hunting, men with an erected phallus, animals (Figure 13).

Deep layers don't contain ceramics, and higher layers have crockery and sculptures (Figures 14-15).

*Figure 12. Excavated part of Çatalhöyük (Dostoyanie planeti. Chatal-Kyyuyuk, n/d)*

*Figure 13. Frescos of Çatalhöyük (Dostoyanie planeti. Chatal-Kyyuyuk, n/d)*

*Figure 14. Baked clay crockery from Çatalhöyük (Dostoyanie planeti. Chatal-Kyyuyuk, n/d)*

Neither houses of special status nor workshops have been discovered till present. Probably, it witnesses that there was weak social stratification and no centralized control existed. Also, houses decorated with sculptures of bulls were used for everyday life i.e. they can't be considered as sanctuaries as we understand it. Multiple bullheads and sculptures of the Great Mother imply that people worshiped the Bull and the Earth Mother [1] - a deity of men and deity of women. Also, no evidence of war was detected – it's obvious that there are no houses of the war chiefs.

The recent research gives the idea that hunting and gathering were the main food sources of Çatalhöyük inhabitants. And agriculture and domesticated animals played an only inferior role.

It follows that in the Neolithic age two society models existed: votaries and war chiefs didn't come forward in those regions where local resources made it possible for people to live and satisfy their needs and there were no external enemies; in those regions which lacked these conditions, society differentiation progressed rapidly. Thus, the end of the Neolithic age witnessed considerable growth of those elements of human activities that build civilization. This age is called the Neolithic revolution for the right reasons.

*Figure 15. Bullheads from Çatalhöyük (Dostoyanie planeti. Chatal-Kyyuyuk, n/d)*

## Dating Issues

Before beginning to analyze how Neolithic pictorial art depicts expanse, it's necessary to clarify which samples origin from a particular period. Different researchers refer to the same drawing to the Mesolithic and Neolithic age or Neolithic and Eneolithic age. If date them according to formal sings, some pictures (Figures 2, 3,7, etc.) will belong to Palaeolithic age.

But it's the incorrect approach: people settled in new areas in a different time and brought their more or less developed culture, faced a favorable or hostile environment, peaceful or aggressive autochthons, and thus Neolithic age started and ended in a different time in different places. In the end, even in the up-to-date age of the Internet and smartphones, tribes are living in remote regions of the planet which still use stone tools. That's why archaeologists embrace all findings related to material, social and cultural legacy – and only then they form an opinion which artifact belongs to a particular epoch.

The task to check estimates on self-organization theory related to the evolution of the conscientious doesn't make such troubles: a condition of conscientious is defined by special aspects of expanse depiction and stages how myth about the World Tree evolved and not specific names of historical periods which is simple to use as they are conventional. Such an approach erases the issue of precise dating and it is correct to use it further on.

# NEOLITHIC PAINTING: THE IMAGE OF SPACE

## Visual Imagery

Rock paintings of the Levante in Spain are the most famous and fully studied samples of the pictorial art of Neolithic age. There's an opinion that they partially belong to the Mesolithic age. There are illustrations of the tendencies at the end of the Stone age (Figures 16-18).

The expanse became flatter: similar paint makes it impossible to create the illusion of the third dimension even for particular characters. It becomes more "tranquil" as if it gives movement to the characters. Still, there are no orientation points – horizon or viewpoint.

*Figure 16. Hunting scene. Cova dels Cavalls) (Spain) (Genofond, n/d)*

*Figure 17. Primeval war of Neolithic age (?). Rock painting of the Levante in Spain (World history vol.1 Stone age, 1955)*

The characters emerge from the stone and disappear in it. Despite that pictures are sketchy and stylized; some details look very realistic.

Agility retains its positions and even reinforces them; the hunting episode changes and became more "pragmatic".

*Figure 18. Group hunting, ritual dance (?), fighters running (World history vol.1 Stone age, 1955)*

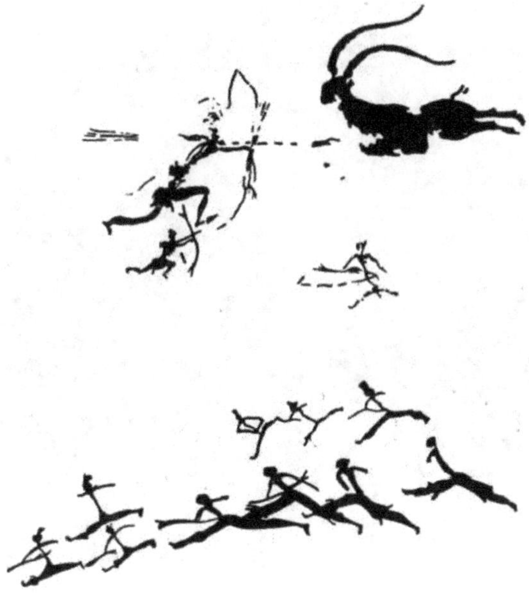

The next drawing shows how social relations became more complicated and aggressiveness spread in Neolithic age – there are groups of fighters and the drawing itself celebrates the victory of one tribe over another (Figure 17).

*But man bade man seek Antiar —*
*One look sufficed...*
*With poison did his henchmen smear*
*The prince's arrows at his orders,*
*And to his neighbors far and near*
*Death sent him forth across the borders. ...*
Antiar by A.S. Pushkin,

Another drawing consisting of several episodes also shows "triumph" of community spirit – "one for all, all for one" (Figure 18).

Let's take a look at several examples – frescos from Tassili upland (Figure 19) in Sahara (Algeria).

The frescos are dated 9 000-2000 years ago and they are some sort of chronicle depicting how Sahara had turned into a desert. They show a striking resemblance to the Levante paintings (Figures 20-23).

There are lots of polymorphic characters (Figure 23).

Other examples are similar enough; it's not necessary to present them.

Let's highlight distinguishing features of the Neolithic pictorial art and compare them with Palaeolithic ones. Thus, it is possible to separate the same 9 special features, as it was in chapter 5, *Paleolithic: Unity*.

Figure 19. The region of Tassili, present time (Rutravel. The region of Tassili, n/d)

Figure 20. Drawings of animals. The region of Tassili (Rutravel. The region of Tassili, n/d)

*Figure 21. Scenes of hunting, battle scenes. The region of Tassili. (Rutravel. The region of Tassili, n/d)*

*Figure 22. Peaceful coexistence. The region of Tassili (Rutravel. The region of Tassili, n/d)*

*Figure 23. Polymorphism. The region of Tassili (Rutravel. The region of Tassili, n/d).*

*Table 1. Special features of Palaeolithic and Neolithic pictorial art comparison*

| Feature | Paleolithic Pictorial Art | Neolithic Pictorial Art |
|---|---|---|
| 1. | It is marked by frankness, vivid colors, a sparkle of plots and expanse itself – dynamic, polymorphic, chaotic in its uncertainty | It is marked by sketchy character, grotesqueness, the monotone of pictures, contrast with the background, dynamics of pictures, the background is inactive |
| 2. | The expanse is depicted as two dimensional and deeply polymorphic. There are no elements of perspective drawing (point of view, horizon, a point of convergence, main beam). But you feel the illusion of depth – due to the relief, texture, and tonality | The expanse is depicted as two dimensional. There are no elements of perspective drawing (point of view, horizon, a point of convergence, main beam). There is no illusion of depth |
| 3. | The design is dynamical, there is no symmetry or center | Design is dynamical, there is no symmetry or center |
| 4. | There are no starting points which would make it possible to fix sky or ground, a motion of time or point of view. Let's make a proviso – some scenes indicate top and bottom | You may distinguish top and bottom, left or right |
| 5. | Natural scale and proportions are distorted, the characters seem to grow out of the walls considering relief and cracks | Natural scale and proportions are distorted, the characters look stylish but sometimes bizarre |
| 6. | Drawings show understanding of the "spirit" of the animals, unity with them, "portrait nature". Although totemism and timing of the animals are dated for the later periods, it is possible to assume that it was this unity which became the condition for taming the animal or they feel it as the ancestor of their bloodline | Images of the animals and people are reflected in a generalized sense, there is no "portrait nature", but many details of the behavior and movement are depicted precisely; violence and battle scenes are presented in large quantities |
| 7. | There are polymorphic drawings | There are polymorphic drawings |
| 8. | There are magic or cult scenes | There are magic or cult scenes |
| 9. | There are abstract pictures. | There are abstract pictures. |

## THE MYTH OF THE WORLD TREE AND THE CULT OF ANCESTORS

The link between the myth of the World Tree and the cult of ancestors is outlined as follows: as it was mentioned in chapter 5, *Paleolithic: unity*, the first stage of the "shaman travel" meant that he should get to the underworld.

The myth developed within Neolithic age to assess how difficult such travel was. Let's examine two plots that were known from the sources of later periods but which have a very archaic origin.

First of it tells us about the departed which helps the living humans. The buried ancestor (under house floor) becomes house spirit which protects it from all unfavorable magic influences if it is treated well. In Slavonic variant, there is a direct link between words "dom" (home of living people) – "domovina" (coffin – home of the dead) – "domovoy" (guardian spirit of the house). In ancient times house as well as the coffin was made of wood, and it turned out that living human as well as the dead enclosed themselves in it, identified with it – here's another link with the myth about the World Tree. The ancient custom of a similar nature was practiced even in Medieval Europe. They put human remains (or even immure a living human) into the base of the castle or bridge pillar. It should have made them indestructible. It is presented in the most refined form in India. Guardian spirit (Purusha) takes care of a house of the human (Purusha) whose soul (Purusha) is in the heart or in the eye (it has the size of a thumb – here you have an idea on Tom Thumb in many fairy tales of different nations). And finally, the whole Universe was created from the parts of a corpse of the giant (Purusha) ritually dismembered by the gods. Moreover, his parts correspond to four varnas (social categories) of Indian society. From the magic point of view, similar names mean the identity of the beings. As it may seem, the myth of Purusha realizes ideas about the unity of micro and macrocosm. But it has considerable amendment – unity is accomplished due to the dead who binds space and human home with the world of the dead.

Such an idea could stem from the hunch of the departed as the relatives, unity with them and even love to them. These are Palaeolithic beliefs as hunch prevailed. Consequently, the transition to the world of the dead, staying there and return to the world of the people or the divine spirits didn't frighten people and it was considered as easy and common.

Another plot presented, for example, in Popol Vuh or Descending of Inanna to the underworld which depicts the spirit world as a dark and frightening place. It is a world of absolute deception, cruel rulers craving to destroy the soul and body of a human who dared to come to their home and sacrifice them. It's impossible to pass the way of the dead and succeed. Even the goddess Inanna could return only after she gave away the soul of her lover instead of hers. And even the great hero Gilgamesh was deceived when he almost accomplished his aim and having lost the chance to gain immortality, he could barely return home (Popol-Vuh, 1993; The Gilgamesh epic, 2015).

And such beliefs developed using a base that matches the spirit of the Neolithic age with its binary opposition, aggressiveness, fear, and mistrust with a weakened feeling of unity in the background. Let's illustrate the above by analyzing the results of the ancient stone sanctuary study (Veselovskiy, 1893; Danylenko, 1986; Mikhailov, 2005; Rudynskyi, 1961; Kifishyn, 2001) (Figure 24).

It is considered that the most ancient painting performed by scratching of the sandstone, is more than 20,000 years old, and the most recent ones belong to the Bronze Age. It contradicts geological dating which claims that deglaciation of the last glacial period in that region started only 17 thousand years ago, and the hill gained its present look only 11 thousand years ago (Geological origin of Melitopol sandstone Kamyana Mohyla (Geological origin, n/d).

*Figure 24. Kamyana Mohyla. View from the nearest hill. Melitopol, Ukraine.*

The object is of great interest because certain symbols of Sumerian wedge writing, as well as an image of Egyptian funeral boat and the goddess of death Meret-Seger, were discovered there. So, in the later periods, Kamyana Mohyla became a global center of the cult of the dead. However, no entombments were found except for the single grave of the Hunnic war chief.

What made an ancient human think that this place was the entrance to the underworld?

Transition to the spirit world is connected with descending – descend to the valley of the river Molochna where Kamyana Mohyla is located is long enough for anybody to feel that they're stepping down into the inevitable.

The entrance to the underworld is in wild and deserted terrain – the hill of the Mohyla with its boulders, cavities, rock shelter (Figure 25) reflects sharp contrast with a surrounding wasteland. Even currently when the area in the vicinity of the Mohyla is thickly settled, the nearest village Myrne is 2 km from the hill.

*Figure 25. Boulders, cavities, rock shelter of the Kamyana Mohyla. The impression of chthonian powers manifestation*

*Figure 26. Sacrifice stone amid shaman stones (in the center). In the distance, you may see the museum of an archaeological site.*

The spirit passing to the underworld crosses the particular border, desert, sea, river (Lethe, Acheron, Styx, Smorodina) – and here a meander of the river Molochna closely approaches the hill. If the river had the same name in the ancient times, it took place another association – with Milky Way. The underworld has its entrance – some bridge, portal, gates. And there is analog of it here – huge boulder having shape similar to the heart. It lies in the very place where the river closely approaches the hill. As for another element of the underworld – snakes are regular beasts of this area even nowadays.

Then the spirit walks along the way to the place of trial or condemnation. And here the only low grade to the top of the hill starts from the heart-shaped boulder. And the place of condemnation – a staff of the archaeological site calls it sacrificial stone (still there is a cavity where the sacrifice was offered and groove where the blood ran) in the midst of five "shaman" stones (Figure 26). Finally, the spirit ascends to the sky (which is in a good light from the hilltop) or descends deep into the depths (across cavities and rock shelters).

The terrain itself serves as a perfect background of the world of the dead (ternary relationship, chapter 2, *Self-organization of complex systems*), so you may either get exposed to its influence or affect it yourself. Drawings made on the walls of the rock shelters give you food for reflection[2]. Let's try to reconstruct it.

Thus, a human ought to walk the way of the dead. He descended to the valley and crossed the river. He felt everything that every dead man felt. When he got to the top, he was ritually sacrificed and became dead "quite literally". For the rest of the ritual, he was in a trance state when illusions replace reality.

A human tried to deceive the rulers of the underworld as he put on the mask and chimerical garment (Figure 28). Supposedly, he tried to defeat them using his strength. To show his victory he stepped on the head of the killed wood snake (Figure 29). Thus, he became immortal in the world of divine spirits and finished his quest across the worlds of the World Tree.

To be fair let`s highlight that there are only a few cult pictures on the walls of the caves. There are fine scenes of everyday life (Figure 30) or mysterious plots (Figure 31), which person of today would consider a sketch map with symbols of rivers, crossings, forests and settlement sites.

However, they don't give any new information for us to analyze special aspects of the way expanse is depicted or plot of the myth on the World Tree.

*Figure 27. A skeleton-like character appearing from the sand shows the scratching technique of pictures*

*Figure 28. The character of the shaman (?) (interpretation of B.D. Mikhailov)*

*Figure 29. Ahead of the wood snake with a symbol of the foot.*

*Figure 30. The scene of everyday life. Presumably, it belongs to later periods*

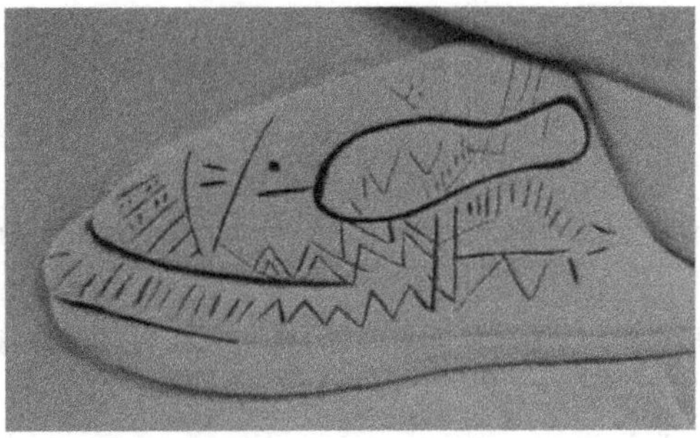

*Figure 31. Images, signs, and ornaments which may be interpreted in different ways*

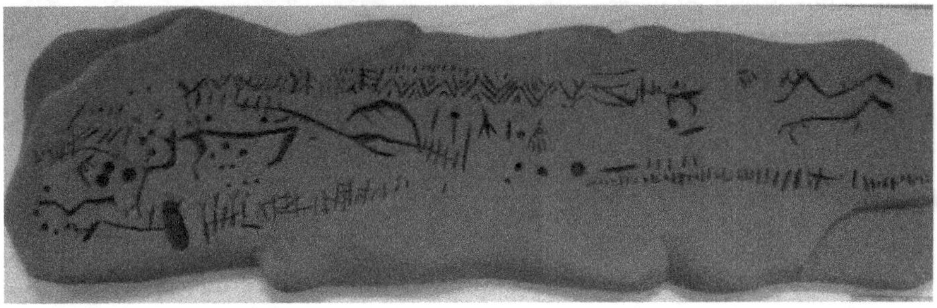

The results of these assumptions are recapped in Table 2 which compares the feature of the Palaeolithic (chapter 5) and Neolithic variants of the myth.

## MARKERS ANALYSIS

It's time to analyze markers.

*Table 2. Features of the Palaeolithic and Neolithic variants of the World Tree myth comparison*

| Feature | Paleolithic Variant | Neolithic Variant |
|---|---|---|
| 1. | The Tree patterns expanse – time and it is a symbol of their unity; canopy, trunk, and roots correspond to the overworld, earth и underworld, the Tree is usually identified with the celestial axis; the Tree is surrounded by the cardinal directions and seasons | The Tree patterns expanse –time and it is a symbol of their unity; however, favorable overworld and earth are opposed to the hostile underworld, the Tree is usually identified with the celestial axis; the Tree is surrounded by the cardinal directions and seasons |
| 2. | A human is identified with the Tree | A human is identified with the Tree |
| 3. | Link with the cult of the dead; their attitude to the living people – well-intentioned | Link with the cult of the dead; their attitude to the living people – hostile |
| 4. | It's possible to move from one world to another along the Tree to gain eternal life in the overworld or reincarnation on the earth | Only the great heroes and shamans are capable of moving from one world to another along the Tree to gain eternal life in the overworld or reincarnation on the earth unless they conduct rituals in special places |

## A Depiction of Expanse

As compared to Palaeolithic age (Table 1), contrast became sharper (binarity, reason), there is dynamics of the characters and inactive expanse (binarity, reason), quantity of violence and battle scenes (ego and will), expanse orientation is improved (nous), feeling of unity with the image (intuition) and vividness (feelings) diminish. Other special aspects – first of all, the two-dimensional expanse – retain.

## The myth of the World Tree and its Relation with the World of The dead

Earth and underworld are forcefully opposed to each other and there's their contrast (binarity), they felt hostility and danger of the underworld (mistrust, weakened hunch) and how hard the transition was (stratification is in progress: shamans and heroes – common people), and they kept faith that it's possible to pass it (weakened hunch still worked).

Let`s move to the conclusions on the state of mind using the diagram of the same scale and definitions as it is on Figure. 37 in chapter 5, *Paleolithic: Unity.*

## NEOLITHIC MIND

It is necessary to underline that it is a general portrait as a mind of a certain person may differ from it.

*Figure 32. Neolithic mind*

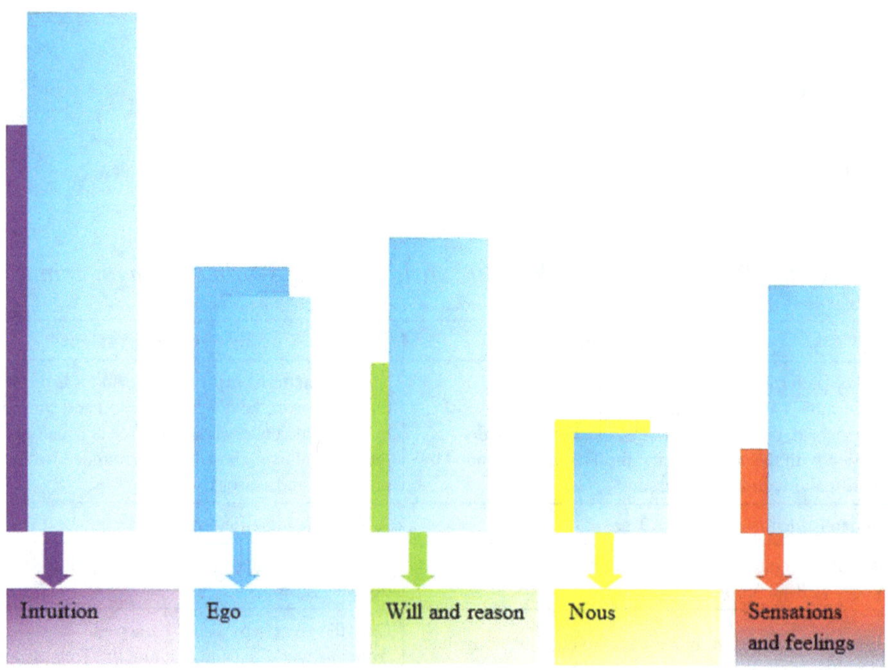

## BEHAVIOR PATTERNS

Let's use Table 7 from chapter *Self-organization: evolution and behavior,* which links perception channels with estate groups and patterns of their behavior.

The hunch is weakened but it continues to prevail. It makes a base to keep "Palaeolithic" relations with the members of the community - love, help, and willingness to make sacrifices (transition to the overworld via underworld to gain wellness for the community) under favorable conditions. People still bury their ancestors under the house floor and within long term settlements sites and such practice became even more widespread. At the same time, people arranged special burial ground outside settlements. The women having excellent hunch were still respected members of the community but they lost monopoly over the gods.

Ego became stronger but it was not enough to get absolute priority of the property and inequality. It's highly unlikely that nobody ruled such big settlements as Çatalhöyük but it's probable that it was an authority of the elders but there was not descendible right to rule.

War conflicts were all-around (they became less violent only by the end of the epoch), as well as a need for hunting. It created conditions for will and mind related to it to develop. They used binary categories which both markers confirm.

It seems unbelievable but it was wars that make it possible to form binary thinking - a base for mathematics logic and it happened due to the potential of the hunch!

Nous became sharper and people needed to streamline subjective expanse which was reflected in the pictorial art and differentiation of the worlds of the World Tree. People built large structures and settlements, improved techniques to process stones, make crockery and burn it. These achievements of the Neolithic revolution and many others – all of them witness strengthened nous.

Sensations, emotions, and feelings became less important. It created psychological pre-condition for people to live in large settlements; in turn, such living weakened sense organs and emotions faded away – other-wise people would slaughter each other.

Thus, above behavior models – "military" related with speeded up differentiation of the society and external expansion and "peaceful" which is reflected by the artifacts of Çatalhöyük find an explanation in channels developed to a different extent which is typical for certain communities. Generally speaking, it will prevail in the first one and mind – in the second one. And if Palaeolithic hunters (traditionally male activity) evolved into aggressive cattle breeders, Palaeolithic gatherers (traditionally female activity) – into peaceful grain growers.

And here it's advisable to return to the ratio of matriarchate and patriarchate elements of the Neolithic society. Perhaps, co-existence of male and female gods which were considered as a spouse (Çatalhöyük) has psychological pre-conditions – weakened female intuition and feelings were balanced with male ego, will, and reason which created a base for harmony between matriarchate and patriarchate elements.

## CONCLUSION

Special features of the Neolithic pictorial art and were defined, in particular, the depiction of space and comparison with the Palaeolithic ones was done.

It's possible to explain the changes occurred within the approach presented in chapter 4, Self-organization: evolution and behavior, as a consequence of the evolution of the human-environment interaction channels.

The additional materials to reconstruct mind were obtained when remodel and analyze the Neolithic variant of the myth of the World Tree was done. They confirm conclusions made.

The diagram on Figure 32 shows the state of the Neolithic human mind but it may be assessed only as a qualitative but not quantitative.

If in the Palaeolithic age weakened hunch and loss of gods related to it invoked discomfort in the spiritual sphere. And palliatives in form of sacral pictorial art and particular ritual and magic practice were quite enough to overcome this crisis. But in the Neolithic Age, the crisis showed up more acutely and in many fields. And it took all the achievements of the Neolithic revolution to restore peace of mind.

## FUTURE RESEARCH DIRECTIONS

The same approach will be applied to analyze the next historical epoch – Eneolithic age presented with the structures defined by the term "proto-civilization".

## REFERENCES

Abramova, Z. A. (1971). *Primitive Art*. Novosibirsk, Russia: NSU.

Anropogenez. (n.d.). Retrieved from http://antropogenez.ru/article/771/

Blednova, N. S. (1998). Primitive Art. Problem of Origin. Kemerovo, Russia: KPI. (Dostoyanie planeti.

Chatal-Kyyuyuk. (n.d.). Retrieved from http://dostoyanieplaneti.ru/2732-chatal-khyuyuk

Civilka. Samurskiy medved. (n.d.). Retrieved from http://civilka.ru/wp-content/uploads/2015/08/samuskiy_medved.jpg

Civilka. Sosud-neolit. (n.d.). Retrieved from http://civilka.ru/wp-content/uploads/2015/08/sosud_neolit.jpg

Danylenko, V. M. (1986). *Kamyana Mohyla*. Kyiv, Ukraine: Naukova Dumka.

Gat, A. (2006). *War in Human Civilization*. New York: Oxford University Press.

Genofond. (n.d.). Retrieved from http://генофонд.рф/?page_id=7277

Geological origin of Melitopol sandstone Kamyana Mohyla. (n.d.). Retrieved from http://www.arabatka.in.ua/?p=1441

Golan, A. (1993). *Myth and Symbol*. Moscow, Russia: Russlit.

Kifishyn, A. G. (2001). The ancient sanctuary of Kamyana Mohyla. Deciphering experience of a proto-Sumerian archive of XII-III millennium BC. In 7 volumes. Kiev, Ukraine: Aratta.

Kovalyov, N. (2009). The Image of Space in the Primeval Epoch. Technical aesthetics and design, 6. pp. 27–33.

Lawrence, H. K. (1996). War before Civilization. Oxford University Press edition.

Leroi-Gourhan, A. (1978). Signs in Prehistoric Art: Mazeno.

Lhote, H. (1962). The Search for the Tassilli Frescoes. Moscow, Russia: Publisher of Oriental literature, p. 140.

Lsvsx-livejornal. (n.d.). Retrieved from https://lsvsx.livejournal.com/482239.html

Megalitica. Naskalnaya zhivopis. (n.d.). Retrieved from https://megalithica.ru/naskalnaya-zhivopis-nourlangie-kakadu-rock-art.html

Mets, F. I. (2001). Samuskiy burial ground. Nations and cultures of Tomsk-Narym Priobye: Materials for the encyclopedia of Tomsk region. Tomsk, 129-130.

Mykhailov, B. D. (2005). Petroglyphs of Kamyana Mohyla: Semantics. Chronology. Interpretation. Kiev, Ukraine: MAUP.

*Popol-Vuh*. (1993). St. Petersburg, Russia: Nauka.

Proistoria. Assets-galleries. (n.d.). Retrieved from http://www.proistoria.org/assets/galleries/184/cueva-de-las-manos-4.jpg

Propp, V. Ya. (2000). Historical Backgrounds of the Fairy Tale. Moscow, Russia: Labyrinth.

Rauschenbach, B. V. (2001). The geometry of Picture and Visual Perception. St. Petersburg, Russia: Azbuka-Classic.

Rudynskyi, M. Ya. (1961). *Kamyana Mohyla (framework of rock paintings)*. Kyiv, Ukraine: Publishing House of Academy of Sciences of USSR.

Rutravel. The region of Tassili. (n.d.). Retrieved from https://www.rutraveller.ru/place/13545

Schmidt, K. (2015). *Premier temple. Göbekli tepe (Le)*. Göbelki Tepe. CNRS Editions.

Stolyar, A. (1985). The origin of the visual arts. Moscow, Russia: Art.

(1955). World history: Vol. 1. *Stone age*. Moscow, Russia: Gospolitizdat.

*The Gilgamesh epic ("On the one who sees all")*. (2015). St. Petersburg, Russia: Nauka.

Veselovskiy, N. I. (1893). Report of Imperial Archaeological commission for the year 1890, Saint Petersburg.

Warspot. Pervobytnaya-voyna. (n.d.). Retrieved from https://warspot.ru/5116-pervobytnaya-voyna-masshtab-tragedii

Wikimedia. Neolithique-001. (n.d.). Retrieved from https://commons.wikimedia.org/wiki/File:Néolithique_0001.jpg?uselang=ru

Wikipedia. Chatal-Hyyk. (n.d.). Retrieved from https://ru.wikipedia.org/wiki/Чатал-Хююк

Wikiway. Bolgaria-peshera. (n.d.). Retrieved from https://wikiway.com/bolgaria/peshchera-magura/

World-rock-art. (n.d.). Retrieved from http://worldrockart.ru/risunki/serra-da-kapivara-braziliya/

World-rock-art. (n.d.). Retrieved from http://worldrockart.ru/risunki/tadrart-akakus-liviya/

World-rock-art. (n.d.). Retrieved from http://worldrockart.ru/risunki/laas-gaal-somaliland/

## ENDNOTES

1 Word "Zemlya" (Earth) in Russian originates from the name of the goddess Semele depicted in form of the cow.

2 To avoid property, abuse the rock shelters are sanded and a museum of National historical and archeological site "Kamyana Mohyla" shows their identical copies. The museum staff kindly permitted us to take photos of them. Sometimes the sand is wind swept and it's possible to take photos of the originals, as, for example, skeleton-like character (Meret-Seger?) on Figure 27.

# Chapter 7
# Eneolith:
## The Search for Balance


## ABSTRACT

*Since the number of Eneolithic cultures has grown considerably in contrast with the previous epochs, it was necessary to determine the criteria and select the cultural content. The choice of the Cucuteni-Tripolye and the Harappa-Mohenjo-Daro, as well as of the Eurasian steppe cultures has been substantiated, and the artifacts for analysis have been selected. Comparisons of the space depicting features with the characteristics of Neolithic painting have been performed. A detailed reconstruction of the World Tree myth has been carried out, and features of the Eneolithic version have been compared with the Neolithic ones. The correlations of markers and the state of levels and channels of subjective space are determined. A generalized psychological portrait of a man of the Eneolithic era is compiled, and his behavioral patterns are described.*


## BACKGROUND

Relatively short in length (from the end of 4000 up to the beginning of 3000 B.C.), the Eneolithic period is interesting, first and foremost, as a transitional period encompassing the prehistoric times of the Stone Age when there was no writing yet and the historic times for which written evidence already exists which one can to rely on. A possibility to trace changes in the human mind over time, in between the "barbarism" and the "culture" is also thought valuable.

Among the Eneolithic painting monuments are rock paintings, reliefs, paintings on rock steles and slabs, painted pottery and certain seals. Art of sculpture is perfected, whilst artistic images continue to be used in embellishing the tools of labor and articles of a household.

Sources for Eneolithic painting are very diverse and are represented both by the results of archeological research of the cultures in question and by theoretical generalizations made both within the framework of history and history of arts.

Accordingly, it is necessary to examine exhibits from museums, reproductions of the pieces of art represented both in the albums and electronically.

DOI: 10.4018/978-1-7998-1706-2.ch007

Historical and archeological publications made it possible to adjust the chronology of Eneolithic cultures, the dating and the identification of paintings (World History, 1955), as well as to consider the images in the context of other achievements of that epoch (Griaznov, 1999); Gromov, 1997). It is this that enabled us to substantiate an approach to the choice of cultures that are viewed as those containing the required content.

There is a number of publications which are specifically dedicated to the culture of the Cucuteni-Trypillia (Gerasimov, 1960; Passek, 1961; Chernish, 1982; Ellis, 1984; Mantu, 2000; Videiko, 2010; Menotti, Korvin-Piotrovsky (Ed.), 2012) whose authors express varying views as to their dating, origination, rise and decline.

The same publications devoted to Harappa-Mohenjo-Daro culture (Hunter, 1934; Bongard-Levin, Ilyin, 1971; Albedil, 1991; Farmer, Steve, et al., 2004; Parpola, et al., 1987-2010; 2010; Rao, 2010), especially to proto-writing problem. Some authors propose that the symbols should be treated as the signs of writing, and make attempts to decipher it, while the others deny this.

Other publications represent the Eurasian steppe cultures (Vadetskaya, 1980, 1986; Leontiev, Kapelko & Yesin, 2006). The results of recent studies, accompanied by numerous and high-quality maps and illustrations, are given on site (European steppe-eneolithic, n/d).

Mindful of the tasks of this research, it would be advisable to focus on those specific features of the artifacts which make it possible to reconstruct the changes in the condition of Eneolithic human while setting aside lots of most interesting materials.

Therefore, of primary interest are the reconstructions of visual representation techniques, of peculiar features for expressing spatial relations, as well as the interpretations of their meaning and incentives of their authors.

It is advisable, remembering the goals of the study, to focus on those artifacts that make it possible to reconstruct the change in the human mind during the Eneolithic era, leaving aside many interesting materials.

First of all, it is of interest to reconstruct the technique of images, the features of the transfer of spatial relations, as well as the interpretation of the meaning of images and the motives of the authors.

Among the many plots traced in the mythology of this period (Rybakov, 1965; Golan, 1993), the myth of the world tree stands out, reflecting ideas about the space of the Universe, acting forces, human interaction with them, as well as the goals of such interaction. Therefore, reconstructions of this myth are important.

## CHOICE OF ENEOLITHIC CULTURES

The Eneolithic period, otherwise known as Chalcolithic or Paleo Metal Period, was marked by the formation of proto-civilizations, whose sites would appear and vanish in the spacious areas of the Middle East, Egypt, Mediterranean, Hindustan and China, as well as on the American continent. In addition to those mentioned, the centers of such cultures are Uruk in Lower Mesopotamia, early Aegean culture in the Eastern Mediterranean, the culture of painted ceramics in the territory of the Central China Plain that triggered the appearance of such early state formations as Xia, Shang-Yin, and Shu, and Olmec culture in Mesoamerica. Their chronological frames varied across different parts of the Earth. It is acceptable to consider even the Egyptian civilization, especially its oldest periods, to be Eneolithic.

Since the number of the proto-civilization sites is large, and their cultural heritage differs, a problem arises of which of them to take as the basis for choosing the images.

Such a choice should comply with the following conditions:

1) A culture should conform to the trend of subdivision into nomadic herders and farmers which existed even in the Neolithic period, i.e., it should belong with one of these two types;
2) A farming culture should be a developed one by the Eneolithic standards, i.e., it should demonstrate all or substantially all of the following attributes: metal (bronze, copper) processing, invention of wheel, use of hoes, draught animals, construction of irrigation structures, construction of megaliths, availability of specific images on rocks and ceramics that may indicate a substantially changed worldview, formation of large communities and settlements that are managed centrally, chieftains and priests being separated into special classes, and an increased trend for patriarchy;
3) A nomadic herding culture should likewise demonstrate the availability of metal tools, wheels, ceramics, change in the plots and techniques of images;
4) A culture should be deeply researched and documented.

There is one more subjective, but still, important criterion, in that authors would like to acquaint the American readers with the cultures which are geographically close to Ukraine (they are known to specialists).

Following the chosen criteria it is advisable to limit the considerable scope to three cultures:

- Tripolye which is represented by the monuments in the territories of Romania, Moldavia and Ukraine and which demonstrates a majority of the attributes of a developed farming culture;
- Harappa which is represented throughout almost the entire catchment area of the Indus River is geographically the largest of the proto-civilizations and demonstrates all the attributes of a developed farming culture;
- the herding cultures of the Eurasia steppes and the Altai Mountains (the Afanasievo and the Okunev culture) situated in the very heart of this great continent.

## THE ARTIFACTS OF THE CUCUTENI- TRYPILLIA CULTURE

The Cucuteni-Trypillia culture (6 000-4 350 years ago) encompassed the territory of Right-Bank Ukraine, Moldova, and Romania (Figure 1).

A set of household articles is represented in Figure 2.

Tools were made of silicon, stone, antlers, and bones; there were few copper articles (awls, fishhooks, decorations; lots of female figurines, fewer zoomorphic figurines; clay armchairs, models of dwellings, ornaments, painted ceramics and cooking ceramics are known.

The main industries were farming and herding. Dwellings were one- or two-story (Figure 3), while settlements could be large, either being rounded in a plan or being in line with the relief (Figure 4).

The origin of the culture is disputable; its basis was formed by the Neolithic tribes of whom the Balkan ones stand out (being possibly the descendants of resettlers from the Middle East). The external appearance of the Tripoleans is shown in Figure 5; however, the remains based on which one could carry out any statistically important reconstruction are not sufficient.

*Figure 1. Sites of the main finds of the Trypillia-Cucuteni culture (Astra-lit. Ukr-ment, n/d).*

The Trypillia culture is known, first of all, by its painted ceramics that offers a wealth of material to study the subjective space of the culture carriers (Figures 6-10).

There are schematic images of animals and plants, as well as the World Tree (Figure 8).

The images are characterized by:

- two-dimensionality;
- contrast – a figure against the background;
- dynamic use of space, plenty of curves, with spirals used most frequently;
- availability of patterns, symbols, animal images, stylized anthropomorphic Figures; there are also images and ceramic Figures of carts and wheels, etc.;
- an image of the World Tree, sometimes replaced with that of a sprouted spike, epitomizing the same very idea of the unity of the three worlds;
- abstractization of the idea of segmenting the Universe either into three or into two parts; this scheme is not only related to the World Tree but is transposed to vessels or other items and expressed not only through meaningful images but also with the help of symbols.

Let us mention some more characteristic traits.

Firstly, it is a complication of a myth plot. If you interpret one of the scenes presented in Fig.11 symbolically, rather than as an everyday scenario in that a dog barked at the squirrel, who then jumped off the tree and took to its heels, - you could notice that two animals have appeared by a tree (which may be associated with many Middle East images of the latest period), that one of the animals is a dog (a warden of the passageway to the underworld, like, for example, Cerberus), that the squirrel (a thought, mind, soul of a human being) cannot freely move between the worlds, especially enter and depart from the underworld.

*Figure 2. The Trypillia culture items: 1-3,5-7,9-14 vessels, 4, 6, 15 – figurines, 8,16 – lids, 19 – knife (copper), 20 – axe. Author: Konstantin (Istorii zemli-tripolskaya, n/d).*

*Figure 3. Reconstructed dwellings of the Trypillia-Cucuten culture: 1 – Lenkovtsy, Ukraine; 2 – Rakovets, Moldova (according to Chernysh, Masson 1982). (Archaic-total-arch, n/d).*

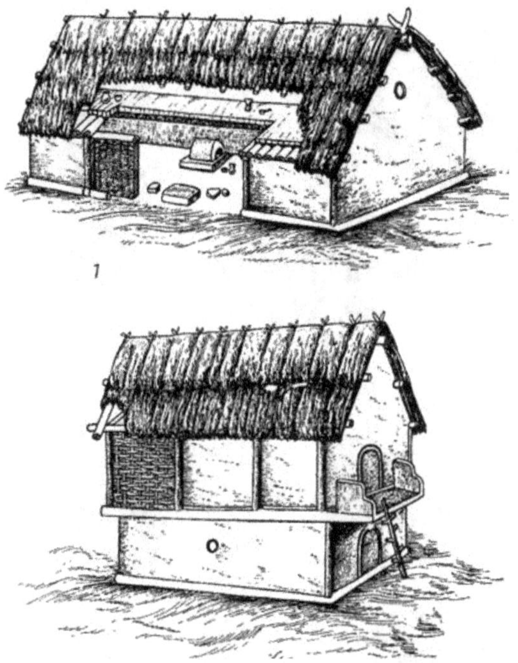


*Figure 4. Reconstructions and plans of fortified settlements of the Trypillia-Cucuteni culture: 1 - Habasesti; 2 - Trusesti, Romania (according to Dumitrescu et al. 1954; Petrescu-Dimbovita et al. 1999). (Archaic-total-arch, n/d).*

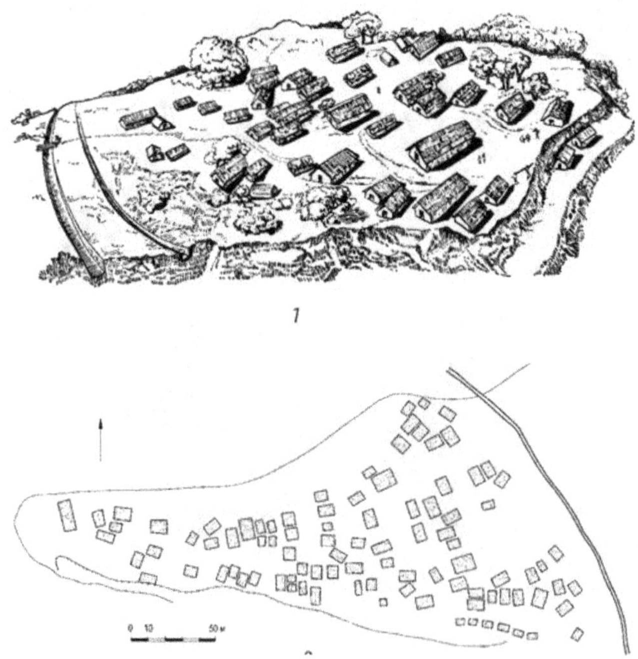

*Figure 5. The reconstructed external appearance of Trypilliaans according to M. M. Gerasimov (Gerasimov, 1960)*

*Figure 6. Pot. End of V - the beginning of IV millennium B.C. Ceramics height. 33 cm. Ident. No. PKP 4 (Antikvar. Tripolskie ornamenty, n/d).*

*Figure 7. Oval mask. Approx 4000 B.C. Ceramics height 14.5 cm. Ident. No. PKP 65 (Antikvar. Tripolskie ornamenty, n/d).*

*Figure 8. Binocular-shaped vessel. 4400–4200 B.C. Ceramics height. 26.2 cm. Ident. No. PKP 47 (Antikvar. Tripolskie ornamenty, n/d).*

*Figure 9. Biconically-shaped pot. 2nd half of IV millennium B.C. Ceramics height. 22 cm. Ident. No. PKP 68 (Antikvar. Tripolskie ornamenty, n/d).*

*Figure 10. Pot depicting sprouted grains. 3850–3500 B.C. Ceramics height of 27.7 cm. Ident. No. PKP 5707(Antikvar. Tripolskie ornamenty, n/d).*

*Figure 11. Elements of paintings (Oringo. Tripolskie Ornamenty, n.d.).*

Secondly, in addition to the binary contrast, the idea of twoness is evident (Fig. 8) – possibly, such vessels became the prototypes of an image of twin brothers which is also known in mythology (for example, Dioscuri)? Might it be another connection with the World Tree?

Thirdly, let us note some methods of composition: self-similarity, fixed forms, the positioning of symbols and forms in a three-world system or as binary pairs.

A rich set of symbols, as well as fixed forms and definite placement of symbols, evidences the emergence of pictographic proto-writing. According to some researchers, some of the paintings depict a calendar.

## THE HERITAGE OF THE HARAPPA CULTURE

The Harappa culture (5 300-3 300 years ago) was formed in the western part of Hindustan, mainly, in the valley of the Indus River (Figure 12). It is the largest of the proto-civilizations in terms of territory and population size (by estimations – up to 5 mln. people).

*Figure 12. Spread of the Harappa civilization at the time of its rise (Pikabu. Gibel Harappskoy, n.d.).*

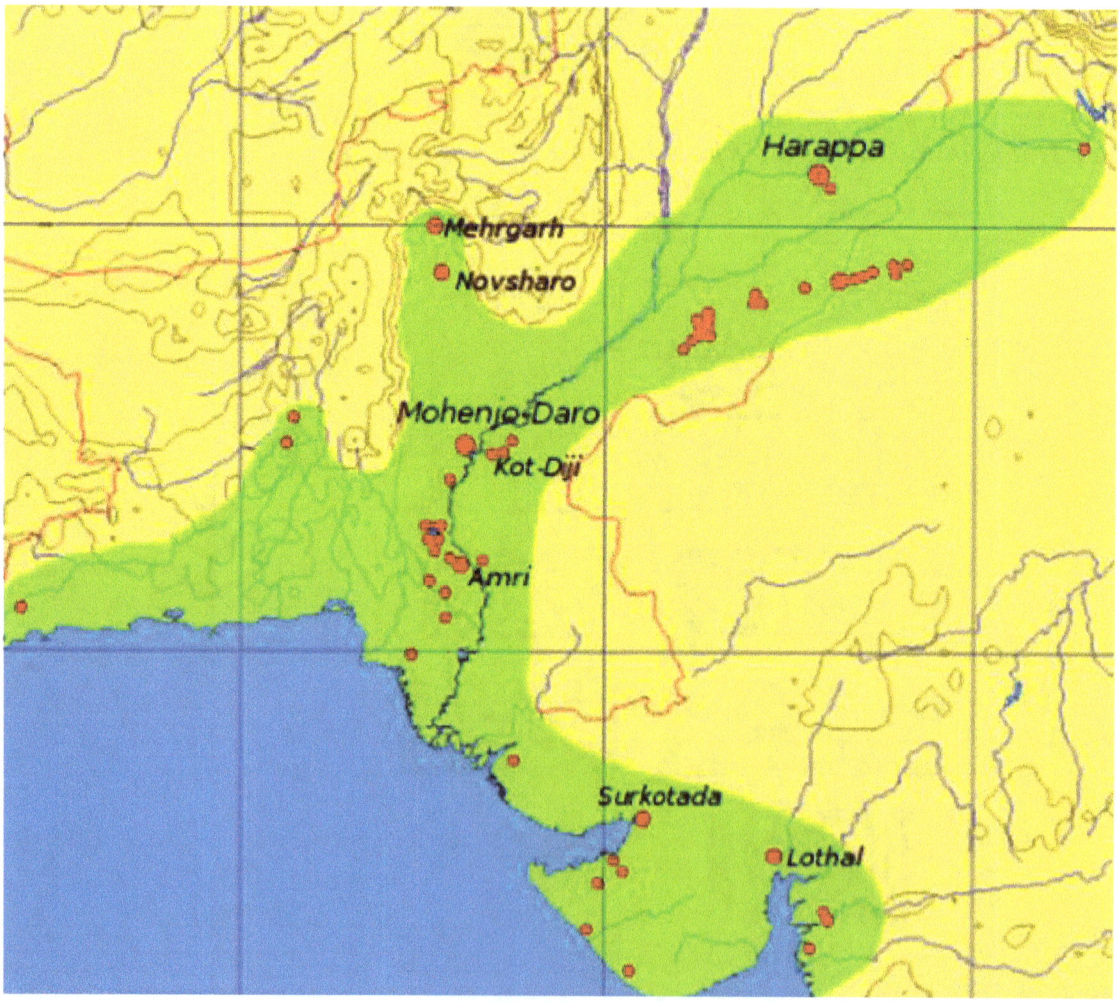

*Figure 13. Mohenjo-Daro (City plan kannelura, n.d.).*

The largest settlements – Mohenjo-Daro (5 300-3 500 years ago) and Harappa – were complexes with broad and straight streets, water supply and sewerage systems, palaces, temples, and pools, that is, they were undoubtedly cities (Figures 13-15).

Even the city's ruins are very impressive. A set of typical articles is shown in Figure 16.

Some metal and stone figurines are masterfully made (Figure 17) and indicate what the Harappians looked like.

Labor tools were primarily made of copper and bronze. A well-organized power institute was necessitated by an important role the irrigation system had in the community's life. The Harappan civilization had trade communications with Mesopotamia and Middle Asia. It is considered that real writing appeared already at that time (Figure 18), for which there is no universally acknowledged deciphering so far.

The issue of the ethnicity of the Indus Valley population has not been settled, with the majority of researchers being of the view that it consisted of Dravidian peoples, whilst the others think that the cities were created by Proto-Aryans.

The images are most commonly found on seals (Figure 9).

Let`s consider the specific features of the images. Against the adjustment of the properties of a material, the images are flat. Space is also lacking depth in that there is no effect of items getting smaller in size as the distance grows. Any directions and viewpoints are missing altogether. It should be noted that the compositions are built to avoid any overlapping of figures and symbols. However, space cannot

*Figure 14. Mohenjo-Daro drainage system used also as a sewerage system. (City plan kannelura, n.d.).*

*Figure 15. The contemporary state of the central part of Mohenjo-Daro (Our civilization. Mohenjo, n.d.).*

*Figure 16. Items from Harappa and Mohenjo-Daro and Harappa: seals, ornaments, labor tools, ceramics, ceremonial articles, figurines (World History, 1955)*

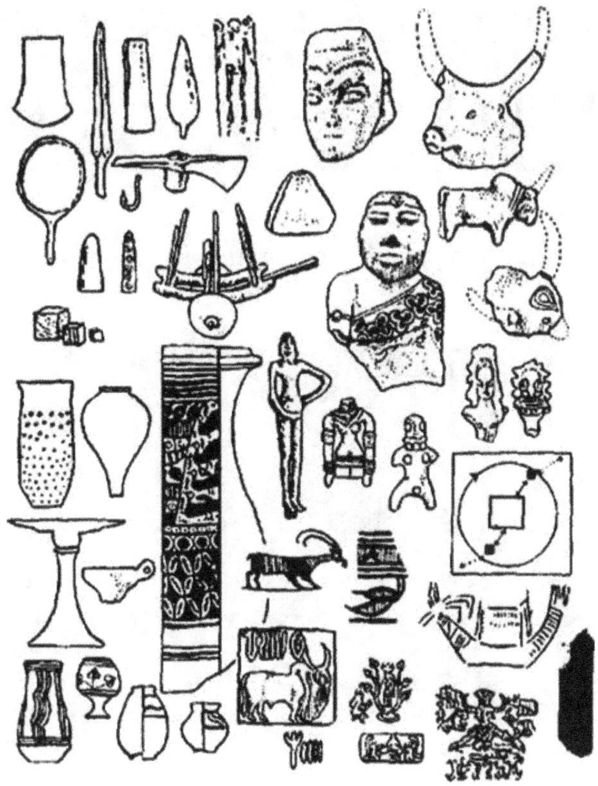

*Figure 17. Samples of the Harappan civilization sculpture (World History, 1955)*

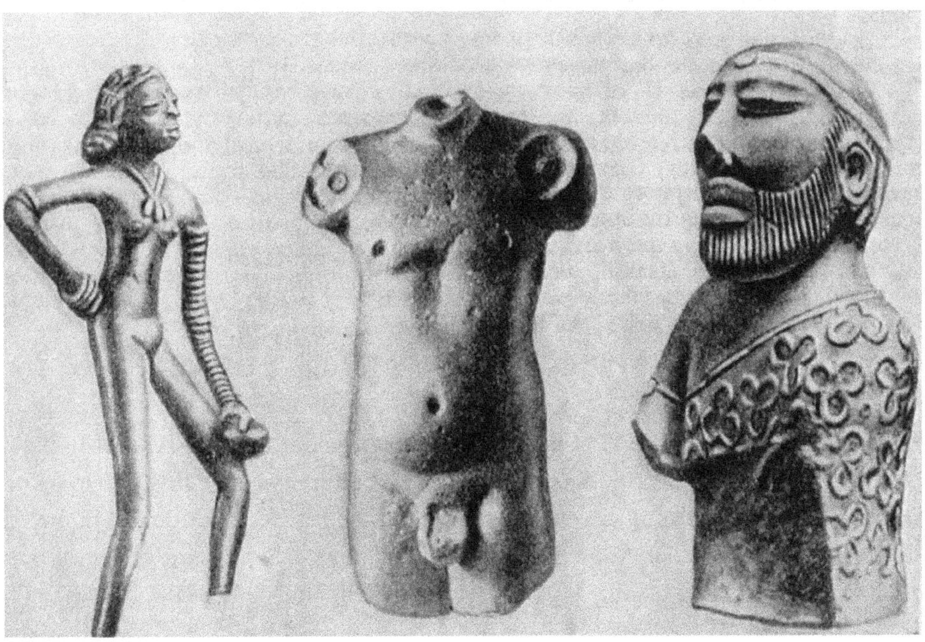

*Figure 18. The longest of the Harappan "texts" (World History, 1955)*

*Figure 19. Samples of seals. The Harappa civilization Author: Sergei Briushinkin (Mail. Community-catastrof, n.d.).*

be called devoid of any orientation, either, as the top, bottom, left and right side plays an important role in constructing the compositions.

Along with quite naturalistic images, there are polymorphic figures of animals, as well as anthropomorphic images, often depicted in ritual postures.

Virtually all seals are accompanied by abstract symbols; many of them look like signatures or even small texts.

Thus, the images combine both archaic (flatness of form, polymorphy) and new traits (signatures and texts). The idea of replicating images by impressing them is, in itself, innovative.

The images of the World Tree and the cultural scenes that will be discussed below are created in the same manner. Meanwhile, let's look at the painting of nomadic peoples.

## ENEOLITHIC PAINTING OF THE CENTRAL PART OF EURASIA

The Afanasievo culture (Griaznov, 1999) occupied the mountainous areas of Altai, Minusinsk Hollow, Mongolian Altai, and Xinjiang 6000 – 4500 years ago. The culture is interesting due to its origin which is connected with the resettlement of the representatives of the Pit Grave Culture who belong with the Proto-Indo-Europeans. The resettlers were nomadic herders that mastered some copper processing technologies (for example, by adding arsenic to obtain bronze). They made needles, knives, awls, hooks of

*Figure 20. Vessels. Afanasievo Culture (Genofond, n.d.).*

copper and bronze, however, stone items were not completely ousted, which gives grounds to attribute this culture either to the Eneolithic or to the Early Bronze epoch.

The most well-researched monuments of the Afanasievo culture are burial mounds grouped into burial sites of 5-20 burial mounds each. Graves constituted rectangular 1.5m deep pits marked by larchen logs, or sometimes by stone slabs. The space around a grave was fenced circle-wise with sandstones, after which a rockfill was erected. 1-2 ceramic vessels were placed within a grave (Figure 20), less frequently, household items, and even less frequently, weapons. The elders' graves, larger in size and more complex in structure, are located away from common burial sites.

The Afanasievo culture had been replaced by the Okunev culture (Gromov, 1997); Vadetskaya, 1980, 1986) that partially occupied the geographical range of the former. The period of existence is dated back 4,500-2,900 years ago. Ethnically, the Okunevans had a mixed (Europoid-Mongoloid) race; they were not descendants of the Athanasian. This culture is also categorized as Eneolithic and Early Bronze. The Okunevans were primarily herders (sheep and cattle), with oxen harnessed to two- and four-wheel carts, which enabled required mobility for a seasonal change of pastures. The settlements were not very long-lasting, with a thin cultural stratum, and are seldom found; small-size stone fortifications are sitting on higher grounds. The Okunevans were familiar with farming (stone hoes, milling stones, stamp mills, reaping sickles with copper blades were found).

The culture is mainly represented by burial monuments. Burial sites consist of one or several burial mounds, whilst a burial pit is not large. Burial mounds are enclosed with a square fence made of slabs put on edge. A deceased was buried in a small case made of sandstone slabs; the bottom was lined with small slabs and the ceiling was made of slabs, too. A deceased lay on their back, legs flexed in knees, with a stone slab placed under their head. The face and body of the deceased were covered with red ochre. Into the grave, they placed clay vessels containing food and other articles, as well as bones of animals and birds, stone and bone pendants (figures of a bear, heads of elk, bird, moon-shaped pendants), bone rods, rhytons relating to the cult items.

The Okunev culture is famous for petroglyphs and anthropomorphic stone sculptures up to 4 m high. It is argued that some sculptures are dated back as IV millennium B.C.E., and that they are even older than the Afanasievo culture monuments. Some of them have bull-shaped three-eyed faces, the others are sun-faced. The first style is bas-relief, and the other is sculpture-graphical.

The sketches of the Okunev culture monuments are shown in Figure 21; Figure 22 shows the "Okunev pebble" which is interpreted as a certain vision of the structure of the Universe; Figure 23 shows the Okunev stela.

The Okunev culture is replaced by the Andronikov one that already relates to the Bronze epoch and is entirely Mongoloid from the ethnical point of view (Aryans of the Eurasian Steppes: Bronze and Early Iron Epoch in the Eurasian Steppes and Contiguous Lands, 2014).

*Figure 21. The Okunev archeological culture. Author: AskizON INFO (Askizon. Okunevskaya kultura, n.d.).*

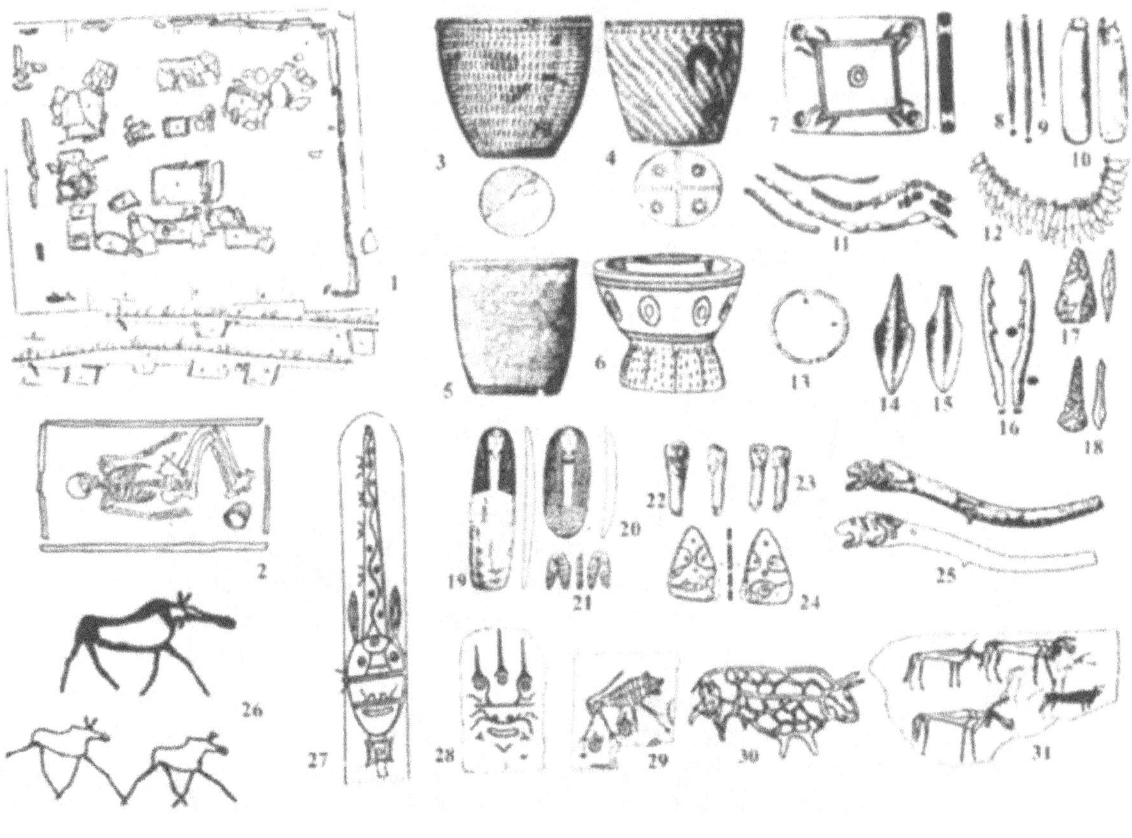

*Figure 22. "Okunev pebble". (Krskstate. Okunevskaya kultura, n.d.).*

*Figure 23. A monument of the Okunev Art as sketched by the Aspelin expedition. Author Aspelin, Johann Reingold and Appelgren (A monument Okunevo, n.d.).*

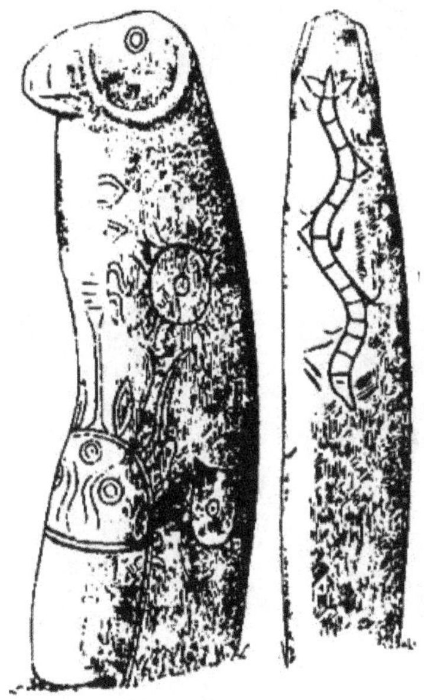

Special attention will be paid to petroglyphs, which are extant in abundance (tens of thousands) in the Mountainous Altai (Karako, Kalbak-Tash, Yelangash, Ak-Yziuk, Turochak Pisanitsa, etc.). As a rule, such sites were utilized by several cultures, from Mesolithic to Medieval, with the images created by the Athanasian and Okunevans accounting only for a certain share. The images can only be dated based on indirect indications, such as plots, depicting extinct animals, cart structures.

Here some examples of images characteristic of Kalbak-Tash (Fig. 24-25) and Turochak Pisanitsa (Figure 26) are given.

*Figure 24. Images of animals. Kalbak-Tash, Altai, Russia (World rock art, n.d.).*

*Figure 25. Anthropomorphic Figures and complex compositions. Калбак-Таш, Altai, Russia. Author: Sergei (Urochishe Kalbak-Tash, n/d).*

*Figure 26. Image of an elk (a sketch of the image is shown in Figure 21). Turochak Pisanitsa. Altai, Russia. (Altai-go. Turochak, n.d.).*

While analyzing these and similar images one may state that the method of their execution and many of specific features are no different from the Neolithic ones, namely:

- space lacks depth;
- the images are structured according to the Fig.-against-background principle;
- the majority of the images are oriented along the top-down or right-left axes;
- there both relatively realistic figures of people, animals, carts and styled anthropomorphic or polymorphic figures, as well as symbols and signs.

The distinctions are:

- images being somewhat less dynamic and more static;
- a lesser number of scenes of war and violence;
- emergence of new plots, figures (sun-shaped characters), symbols (circles, swastikas, etc.).

Comparing the artifacts of the Afanasievo and Okunev cultures with those of the Trypillia and Harappa cultures it is necessary to state that the former is more archaic, demonstrates a lesser level of craftsmanship, and, of course, a primitive state of urban development.

Now it is necessary to discuss whether all of these cultures exhibit certain common changes that would enable to suggest that there was the worldwide process of the human mind evolution. To get an answer the already known markers need to be analyzed.

## AENEOLITHIC PAINTING: TWO-DIMENSIONAL SPACE, NEW STORIES, AND SIGNS

Along with that, farmer and herder petroglyphs and paintings have characteristics in common, too:

- space remains two-dimensional and figures continue to be stand-alone;
- abstractization of the concept of the segmentation and connection of the three worlds and break away from the image of the World Tree;
- abstractization of the concept of the cyclicality and its breakaway from the World Tree; transference of the motion concept from specific forms (running people and animals) into abstract forms (spirals, swastikas, curves) also epitomizing the cyclicality idea;
- clearly expressed contrasting of the world of the living and the world of the dead;
- a contrast of figures with the background;
- availability of both realistic and polymorphic Figures, their sketchiness and grotesqueness;
- the orientation of images;
- more static compositions;
- new signs, Figures, symbols, patterns;
- new plots;
- less violence.

Here is a comparison with the Neolithic painting (Table 1).

*Table 1. Comparison of the Neolithic and Aeneolithic painting features*

| Feature | Neolithic Painting | Aeneolithic Painting |
|---|---|---|
| 1. | It is characterized by sketchiness, one-colored images, contrast with the background, dynamic images, inactivity of background. | It is characterized by sketchiness, one-colored images, contrast with the background, dynamic images, inactivity of background. |
| 2. | Space is depicted as two-dimensional. No elements of perspective images are present (viewpoints, skylines, vanishing points, line of sight). No illusion of depth is created. | Space has a complex structure and is depicted as two or three segments, each filled with different patterns and signs and divided by borders, though sometimes their unity is emphasized with an image of sprouting the grain, etc. |
| 3. | Compositions are dynamic, having no symmetry or center. | Compositions are more static; symmetry and center are sometimes present; dynamics and cycle are expressed by patterns and signs. |
| 4. | Top and bottom, right and left side are normally discernible. | Top and bottom, right and left side are discernible. |
| 5. | Natural scale and proportions are violated; Figures look stylish, sometimes grotesque. | Natural scale and proportions are violated; Figures look stylish, sometimes grotesque. |
| 6. | Images of animals and people are not expressed explicitly, though many details of conduct and motion are reflected very accurately; scenes of war and violence are large in number. | Images of animals and people are not expressed explicitly, details of conduct and motion are rather schematic; a relatively little number of scenes of war and violence. |
| 7. | There are polymorphic images. | There are polymorphic images, in particular, of sun-shaped Figures |
| 8. | There are magic or cult-related scenes. | There are complex magic or cult-related scenes |
| 9. | There are abstract images. | Abstract images, symbols, and signs are large in number and diversity. The emergence of pictographic proto-writing or writing. There appeared and developed image replication through their impression. |

## THE MYTH OF THE WORLD TREE AND THE GREAT GODDESS

The evolution of the World Tree myth is most clearly visible on the Harappa culture seals. Let's look at some seals relevant to the myth (Fig. 27-30). Note that despite the good condition of the seals (Figure 27), it is more convenient to analyze the drawings presented (Albedil, 1991).

The most interesting of those sketches is the image of the World Tree as shown in Figure 28.

*Figure 27. Seal from Mohenjo-Daro (Evpatori Pechat Mohendzo, n.d.).*

*Figure 28. World Tree. M.F. Albedil (Albedil, 1991)*

This is how this image is described and interpreted by M. F. Albedil (Albedil, 1991): "You can see a stylized image of an asvattha tree with a distinctive form of the leaves. The tree crown is represented by seven branches, including the end of the trunk, and by nine leaves, three of which are marked with dots (the central one and two outermost of the five upper leaves), and four of which are pointed downwards. Under the lower branches is a circle made up of two central dots, which are encircled by eight dots. Sitting on necks shaped like buffalo horns, two heads of turs stem from the circle. Their manes are shown as 12 curvy strokes, while the horns on each head are covered with 6-year rings. Under the circle, the roots are visible, below which the beginning of the upturned tree shows. The tree crown personifies the vault of heaven. The central branch points to the zenith. It ends with a small leaf; whose end indicates a direction to the point of intersection of the meridian with the celestial equator that is declined southwards from the zenith in India's sky. Through this point pass the constellations of the celestial equator that rise in the east, go near the zenith with a declination to the south, and set in the west. Tilted at a 45° angle, two middle branches sprout up from the tree trunk, pointing at intermediate directions. Through these points pass the constellations that rise in the north-east and southeast and cross the celestial meridian at the height of 45 arc degrees. The points that mark three leaves may indicate to the main celestial directions. Similarly, other branches and trees point to additional directions. Thus, an arc between the leaves of the upper branches indicates the points of the sunrise over a year, whilst the upper branches and their elongation downwards form a hexagonal Fig. that vividly demonstrates the division of a year into six seasons.

Sprouting from the branch, below the leaved branches are two curved branches with circles on their ends that, most likely, denote the sun and the moon. To all appearances, the circle placed at the bottom of the trunk indicates the "circle of the earth", while eight dots placed on it correspond to eight directions – 4 main and 4 intermediates.

In the Proto-Indian calendar, tur symbolized a year, as well as one of the six seasons starting from the winter solstice. The same semantics applies to the two heads of turs that stem from the tree on their long-arched necks, facing each other. They symbolize two half-years in between the solstices, during which the point of the sunrise is shifting northwards and southwards from the west. The necks of the turs' heads are arched in the way which exactly replicates the bend of the buffalo horns with which the head of the supreme deity is crowned, whose image we considered in the previous Chapter. That is why 12 strokes indicating the turs' manes do not only mean 24 half-months, into which a human year is divided, but also a 12-year cycle of the Jupiter planet, which was thought as the "year of gods". The same idea is represented by six-year rings on the horn of each tur head, which add up to the same number of 12".

If this interpretation is correct, then there is a complex and structured idea of space-time, very far from the direct intuitive sensation of the Paleolithic and even Neolithic people. Such understanding must be underpinned by long-term observations and a certain "school" of their interpretation, albeit symbolically expressed.

Among other specific features of this seal, it is possible to note its tripartiteness, polymorphy, and, probably, the availability of the writing signs (Parpola, Asko, et al. 1987-2010; Farmer, Steve, et al. 2004; Rao, 2010).

Figure 29 shows a certain ceremony of worshipping the Great Goddess living inside a tree, which thus becomes the Tree of Life. The ceremony is done by women only, no men are present. A priestess is bowing to the tree, and women clad in ceremonial garments are marching by and, probably, singing. The animal may be the Bull-husband of the Goddess (by analogy from, for example, the figurines of the

*Figure 29. Admiration of the Great Goddess. M. F. Albedil (Albedil, 1991)*

*Figure 30. Expulsion of the Tree Keeper (?). M. F. Albedil (Albedil, 1991)*

goddess and the bull's heads in Chatal Huyuk), or it may be being sacrificed. As a result of the ritual, the Goddess reveals itself on earth and is ready to heed the prayers of those participating in the ritual.

Figure 30 probably shows the purpose of such ceremony: attracted either by the husband or by the victim, the Goddess is ousting a certain predator (a keeper of the passage?), thus opening a way to the world of the dead, as the first stage of transition to the worlds of eternal life. At the same time, this

*Table 2. Comparison of the Neolithic and Eneolithic versions of the world tree myth features*

| Feature | Neolithic Version | Eneolithic Version |
|---|---|---|
| 1. | The tree structures space and time and symbolizes their unity; however, benevolent worlds of heaven and earth are set against the hostile underworld, with the identification of the Tree with the World Axis being quite common; parts of the world and seasons are grouped around the Tree. | The structuring of worlds is not associated solely with the World Tree. The Tree becomes a certain soteriological symbol required for opening the path to eternal life. |
| 2. | Identification of human with the Tree. | Identification of the Goddess with the Tree. Only women, being the keepers of the Great Goddess cult and having the necessary intuition can induce the appearance of the Goddess in the world. |
| 3. | Connection with the cult of the dead, whose relation to the living is hostile. | Connection with the cult of the dead: the worlds are divided by the keeper and to enter the world of the dead and to start the search for eternal life one needs to overcome his resistance. |
| 4. | Only great shamans and heroes are able, with the use of special rituals in specially dedicated places, to move from one world to another across the Tree in the search of either eternal life in the world of the heavenly spirits or reincarnation on earth. | Even priests are unable to move around the world of the dead unassisted, they need the help of the Great Goddess, as a minimum, to open the passageway. |
| 5. | The myth plot is simple enough to be limited to an intuitively understandable symbol and not to require to be fixed in writing. | The myth plot gained a soteriological connotation, it became multidimensional, intertwined with other myths, got more complicated and, to be more adequately articulated, it required a complex set of symbols, ceremonies (mysteries) fixed in writing. |

might be an expression of fear against the keeper which underlies the taboo on the journeys throughout the world tree.

A summarized result of these speculations is given in Table 2 that compares the specific features of the Neolithic and Eneolithic versions of the myth.

## MARKERS ANALYSIS

Based on the resultant Tables 1 and 2, let us analyze the state of the markers.

**Space Representation.** In contrast with the Neolithic period (Table 1), the space became more complex and structured (strengthened nous), compositions became more complicated and lost their dynamics (strengthened nous), the number of scenes of war and violence was reduced (weakened ego and will), a sense of unity with the image decreased (weakened intuition), various 'explanatory' symbolics and writing (strengthened nous), loss of vivacity of impressions (strengthened nous, weakened sensations, and organs of senses); replicability appeared (strengthened nous). Other specific features, such as two-dimensionality of space, persisted.

**The World Tree Myth Connected with the Great Goddess.** The Paleolithic simplicity (fallen leaves (human bodies) received by the Mother Earth for them to be reborn (reincarnated) in spring again had been left far behind. Fear of death (weakened intuition, strengthened ego) had taken root in the hearts of people forever. Dangers of transition started to seem insuperable (weakened will), deceits and illusions of the underworld started to seem beyond the human comprehension (weakened reason, which is

*Figure 31. Eneolithic mind*

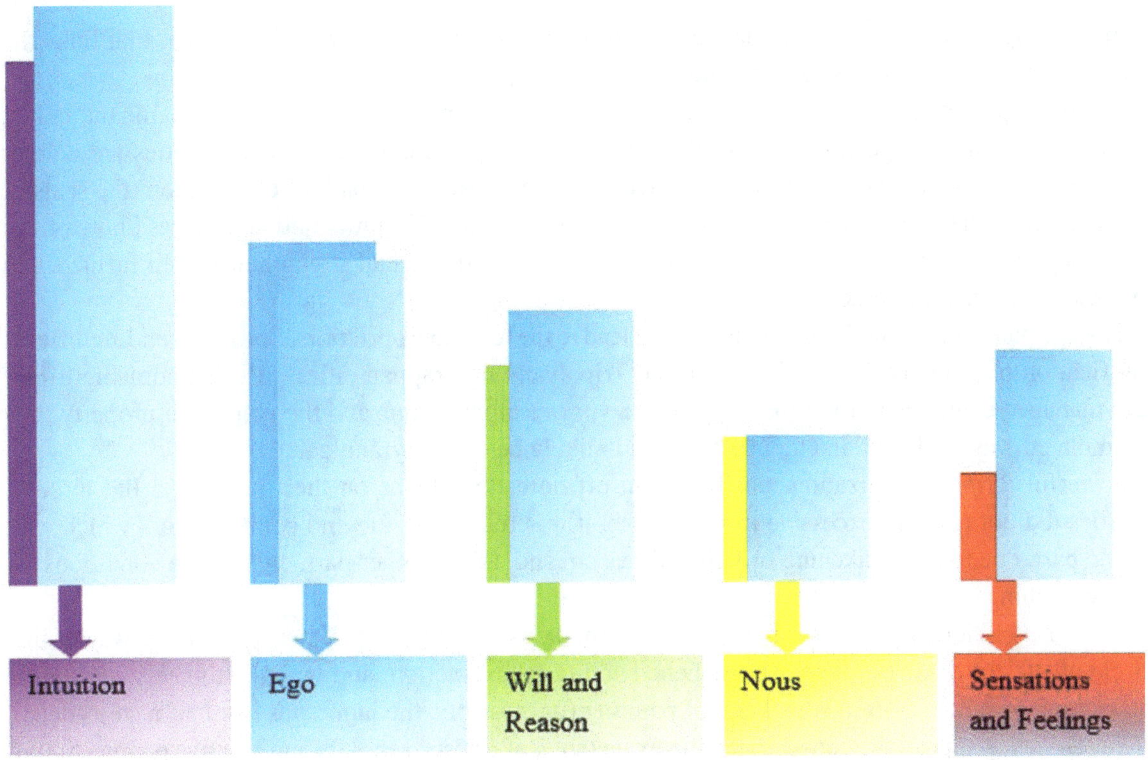


unable to tell truth from lie). Even shamans, priests, and heroes of the countries pampered by civilization were afraid to make self-sacrifice. And since the desire to live eternally (ego) was still there, there was a need in a Soter (savior), whom the formidable Great Goddess, by that time incomprehensible for the men, became. And she needed to be summoned by doing a complicated ritual, to be appeased, and obeyed in every way by accepting her help, which was more natural and more characteristic of women, who were more sensitive and apt to ask, who loved their children and were accustomed to obeying their husbands (Table 2).

Probably, this is related to a new increased role of women in society, and the relative peacefulness of the Eneolithic epoch might serve as an indirect indication of that.

Let us proceed to reconstruct the Eneolithic mind, by expressing its channels once again in the same scope as in previous chapters (Figure 31).

## ENEOLITHIC MIND

Let us once more note that this is a certain generalized qualitative portrait and that the mind of an individual personality may differ significantly from it.

## BEHAVIOR PATTERNS

Let us use Fig. 31 and Table 7 of chapter 4, *Self-organization: evolution and behavior*, that links the perception channels with class groups and their behavior patterns.

The mind potential had decreased. The intuition had weakened and is almost non-dominant, being, due to the ego, in a proportion close to that of a balanced personality. Such personalities are neither characterized as aggressive nor by self-sacrifice, choosing instead the path of observation (the seals of the Harappa civilization portray a large number of figures depicted in meditative postures known from the latest yoga). Women – who possess a higher level of intuition – had strengthened their influence in the society and in sacral practices.

The ego had strengthened a bit but did not yet lead to the total appropriation of property and significant stratification of property. It is definite that the Tripolyan or Harappan cities, just like nomadic tribes, were managed centrally, but in times of peace the power of chieftains and the volume of property they accumulated was not large, and their power can unlikely be called tyrannical.

Powerful farming civilizations easily fought off nomadic attacks on their borders; a life in communities did not result in excessive intransigence, it was relatively safe and comfortable – which was, for one part, due to the weakening of will and reason, and, for the other part, created the conditions for such weakening.

Nous had obviously grown stronger, as shown by the streamlined Universe, space, time, way of life, "circle of life" of the community, strengthened ability to abstraction, and which ultimately resulted in outstanding inventions – of a wheel and of painting (or, possibly, the latter was just half a step ahead). The intelligence appeared – consisting, quite expectably, of priests, scholars and creative people. Stability appeared in the society, but, along with them there also appeared staleness and inability to respond to new challenges. Most of all, this is where the main reason of the death of proto-civilizations: the climate had changed, but they did not find the strength to resettle or create an irrigation system; internal conflicts escalated, but the power did not reposition itself and did not address them; the conquerors had appeared, but forces were not mobilized to fight them off, and so on.

Sensations, emotions, and feelings had weakened again. As with the Neolithic settlements, this was a condition of peaceful coexistence; cohabitation induced the weakening of sensory organs and to the waning of emotions.

Undoubtedly, the processes described varied across the farming and herding cultures. That circumstance has already been mentioned in chapter 6, *Neolithic: separation and war*, concerning the Neolithic cultures. However, though manifesting themselves differently and at a different pace, those processes were nevertheless taking place, driven by the self-organization laws.

On the whole, the Eneolithic epoch may be characterized in two ways: the Neolithic processes and tendencies had reached their peak, but there emerged a new factor which was connected with a gradual development of human nous at the backdrop of the weakened intuition, will and reason. The significance of this factor and its influence upon the painting technique and the World Tree myth development will continue to grow in the subsequent epochs.

## CONCLUSION

Criteria have been determined and cultures and artifacts have been selected representing the Eneolithic epoch.

Specific features of the Eneolithic painting have been identified, in particular, those regarding depicting the space; they have been compared with those of the Neolithic epoch (Table 1).

It has been possible to explain the changes that took place, by using the approach suggested in chapter 4, *Self-organization: evolution and behavior,* as a consequence of the evolution of the condition of human-environment interaction channels.

Additional materials to assess the state of mind have been obtained in the course of analyzing the Eneolithic version of the World Tree myth (Table 2). These materials confirm the conclusions made.

The diagram is shown in Fig. 31 demonstrating the state of mind of an Eneolithic human, can be regarded as qualitative, rather than quantitative.

In the Neolithic era, the weakening of intuition and a sense of spiritual comfort caused numerous wars and brought to life great inventions that changed the way of life of hunters for the life of farmers and pastoralists.

On the contrary, in the Eneolithic era, "longing for a lost paradise" was compensated by complex myths in which only a few priests, thanks to sophisticated rituals, rich sacrifices, and grace of higher powers, as well as heroes, thanks to personal valor, divine origin, and extraordinary merits, allowed a bright world of gods. And the rest of the people could only hope for the intercession of priests and heroes and obey them in life. Nevertheless, they could count on the convenience of urban life, as compensation for the heavenly beauties and their beliefs, as compensation for the direct contemplation of the gods.

## FUTURE RESEARCH DIRECTIONS

The same approach will be used to analyze the following historical epoch of Ancient Civilizations.

## REFERENCES

A monument of the Okunev Art as sketched by the Aspelin expedition (n.a.). Retrieved from Alt-Altaische Kunstdenkmäler: Briefe und Bildermaterial von J.R Aspelins Reisen in Sibirien und der Mongolei 1887-1889 herausgegeben von Hrsg. Von Hjalmar Appelgren-Kivalo. Hfors 1931. 47 s. + 76 kuvas. Authors: Aspelin, Johann Reinhold and Appelgren. https://ru.wikipedia.org/wiki/Файл:Из_Alt-Altaische.png

Albedil, M. F. (1991). *A Forgotten Civilization in the Indus Valley*. Saint Petersburg, Russia: Nauka.

Altai-go. Turochak. (n.d.). Retrieved from http://altai-go.ru/index/turochakskaja_pisanica/0-109

Antikvar. Tripolskie ornament. (n.d.). Retrieved from https://antikvar.ua/tripolskie-ornamenty/

Archaic-total-arch. (n.d.). Retrieved from http://archaic.totalarch.com/dwelling_and_settlement

Aryans of the Eurasian Steppes. (2014). *Bronze and Early Iron Epoch in the Eurasian Steppes and in Contiguous Lands*. Barnaul: Altai State University.

Askizon. Okunevskaya kultura. (n.d.). Retrieved from (https://askizon.ru/item/915-okunevskaya-kultura-konets-3-nachalo-2-tys-do-n-e-epokha-rannej-bronzy-khakasiya-khakassko-minusinskaya-kotlovina

Astra-lit. Ukr-ment. (n.d.). Retrieved from https://www.astra-lit.com/ukr-ment-ru/ukr_ment_3.htm

Bongard-Levin, G. M., & Ilin, G. F. (1971). *Indiya v drevnosti*. Moskow, Russia.

Chernish, E. K. (1982). Eneolithic of the Right-Bank Ukraine and Moldova. Eneolithic USSR. Moscow, Russia, 1982. 166—347.

City plan kannelura. (n.d.). Retrieved from http://kannelura.info/?p=8997

Ellis, L. (1984). The Cucuteni–Trypillia culture: study in technology and the origins of complex society. Volume 217 of the BAR international series; British Archaeological Reports (B.A.R): Oxford, 1984.

European steppe-eneolithic. (n.d.). Retrieved from https://indo-european.eu/tag/steppe-eneolithic/

Evpatori Pechat Mohendzo. (n.d.). Retrieved from http://www.evpatori.ru/pechat-iz-moxendzho-daro.html

Farmer, S., Sproat, R., & Witzel, M. (2004). The collapse of the Indus-script thesis: The myth of a literate Harappan civilization. *Electronic journal of Vedic studies, 11*(2), 19-57.

Genofond. (n.d.). Retrieved from http://генофонд.рф/?page_id=3362

Gerasimov, M. M. (1960). *The appearance of a person from Nezvisko. Materials and research in the field of archeology of the Southwestern USSR and the Romanian People's Republic* (pp. 83–90). Kishinev.

Golan, A. (1993). Myth and Symbol. Moscow, Russia: Russlit.

Griaznov, M. P. (1999). Afanasievo Culture along the Yenisei. St. Petersburg, Russia: Academic Press.

Gromov, A. A. (1997). *Origin and Links of the Okunev Culture Population.* St. Petersburg, Russia: Petro-RIF.

Historic Cultural Heritage MuseumP. L. A. T. A. R. (n.d.). Retrieved from https://antikvar.ua/tripolskie-ornamenty/

History, W. Vol.1 Stone Age. (1955). Moscow, Russia: Gospolitizdat.

Hunter, G. R. (1934). *The Script of Harappa and Mohenjodaro and Its Connection with Other Scripts.* London, UK: Kegan Paul.

Istorii zemli-tripolskaya. (n.d.). Retrieved from http://историиземли.рф/tripolskaya-kultura.html

Krskstate. Okunevskaya kultura. (n.d.). Retrieved from http://my.krskstate.ru/docs/early_history/okunevskaya-kultura/

Leontiev, N. V., Kapelko, V. F., & Yesin, Yu. N. (2006). *Sculptures and Stelas of the Okunev Culture.* Abakan, Russia: Khakassia Research Institute of Language, Literature, and History.

Mail. Community-catastrof. (n.d.). Retrieved from https://m.my.mail.ru/community/catastrof/130BB97B091C716B.html

Mantu, C.-M. (2000). Cucuteni–Trypillia cultural complex: relations and synchronisms with other contemporaneous cultures from the Black Sea area. Studia Antiqua et Archaeologica. Iaşi, Romania: Iaşi University. VII: 267.

Menotti, F., & Korvin-Piotrovsky, A. (Eds.). (2012). *The Trypillia Culture giant-settlements in Ukraine. Formation, Development, and Decline*. Oxford, UK: Oxbow Books.

Oringo. Tripolskie Ornamenty. (n.d.). Retrieved from https://oringo.com.ua/forum/tripolskie-ornamenty

Our civilization. Mohenjo. (n.d.). Retrieved from https://our-civilization.com/civilizacia/arhitektura-drevnih/mohendzo-daro

Parpola, A. (2010). Dravidian solution to the Indus script problem. Central Institute of Classical Tamil, 2010.

Parpola, A. (1987-2010). Corpus of Indus seals and inscriptions, Helsinki, Finland: Suomalainen Tiedeakatemia (Academia scientiarum Fennica).

Passek, T. (1961). Early agricultural (Tripoli) tribes of Transnistria, MIA, n. 84. Moscow, Russia, 1961.

Pikabu. Gibel Harappskoy. (n.d.). Retrieved from https://pikabu.ru/story/gibel_kharappskoy_tsivilizatsii_5265336

Rao, R. P. N. (2010). Probabilistic Analysis of an Ancient Undeciphered Script (PDF). *IEEE Computer.*, *43*(4), 76–80. doi:10.1109/MC.2010.112

Rybakov, B. A. (1965). Cosmogony and mythology of the Eneolithic farmers. Soviet Archeology, 1965, No. 1-2.

Universal History of Architecture. (2008). *Ancient World Architecture* (Vol. I). Moscow, Russia: Arhitektura-S.

Urochishe Kalbak-Tash. (n.d.). Retrieved from http://kp74.ru/urochishhe-kalbak-tash.html

Vadetskaya, E. B. Sculptures of the Okunev Culture. (1980). London, UK: Science.

Vadetskaya, E. B. (1986). *Archeological Monuments in the Middle Yenisei Steppes*. Okunev Culture. London, UK.

Videiko, M. (2010). *Ukraine: from Trypillia to Rus*. Kyiv, Ukraine: Krion.

World Rock Art. (n.d.). Retrieved from http://worldrockart.ru/risunki/petroglify-altaya/

# Chapter 8
# Ancient World:
## Ego, Archaic, Immortality


## ABSTRACT

*The features of depicting space in the reliefs and murals of Ancient Egypt are considered. Attention is drawn to the preservation of the connection of ancient Egyptian art with primitive art in sacred paintings and to the evolution of the ways of depicting space in secular scenes. There is enough material to reconstruct the ancient Egyptian version of the World Tree myth and to establish links with other archaic myths and ideas about the World Tree in the synchronous cultures of the Middle East. When analyzing markers of evolutionary changes, the most active channels were established and the forecast of the self-organization scenario was checked. The results are presented in the form of generalized psychological portraits and behavior patterns of representatives of the main estates.*


## BACKGROUND

The timeframes of the research need to be established.

Archeological finds in the Nile Valley date back to the Paleolithic, Neolithic and Eneolithic Ages. Speaking about a relatively developed urbanistic civilization comparable with the Harappa or Sumer, such one appeared in the so-called predynastic times (approximately 6 000 -5 000 years ago). However, of interest here is the pictorial art of the epoch of the earliest states, as it is somewhat different from the art of the Eneolithic period. Those timeframes encompass the so-called dynastic period of the Egyptian history (5 000-2 650 years ago) which includes the periods of the Old, Middle and New Kingdoms, together with the intermittent periods of disturbances and foreign invasions (Breasted, 1915; Gardiner, 1964; Grimal, 1992; Dodson, Hilton 2004; Wilkinson, 2010). In that period Egypt's influence sometimes spread to the west and east of the Nile Valley, as well as far back southwards, sometimes, it narrowed down to the territory spanning from the Nile delta in the north to approximately the latitude of the present-day Aswan in the south (Figure 1).

DOI: 10.4018/978-1-7998-1706-2.ch008

*Figure 1. Ancient map of Egypt with names of towns also in hieroglyphs. The map is made by Jeff Dahl Hieroglyphic names of towns and nomes. Author: JLG. Arts. (Wikipedia, Ancient Egypt, n/d)*

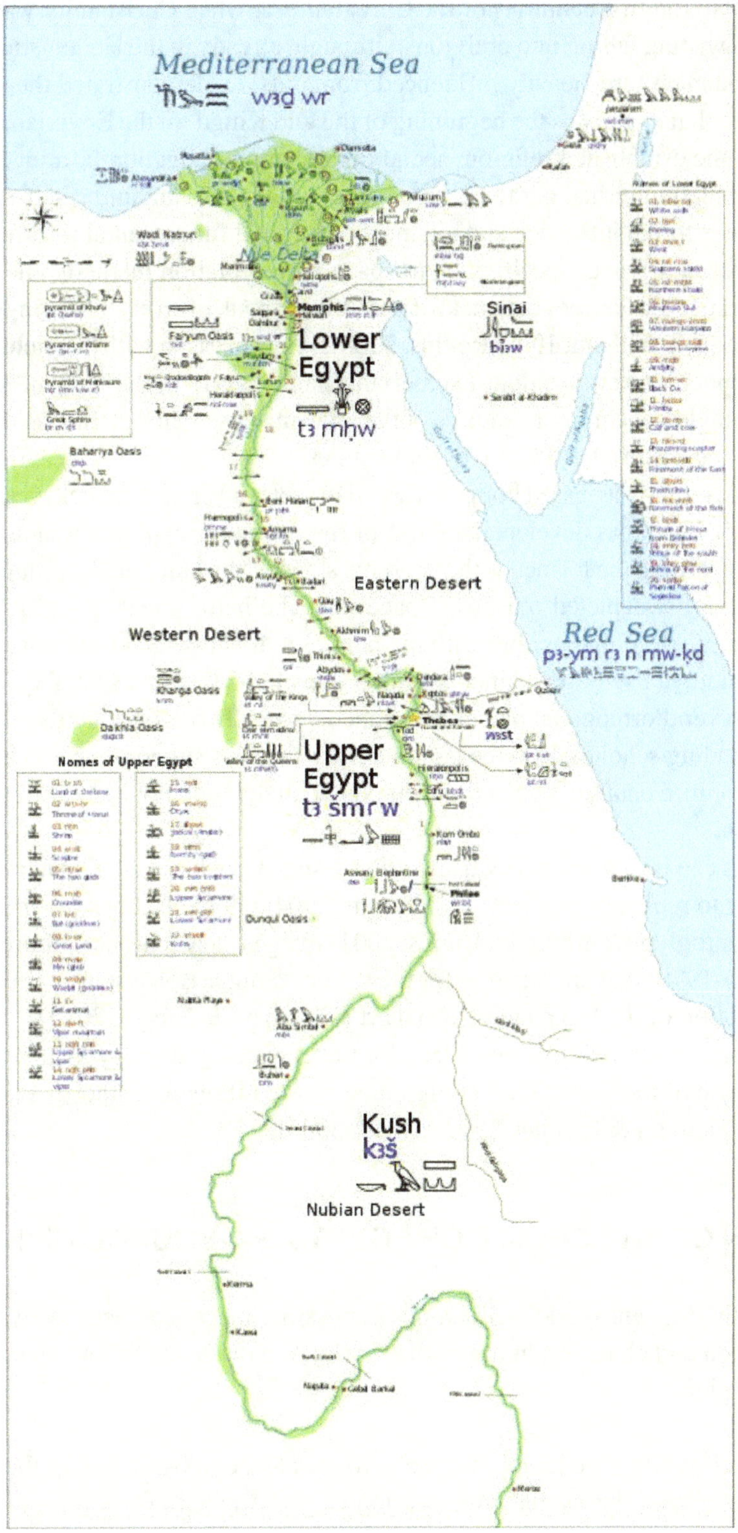

This period ended with, firstly, the Assyrian, then Persian and Macedonian conquests, as a result of which Egypt lost its governmental autonomy; whereas the cessation of the spiritual and cultural tradition should be dated to the first centuries of the Christian Era, when Christianity was adopted, with the ancient religion and writing falling into oblivion. Although, already in the Hellenistic and Roman period the traditional Egyptian culture, heavily influenced from outside, demonstrated the signs of stagnation.

It should be noted that as early as the beginning of the Old Kingdom the Egyptian culture was already rather mature, with the established religion, social organization, irrigation farming, writing, literature, and arts. Such combination turned out to be robust enough while maintaining self-similarity in several respects, despite the external influence and attempts of internal fundamental reformation.

Concerning pictorial art, and, specifically, to space representation methods, the monuments of the time also have some specific features that make it possible to speak of an easily distinguishable "Egyptian style" (Smith, Simpson & William, 1998; Robins, 2008). A comparison with the pictorial art monuments of synchronic Mesopotamian civilizations (Sumerian, Elam, Assyrian, Babylonian) is also necessary as it will enable us to single out common features of the ancient states while setting aside certain particular features inherent in the pictorial art of any of those states.

The researches differently assessed both the Egyptian pictorial art and its link with the art of previous epochs. Matie (1956, 1966) has developed an idea of the Ancient Egypt pictorial art being close to the primeval art while regarding the former as the supreme stage of the latter. In the opinion of Raushenbakh (1980), an Egyptian artist depicted objective space with the help of orthogonal projections; and, the author supposed that it was only possible with the help of a drawing made according to the orthogonal projection rules. Kolotov (1998) maintained a similar view believing that spatial correlations were portrayed by aligning several orthogonal planes positioned vertically or horizontally.

Therefore, the leading scholars advance contradictory theories supported by an analysis of a certain part, rather than the entire data set, whilst the methods of analysis do not demonstrate the required consistency of approach.

These drawbacks can be eliminated by using other sources, such as visual materials on the Egyptian pictorial art in order to replenish the data set (Verman, 1904; Global History of Arts, 1956; Gombrikh, 1988; Hill, 2007); hieroglyphs (Adkins & Adkins, 2001) images of synchronic cultures (Ragozina, 1902; World History, 1955-1970) that are required for the sake of comparisons; self-organization theory – applicable to the evolution of the Egyptian pictorial art (Kovalyov & Nitsyn, 2010).

The history of magic and religion, necessary for understanding the Egyptian version of the cult of ancestors and the myth of the World Tree, is presented by the following authors (Shafer & Byron, (ed.), 1991; Pinch, 1995; Fleming & Lothian, 1997; David, 2002).

## JUSTIFICATION OF THE CHOICE OF EGYPT AS A SOURCE OF FACTS

While considering the Ancient World civilizations as possible sources of facts about changing the methods of space depiction and on the evolution of the World Tree myth, the following conditions should be taken into consideration:

- The civilization should have Eneolithic roots (to enable a comparison with the previous epoch).
- It should have existed during the entire period of the Ancient World, or, better still, before the distribution of Christianity (to enable traceability of the continuous sequence of changes).

- To be sufficiently "vivid", i.e., to feature a large number of outstanding artifacts.
- To be well-researched, i.e., the artifacts should be described and represented in the form of available materials.

Of all the oldest civilizations (which, like the centers of the Eneolithic culture, were situated on almost all of the continents), it is the Ancient Egypt that demonstrates the best correspondence to these conditions.

True, if one considers only one civilization, no matter how brilliant it might be, there is a danger of losing sight of some of the details, and, at the same time, of mistaking a unique feature of a civilization detail for a global pattern. To avoid such mistakes, correlations will be made with synchronic civilizations, primarily, of the Middle East.

## IMAGE OF SPACE IN EGYPTIAN PAINTING

Because of the abundance of materials, ancient Egyptian images need to be classified; peculiarities of space representation methods need to be singled out and compared with the achievements of pre-historic and synchronic cultures; the place of the ancient Egyptian pictorial art in the evolution of the human mind and establishment of the modern understanding of space needs to be determined.

Further, it is necessary to find out which state of the artist's mind his works reflect and for the activation of which of the channels they are designed. That will make it possible to correctly interpret the meaning and technique of the images, as it is obvious that, for example, the "space of will", physical space and "emotional space" have qualitatively different objects being in qualitatively different relations that call for different means to be portrayed.

Let us single out three groups of such "spaces":

- **Sacral Space** will unite what is perceived through the first three channels and is neither perceivable by reason or by feelings. This area customarily fell within the competence of religion and magic.
- **Geometric Space** will unite mind-perceivable abstract relations of objects of the physical world, where interesting, in the first place, by the relations between the forms, i.e., geometry.
- **Emotional Space**, perceivable through sensuous comprehension and constituting the subject matter of psychology and aesthetics.

The proposed classification of the ancient Egyptian images meets the understanding and methods of the research that is outlined in previous chapters. There are six types of images represented by paintings, reliefs, and impressions:

- Universe "models" (the World Tree images will be considered separately).
- Images of gods.
- Scenes of "official chronicles".
- Hieroglyphs.
- "secular chronicles", scenes depicting husbandry and everyday life.
- Images of nature.

## "Universe Models"

Let us compare the images on an Assyrian relief (Fig. 4) and an Egyptian papyrus (Figures 2-3).

The Assyrians maintain an understanding of the "World Tree" that unites the underworld, the earth, and the overworld, such understanding being characteristic of the previous epochs and reflecting, as noted above, the "trueness" of the intuitive perception.

In the Egyptian myth, a novation can be noted in that the Universe becomes binary, as represented by the figures of the spouses – Geb, the god of the earth, and Nut, the goddess of the sky. And these symbols reflect binarity of will/reason. A primeval human, who was not infrequently on the edge of dying, most certainly noticed that it was there where unusual insights would emerge, useful for personal survival and the welfare of the community, - in other words, where the intuitive perception of things is activated. Since ancient times, snakes and fish have symbolized the world of the dead. Hence, a Sumerian myth on the fish-human Oanness who taught people farming, crafts, and urban construction. Here, a human, too, started to separate himself from nature (ego activation) and to endeavor to "conquer" it (activation of will/reason), thus weakening its intuition. So, its stimulation became necessary in the form of an imitation of near-death states, which was monopolized by shamans and priests (Chapter 6, *Neolithic: separation and war*). Thus, the underworld became somewhat identical to the overworld, as reflected in the latest myths about civilizers, for example, about the Plumed Serpent Quetzalcoatl (a bird is a symbol of the overworld) or Triptolemus, who moved in a chariot to which winged serpents

*Figure 2. The Tree of Life and Wardens. Assyrian relief (Tseren, 2007)*

*Figure 3. Geb and Nut (fragment). Henuttawy Papyrus. London, the British Museum. XI century Photo by E. Tsarev (Mail. Community-catastrof, 2007)*

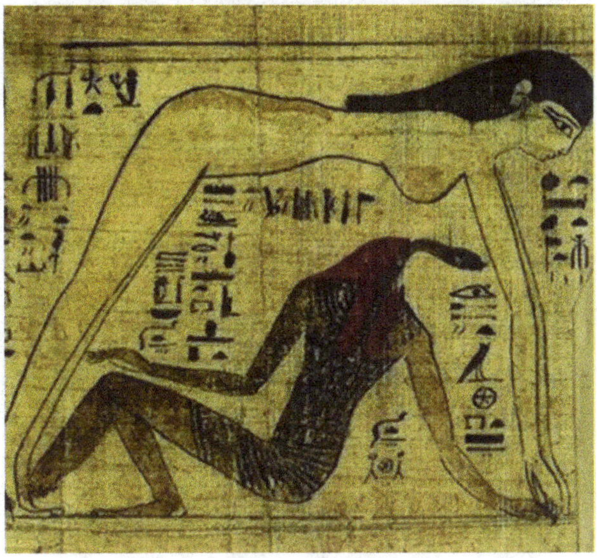

were harnessed. The human world became self-sufficient, the outer world – binary, and the World Tree ceased to correspond to the new reality.

Probably, the minds of Egyptians evolved in the same way (let us draw attention to the snake-like head of Geb, which is indicative of his chthonic nature, and to his spousal relations with Nut, which symbolized the identity with heaven).

Indirectly, the image still epitomizes, as does the World Tree, the idea of the unity of all the worlds – through kinship with the creator god Ra-Atum-Khepri.

Thus, Fig.3 shows relations within the sacral space.

**Images of gods** are represented as human figures and human figures with heads of animals, i.e., are polymorphic (Figure 4). This most remarkable feature is well known from the primeval pictorial art (previous chapters); it evidences the depiction of relations with the sacral spaces.

**Scenes from "Official Chronicles"**: of victories over enemies and of rituals. The pharaoh was their main character (Figure 5).

Since the pharaoh was deemed to be a descendant of gods and a living god, the aforementioned style was used to portray him, whereas a simple method of increasing the size of his figure in contrast with the sizes of his subjects and prisoners accentuated the regal greatness. It is thought that here a triumph of Narmer Pharaoh, the unifier of the Upper and Lower Egypt and the first ruler of the unified kingdom, is portrayed.

However, the scene in Fig.5 may as well be interpreted as a manifestation of the pharaoh's power over the three worlds – that of the dead, of the living and of gods (the scene features their images, as well as the border between the worlds of the living and the dead); that is, here relations within the sacral space are also represented.

One more interpretation is that depicted are not merely pictures, but pictograms, thus, this panel is a record of the emerging writing.

Figure 4. Image of Thoth, the god of wisdom, as a polymorphic figure (Silnaya mplitva Bogu, n/d)

Figure 5. The Narmer Palette 3000 BC ± 100 years (Studblue. Notes-midterm, n/d)

*Figure 6. Egyptian hieroglyphs (History. Egipetskie ieroglify, n/d)*

*Figure 7. Sumerian cuneiform (Symbol. Shumerskaya pismennost, n/d)*

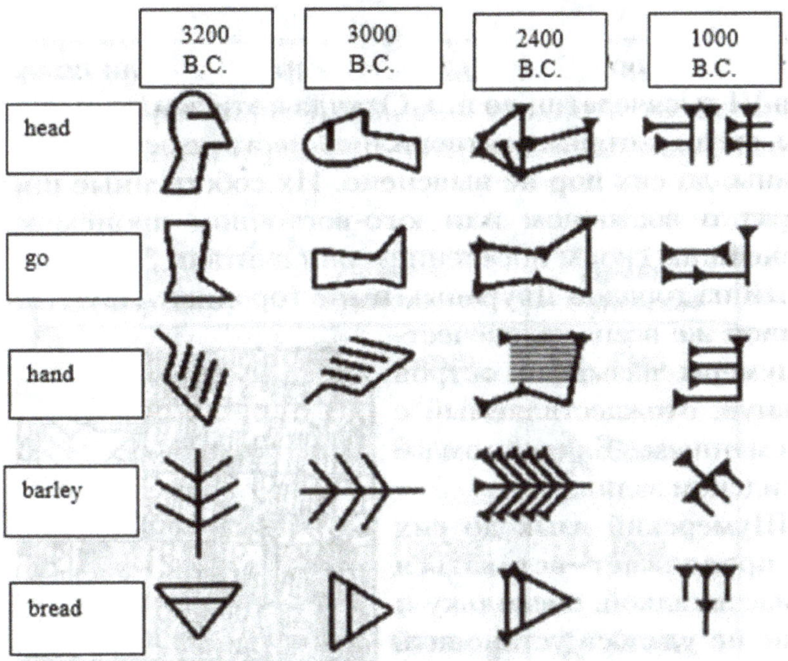


**Hieroglyphs** representing both separate notions and sounds were mainly used for recording ritual texts (simpler hieratic and demotic writing systems existed in Egypt as well), which gives grounds to suppose that their graphemes reflect relations within the sacral space (Adkins, L.; Adkins, R, 2001; Sacred Texts, n/d). This supposition is far from being obvious, more than that, for many hieroglyphs represented in Figure 6 quite "earthly" matches may be easily found. Likewise, many other writing systems evolved as abstractions of certain "earthly" objects (Figure 7).

It should not be forgotten that it is only in the sacral space that the unity exists between the subject, symbol and object, wherefrom the idea of designation – firstly, through a specific image, and then, through an abstract symbol, - itself originates. Such interpretation is confirmed by the manner of drawing hieroglyphs, which is no different from the one considered above.

**Images of "secular chronicles", scenes depicting husbandry and everyday life** illustrate an increasing thrust towards naturalism (Figure 8) reaching its highest in the Amarna period – thus, a degenerative appearance of the reformer pharaoh Echnaton is portrayed without any single "retouch" (Figure 9).

*Figure 8. Architect Khesira. Relief from the tomb from Khesira in Saqqara. Wood. 2 800 B.C. Egyptian Museum. Cairo. (Vselova. Zodchiy Hesira, n/d)*

*Figure 9. Amenhotep IV (Echnaton). Approximately 1360 B.C. Limestone. 14 cm high. Berlin, Egyptian Museum (Guner. Bibliotec Culture, n/d)*

New features may be noted among other samples of this type of images:

- The clearly conscious and even "standardized" set of methods for portraying forms and relations within the three-dimensional space (Figures 10-11).
- A combined representation of the external appearance and interior structure of objects (Figure 12).

Thus, the images of this group start to depart from the ancient canon by representing relations of the sensuously comprehensible three-dimensional space.

*Figure 10. Orchard of Nebamun. Approximately 1400 B.C. Painting from a tomb in Thebes 64 x 74.2 cm, London, the British Museum (Guner. Bibliotec Culture, n/d)*

*Figure 11. Sennefer's Tomb: orchard. Approximately 1410 B.C. Western Thebes (Art. Biblioclub. Grobnitsa-Sen_n, n/d)*

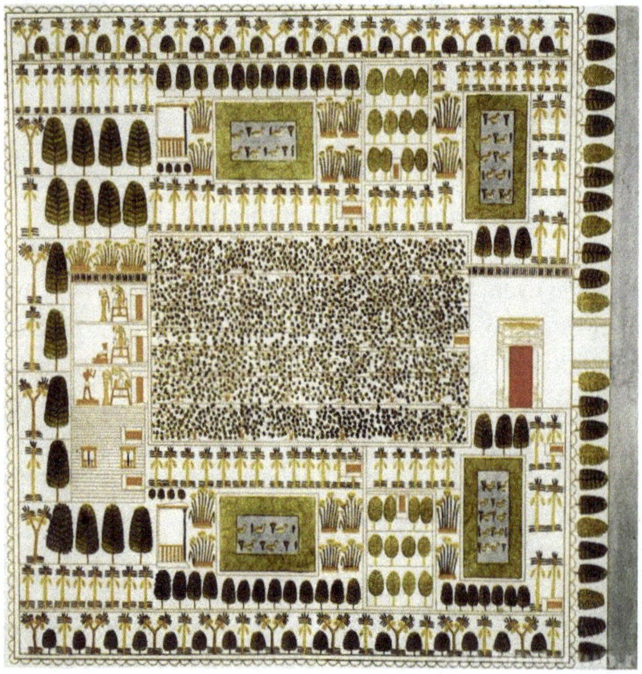

*Figure 12. Guests coming to a village house (after Ros. M. C., 68) (E-reading Club. Ehrman, n/d)*

## Images of Nature

Ancient Egyptian artists left quite many wonderful images of nature (Figures 13-14), which are deeply impressive, i.e. represent relations of the emotional space.

Now, while setting the plot point aside, the specific features of the ancient Egyptian image representation technique should be determined. The following can be singled out:

- **Planeness**: Each figure is delineated with a continuous line of unvarying thickness that is evenly filled with paint of the same color (Figure 15).
- **No Aspect Angle**: Altogether absent is the foreshortening of parts of an object because of their being at different distances from the viewer. The aspect angle is eliminated because the image of an object combines the images of its separate parts created from the points of view and lines of sight which exclude the perspective distortion of a form (Figure 16).
- Demonstration of beauty and expressiveness of a form with the help of special methods. Thus, in portraying a human figure, legs are shown from the viewpoint that best represents the shape of feet, whilst the upper part of the body is painted by the artist along such line of sight which enables its perception without foreshortening. As a consequence, the plane of the thoracic girdle is turned against the plane dividing the body by half, with the angle of a quarter of a circle, - representing a peculiar feature by which one can unmistakably identify a work of the Ancient Egyptian art. Feet are depicted from the side of the big toe because the view of a foot that reveals the internal line of the instep is more expressive. Hands are depicted in the same way. That is why a human is depicted in ancient Egyptian paintings as having two right or two left legs, as well as two right or

*Figure 13. Tomb of Khnumhotep in Beni-Hasan. Fragment of the painting: birds (Egyptopedia. Arkhitektura-srednego-tsarstva, n/d)*

*Figure 14. Painting from a tomb in Thebes. XV-XIV century B.C.E. British Museum, London Author: Alexey Berezin (Alex-aka-jj, n/d)*

*Figure 15. Thebes, the tomb of Menna, XVIII Dynasty (1420-1411 B.C.). Author: Alexey Berezin (Alex-aka-jj, n/d)*

*Figure 16. Beni-Hasan, the tomb of Khnumhotep, XII Dynasty (1920-1900 B.C.). Author: Alexey Berezin (Alex-aka-jj, n/d)*

*Figure 17. Papyrus from Dair al-Madina. Thebes, XIX Dynasty. Brussels, the Museum of Art and History, (E 5043) (Maat-object, n/d)*

two left arms. However, because of this strangeness, the ancient Egyptian pictorial art has a larger aesthetic value than anatomically correct images (Figure 17).

- Alignment with the picture plane of several images that best reveal the form of different objects. To show the orchard (Figure 18), the artist chooses the viewpoint situated at the height of the bird's flight and looks at the earth from above. However, the villa, the pond, the trees growing along its edges, the barns full of supplies are all shown in the direction of the sight that gives the best representation of their form. And, at the same time, the images of the trees are positioned at right angles to the edges of the pond, and that is consistent with how the trees grow in reality.
- Another manner of depicting the depth of the space is also used (Figure 19). If the picture plane is positioned at the right angle to the earth's surface, and the line of the sight – at the right angle to it, the image of the earth will turn into the straight line, i.e., into the picture trace of the object plane. The images of figures resting on the earth are not hanging in the air, they're touching the straight line will be called the line of support. Usually, several lines of support can be found on the ancient Egyptian paintings and reliefs, such lines dividing the image plane into a multitude of spatial belts. In so doing, an artist may erase lines of support from the image plane and return to the space representation method, in which the lowest line of support is featuring, while the objects arbitrarily fill in the image plane and seem to be soaring above the line of support. I.e., in depicting the

*Figure 18. Egyptian villa. According to bass-relief in Tell el-Amarna. Middle Kingdom (Gardens. Istoriua parkovigo, n/d)*

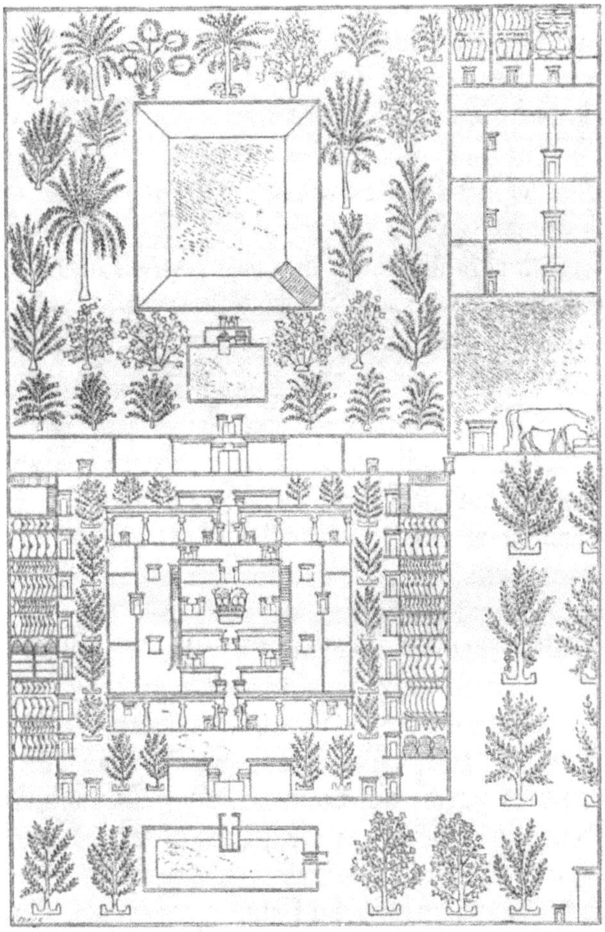

space an artist would employ one of the simplest signs of the depth of space –approximation to the skyline of the visual image of an object as the distance to the object grows; according to such method, the image of the object which is farthest from the viewer, is shown higher than the images of the objects located closer to the viewer. It should be noted that the Assyrian artists used similar methods (Figure 20).

- Images of objects included in one spatial stratum in the direction of the line of support (Figure 1).This method was used when one object was obstructing the others. When a procession is portrayed, figures are shifted to ensure that those marching nearest to the viewer should not hinder the correct perception of those marching at the end of the column. Thus, an ancient Egyptian artist was aware of the existence of one more monocular sign of depth, that of superimposition, when an image of an object positioned nearer is covering by itself images of objects positioned farther. Although this peculiarity may be creating an impression of a rudimental linear perspective, the image is, in fact, devoid of the depth of space, which can be seen, for example, from the fact that the images of warriors positioned along the line of support are of the same size.

*Figure 19. The attack on the Hittite Fortress of Dapur (according to L. D., iii. 166) (E-reading_club. Ehrman, n/d)*

*Figure 20. Bass-relief of the times of reign of Tiglath-Pileser III (VIII century) (Xlegio. Siege machinery, n/d)*

- A combination in one picture plane, in a certain sequence, or related events taking part at different times (Fig.22), through which the plot could be deployed through the time perspective.

*Figure 21. Warriors. Painting from the tomb of Januni in Thebes (Egyptopedia. Voennaya politika, n/d).*

*Figure 22. Farmers at fieldwork. Painting at the tomb of Ty (Egyptopedia. Selskoe-khozyajstvo, n/d)*

Attention is drawn to one more circumstance. Whilst in the Eneolithic and Neolithic ages, the pictorial art remained relatively homogeneous, Ancient Egypt experienced one more very vivid break from tradition. This is the so-called Amarna Period related to the reign of Echnaton. In the first place, it is related to a rapid increase of interest in depicting the relations within the emotional space, with a thrust towards naturalism. Even after the reign of Echnaton had ended and his dynasty had ceased to exist after the departure from the canon (which was preeminently represented by the sacral and "official" art) had become less noticeable, this tendency survived.

During the period of the Old Kingdom (it is chosen for the sake of control) the main characters of the images were gods and pharaohs, who were depicted very strictly and without excessive dynamics, with a limited amount of color and, mainly, against a grey background (Figure 23). At the times of the Middle Kingdom the dynamics, complexity of composition and interest to nature enhanced (Figure 24, images were often depicted against a warm yellow or beige background). At the times of the New Kingdom images grew much more expressive (Figure 25), scenes – dramatic (Figure 26), colors and hues grew in number, while the background often turned out to be white (Figure 27). To make those changes more illustrative, Figures 23-27 were selected as two-dimensional images portrayed in the "Neolithic" style.

Let us have a look at which correlations in the minds of the Egyptians those changes were accompanied.

The Old Kingdom brought an image of the God-Pharaoh, and omnipotent descendant of Osiris, Isis, and Horus. No doubt, the soteriological sentiments noted at the end of the Eneolithic Age greatly

*Figure 23. Image of the Old Kingdom period (to the right you can see pictograms turning into hieroglyphs). Ritual posture, minimum of the color, grey background (Egyptopedia. Iskusstvo Drevnego tsarstva, n/d)*

*Figure 24. Image of the Middle Kingdom period. An everyday dynamic scene deployed in space and time against the backdrop of warm hues (Egyptopedia. Kultura Srednego tsarstva, n/d).*

*Figure 25. Image of the New Kingdom period. Wailers, maximum of emotions, light background (Egyptopedia. Iskusstvo Novigo tsarstva, n/d)*

*Figure 26. Battle of Kadesh. Turning point: the Great Ramses II is breaking out of the Hittite encirclement. Modern trace drawing (Balagan. Battle of Kadesh, n/a)*

*Figure 27. Fresco paintings of the tomb of Nefertari. Light background, vivacious hues. XIX Dynasty. The first half of the 13 century B.C. Valley of the Queens, Thebes (Megabook-media. Fivy, n/d)*

contributed to this. And the task of going along the path of the dead and coming to light - the pharaoh's task – was becoming public as well: if the pharaoh cannot do it, who will help us, ordinary mortals? Both the pharaoh believed himself to be a god, and his subjects were mobilized, too, and we're ready to devote all their forces to constructing temples and pyramids. But while such a state of mind in the pharaoh and priests was maintained by continuous practices and rituals (as for the priests – by undergoing rigorous selection as well) enhancing intuition with its access to "extra-systemic forces", the nobility and the people, who sporadically participated in ceremonies, were devoid of such support and their minds were gradually moving to lower levels, among which was the level of reason, as is seen based on the increased interest to realism and three-dimensional space.

The unreadiness of the society is to permanent mobilization and the exhaustion of resources consumed by the "Megalithic" construction resulted in the decline of the Old Kingdom.

The Middle Kingdom was a certain period of equilibrium, when the priests popularized a path to light not linked with a certain pharaoh (complex mourning rituals, mummification, the "Book of the Dead", etc.), thus proving their competence. Senusret I can be regarded as one of the "exemplary" pharaohs of that epoch, but even he, while fighting the enemies on the borders, was characterized by peaceful reign. The pharaoh remained as a certain "invisible locomotor" activating the entire religious and social order. The nobility and common people, being stripped of the great goal, dissipated their potential by enjoying life.

As the potential decreased, that equilibrium became disrupted, and the Middle Kingdom fell into decline and was conquered by the Hyksos; foreign pharaohs, though retaining a formal status of living gods, were not certainly regarded as such.

Reconstructed the Middle Kingdom had long searched for its path: women pharaohs appeared (the most famous, Hatshepsut, pushed such a great warrior as Thothmes III away from power for about 20 years), then Thuthmos III, likewise his son Amenhotep II, and, subsequently, Ramses II thought that their mission consisted of external conquests, whereas Echnaton set about reforming religion. Therefore, they already had neither the mind, nor the modus operandi of gods, though they possibly themselves believed in their divinity, and, certainly, retained references to it in their titles. Other Middle East kings acted similarly.

Documents of that epoch evidence the growing hedonism of social life. The so-called Turin Erotic Papyrus created at the times of the successors of Ramses II, may serve as an example. The reconstructors of that badly preserved document count that it features 12 sexual scenes, as well as the scenes that depict mice besieging fortresses, and other animals performing human tasks. The papyrus possibly constituted the first satire on the public moral, and it, undoubtedly, served as the primary source for the "Planet of the Apes"! It is beyond all doubt that, seeing on the walls of a temple the manipulations of gods with the source of life – the semen of men – the Egyptians started, over time, to understand them literally, merely as sexual scenes. No wonder, that after experiencing a short period of flourishing (a quick "release of potential" during the transition to the next level of mind), Egypt was never able again to reach its former height. It should be noted that it was the priesthood – and one cannot fail to notice that it is an effect of practices – that managed to last longest; however, during the Roman rule it degraded too (to illustrate this, one may recall the Golden Ass novel by Apuleius), and, as early as the first centuries of the Christian era, even the writing was forgotten.

Now let`s compare the Egyptian pictorial art with that of the Eneolithic period (Table 1).

*Table 1. Comparison of the Eneolithic and ancient Egyptian pictorial art features*

| Feature | Eneolithic Pictorial Art | Ancient Egyptian Pictorial Art |
|---|---|---|
| 1. | It is characterized by sketchiness, one-colored images, contrast with the background, less dynamic images, inactivity of background. | The sacral painting was characterized by canonicity, a monotony of color in images, contrast with the background, lesser dynamics, inactivity of background.<br>Genre painting is characterized by the violations of the canon, polytonality of images, contrast against the background (with changes in the color of the background), dynamics |
| 2. | Space is depicted as two-dimensional. No elements of perspective images are present (viewpoints, skylines, vanishing points, a line of sight). No illusion of depth is created. Space has a complex structure and is depicted as two or three segments, each filled with different patterns and signs and divided by borders, though sometimes their unity is emphasized with an image of sprouting the grain, etc. | Sacral painting: space is two-dimensional, no elements of perspective, no illusion of depth.<br>Space structurization is not related to the World Tree image and is communicated through a complex semantics of symbols and writing signs.<br>Secular painting: along with two-dimensional images, methods appear to reflect the depth of space and deploy action in the course of time. The action takes place on earth.<br>A rather expressive "emotional space" is being organized.<br>Many new plots appear. |
| 3. | As compared with the Neolithic paintings, compositions are more static; symmetry and center are sometimes present; dynamics and cycle are expressed by patterns and signs. | Most often, a center exists, but not in the geometric sense – the pharaoh's or noblemen are made prominent (including, in size), the layout of a villa, lake contour, etc. Cyclicity is expressed to a lesser extent. |
| 4. | Top and bottom, right and left side are discernible. | Top and bottom, right and left side are discernible, as are the past and the present. |
| 5. | Natural scale and proportions are distorted, figures look stylish, sometimes grotesque. | The natural scale is distorted (especially, when it comes to the figures of gods and rulers), while proportions and figures sometimes look grotesque (the Amarna period). |
| 6. | Images of animals and people are not expressed explicitly, details of conduct and motion are rather schematic; a relatively little number of scenes of war and violence. | Images of animals and people are represented canonically; details of conduct and movements are schematic; scenes of war and violence are relatively small in number; except the Amarna period and battle scenes, especially of the New Kingdom |
| 7. | There are polymorphic images, in particular, of sun-shaped figures. | Polymorphic gods are large in number. |
| 8. | There are complex magic or cult-related scenes. | There are complex magic or cult-related scenes. |
| 9. | Abstract images, symbols, and signs are large in number and diversity. The emergence of pictographic proto-writing or writing. There appeared and developed replication of images through their impressions. | Pictograms quickly turn into hieroglyphic writing, which, which, as signs gradually lose similarity with the object they denote, evolves first into the hieratic, and then into the demotic writing. |

## TIGHT GATES: WORLD TREE AND THE BOOK OF THE DEAD

The culture of Ancient Egypt remained largely a culture of the Stone Age – even if such which discovered writing and learned how to smelt metals. The sacral painting vividly illustrates that.

But the sacral painting evidences a deep crisis of the worldview of the Stone Age. The emergence of a canon and standard methods of depicting figures and space shows that inspiration vanishes from the creative workers who worked on the priests' instructions. Instead of vivid intuitive impressions of the subjective space, where no figure resembles the other, there are illustrations to myths and rituals, with only formal characteristics of the 2 and 3 levels of self-organization remaining.

*Figure 28. The God of the Sun Ra in the form of a cat wounding the Apophis snake (the personification of darkness and chaos) by the roots of the World Tree. Dair al-Madina, the tomb of Inher-ha, Thebes, XIX Dynasty (Bestiary. Bog-Ra, n/d)*

*Figure 29. Meretseger in the form of a snake. Dair al-Madina, Thebes. (Bestiary. Zmienoobraznaya-Boginja, n/d)*

The same crisis is evidenced by the transformation of the World Tree Myth, notably, its binarization which means "struggle for unity" instead of "unity". The goal – immortality – is preserved, but the path to it means that Horus needs to defeat Seth, Ra - Apophis (Figure 28)[1], human – to undergo severe tests and, either when alive or after demise, to defeat the death itself (the goddess of death – Meret-Seger – was also portrayed as a snake, Figure 29).

Thus, the World Tree becomes the "tree of good and evil" as well, and is, in such a capacity, tabooed (protected by the snake). This tabooing, also typical of the Mesopotamian cultures (already mentioned myths about Inanna's descent to the underworld, and about Utnapishtim and Gilgamesh – only Utnapishtim managed to reach immortality) was subsequently included into the Biblical myths.

The "Book of the Dead" (Anonymous/ Budge, 2014) represents the idea of revival by using an entirely different set of symbols; it relates to the World Tree myth only indirectly, such relation dating back to the Eneolithic times.

According to the "Book of the Dead" (the authentic name is the "Book of Coming Forth by Day"), after death, the soul undergoes the judgment of gods (Figure 30) and may irrevocably perish by their sentence or as a result of numerous perils faced during after-death wanderings. And only after passing all trials, and after realizing that the antipodes – Horus and Seth – are, in fact, the same being, and after becoming not quite human anymore, the soul comes to light and finds eternal life.

It should be noted that the "topography" of the underworld is described in sufficient detail, with the specification of road forks, fire lakes, dangerous places where snakes, scorpions, and ghosts dwell, places where one can and need appeal to gods. The appeals themselves are described in detail, and the milestones of the realization of unity are illustrated – through the understanding of kinship among gods and their functions in various episodes of the creation of the world (Figure 31). It is those milestones that constitute the stages of the soul's transformation; figuratively speaking, the ontogenesis of the soul is linked to the orthogenesis of the world in the process of the soul coming forth to light. The elaborate language, a system of illustrations, rituals conducted over a dead body, all may be compared with an

*Figure 30. The judgment of the dead in the presence of Osiris. Papyrus of Khanufer. From Thebes, Egypt. 19th Dynasty, around 1275 BC. British Museum. (Webcitation, n/d)*

*Figure 31. A drawing from the papyrus of a Henuttaui priestess. London, the British Museum. XI century Author: E. Tsariova (Mail. Community-catastrof, 2007)*

"enhanced" neurolinguistics programming, whose goal is to structure the post-mortem visions, to make it possible to discern their meaning and to steer the transformations of the soul in the right direction.

Such an interpretation indicates that the Egyptians remembered that binarity and juxtapositions led to death, whereas intuition leads to revival, through the sensation of the unity of gods and with gods, but that the path to acquiring it seemed incredibly difficult to them and possible, for most of them, only after death.

After interpreting from the position of the self-organization theory and of reconstructions performed in previous chapters, the following inference can be made: a scenario of folding up the C-space from level 3. Such folding up is expressed in the following manner: with the help of gods (soteriological Eneolithic idea, chapter 7, *Eneolith: The Search for Balance*) the soul overcomes binary juxtapositions (Neolithic

*Table 2. Comparison of the Eneolithic and Ancient Egyptian Versions of the World Tree Myth Features*

| Feature | Eneolithic Version | Ancient Egyptian Version |
|---|---|---|
| 1. | The structuring of worlds is not associated solely with the World Tree. The Tree becomes a certain soteriological symbol required for opening the path to eternal life. | The structuring of worlds is not associated solely with the World Tree. |
| 2. | Identification of the Goddess with the Tree. Only women, being the keepers of the Great Goddess cult and having the necessary intuition can facilitate the appearance of the Goddess in the world. | This motif is absent. |
| 3. | Connection with the cult of the dead: the worlds are divided by the keeper and to enter the underworld and to start the search for eternal life one needs to overcome his resistance. | Connection with the cult of the dead: direct one – is absent; indirect one – to access the tree one needs to overcome the resistance of Apophys. But the serpent is not a "borderline" deity, it is a personification of chaos and darkness, and even gods are incapable of killing him. |
| 4. | Even priests are unable to move around the underworld unassisted, they need the help of the Great Goddess, as a minimum, to open the passageway. | No data is available as to how the information from the "Book of the Dead" is related to the movements around the World Tree |
| 5. | The myth plot gained a soteriological connotation, it became multidimensional, intertwined with other myths, got more complicated and, to be more adequately articulated, it required a complex set of symbols, ceremonies (mysteries?) fixed in writing. | The myth is tabooed, but though slipping to the periphery of the mind, it is still not forgotten. |

mind), acquires integrity (the state of mind in the Paleolithic age), and, overcoming that stage, unites with the light (while restoring the intuitive state of the pre-Paleolithic human)[2]. A summarized result of these speculations is given in Table 2 that compares the features of the Eneolithic (chapter *Eneolith: The Search for Balance*) and Ancient Egyptian versions of the myth.

## MARKERS ANALYSIS

### Space Representation

In contrast with the Eneolithic period (Table 1), the space became more complex and structured, its depth appeared, as well as a linear deployment of events during the time period (strengthened nous), compositions became more complicated (strengthened nous), the number of scenes of war and violence, with the exception of short periods, can in no way be comparable with the permanent war of Assyria against the entire world (weakened ego and will), sensations of unity with the image grew lesser in size (weakened intuition), explanatory pictograms and texts appeared (strengthened nous, and weakened sense[3]), there is canon in sacral and official scenes (strengthens reason, weakened intuition), a departure from canons, diversity of plots, the brightness of colors and well-pronounced emotionality in everyday scenes (exaltation of feelings). Two-dimensionality of space was mainly preserved in sacral painting.

### The World Tree Myth

Fear of death (weakened intuition, strengthened ego) increased to such an extent that even thought about going through the path of the dead and coming back alive, became tabooed. Even gods could not kill Apophys; defeated, he attacks the sailboat of gods again and again. Only an already dead person who has nothing to lose may try to pass from darkness to light, after vindicating himself during the judgment and gaining support from gods (weakened will) and resorting to the prepared prayers and incantations from the Book of the Dead. Even pharaohs seem to have ultimately stopped setting such a goal, being content with the victories over the earthly enemies, even more so, over their wives and concubines. But the myth was not forgotten altogether, and the following chapter will disclose an unexpected turn of its plot.

## MIND OF THE ANCIENT EGYPTIAN HUMAN

In contrast with all cultures examined in previous chapters, undisputable and explicit class distinctions are manifested only in Ancient Egypt. That is why it is no more possible to compile a certain generalized psychological portrait, but a certain "cross-chapter" of the society is required, by examining the representatives of each of the classes – pharaohs, priests, aristocracy, commoners (Figures 32-35), which it is logical to limit to the timeframes when the Egyptian culture flourished, or, more exactly, by the periods of the XVIII-XIX Dynasties of the New Kingdom.

It should be noted that these are certain generalized qualitative portrayals, whereas the mind of an individual personality may be significantly different.

*Figure 32. State of mind of the Pharaohs of the New Kingdom period*

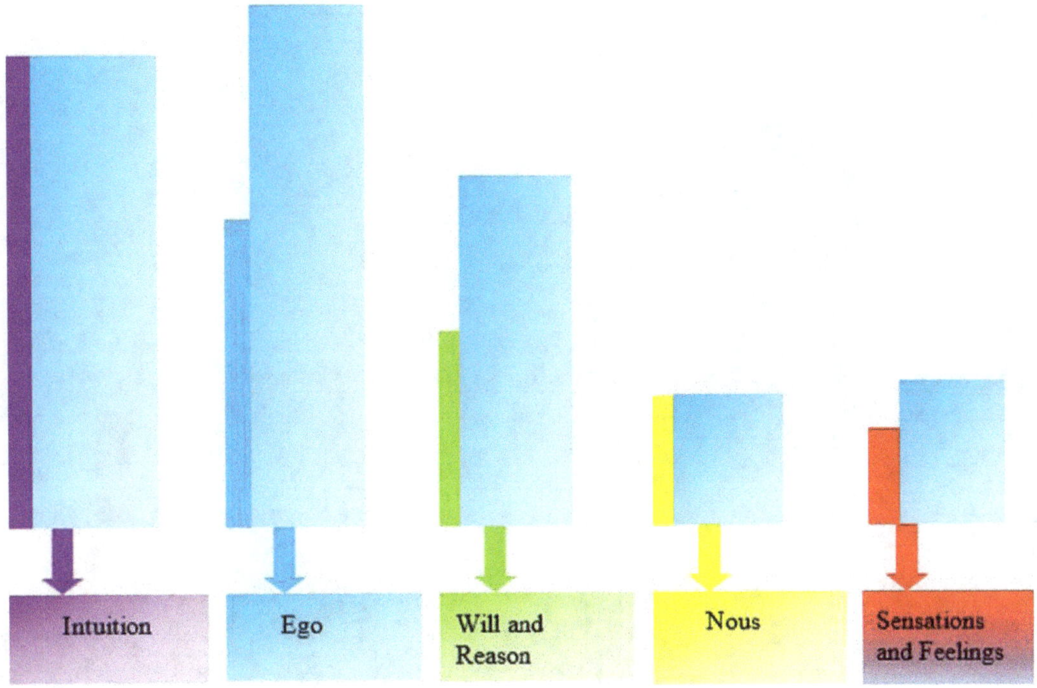


*Figure 33. State of mind of the priests of the New Kingdom period*

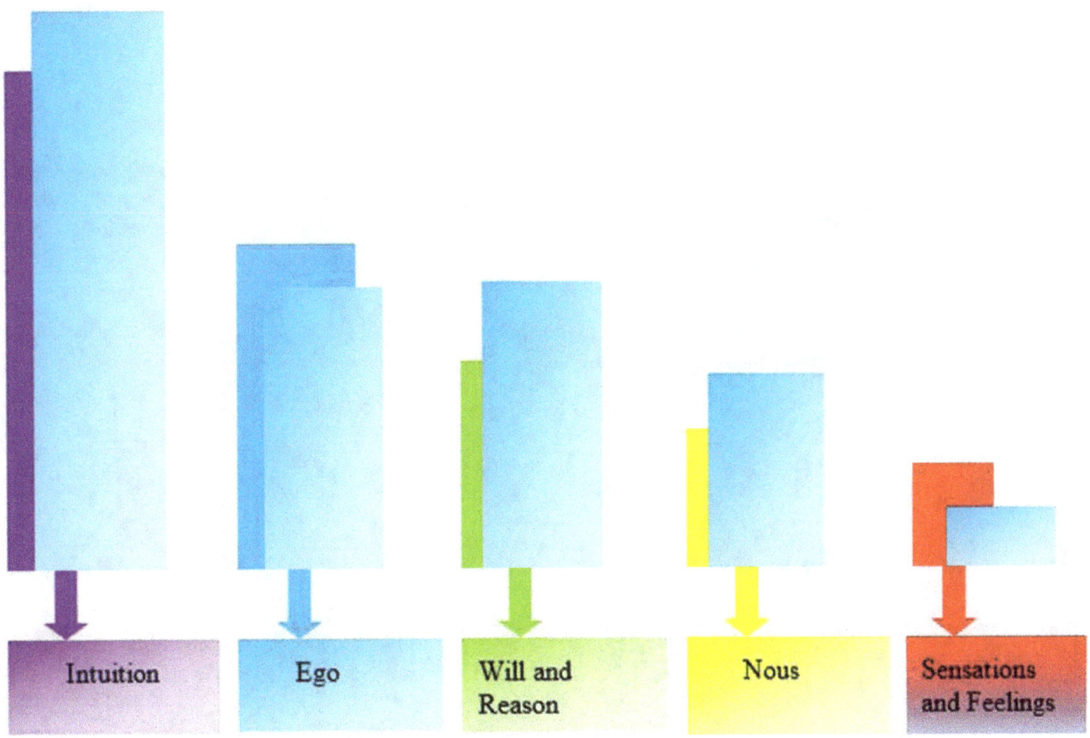

*Figure 34. State of mind of the aristocracy in the period of the New Kingdom*

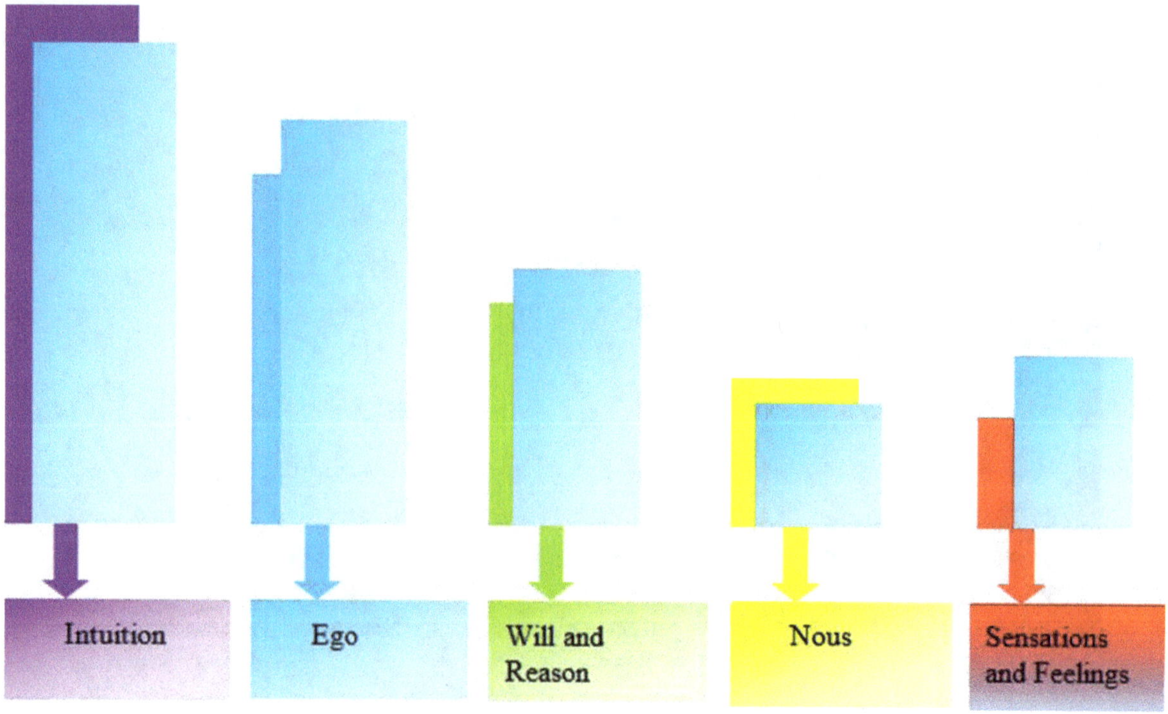


*Figure 35. State of mind of common people in the period of the New Kingdom*

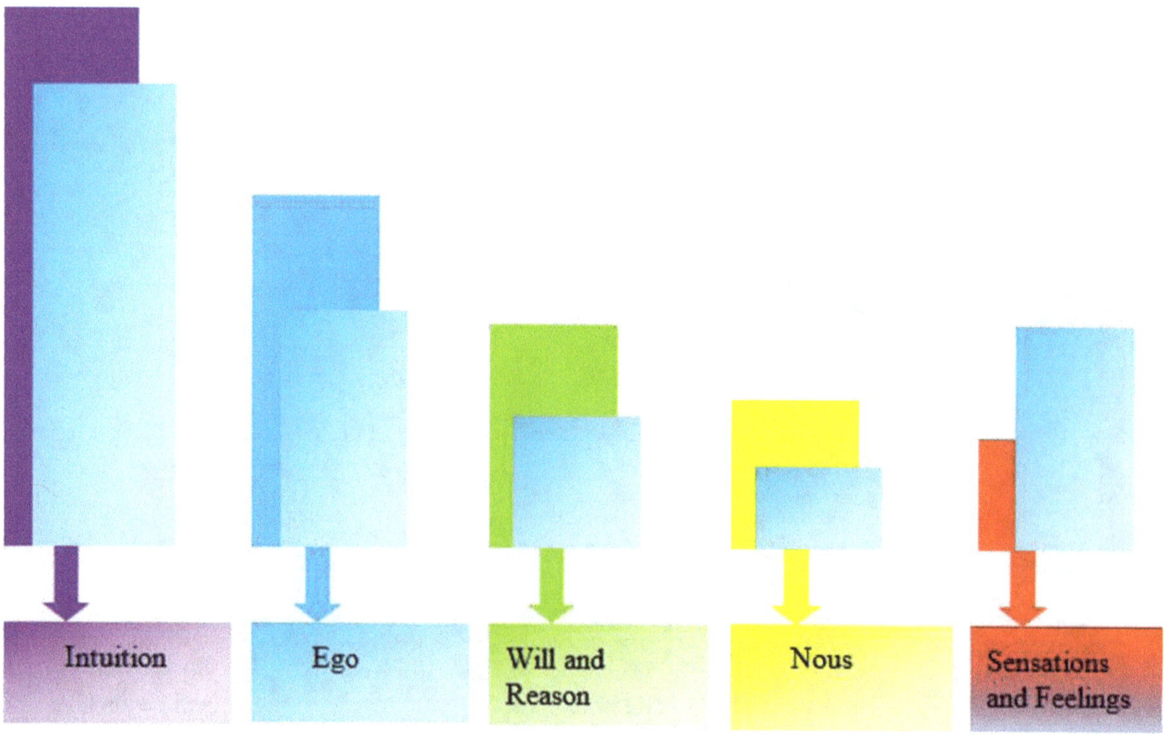

## BEHAVIOR PATTERNS

Further Fig.31 and Table 7 of Chapter 4 that links the perception channels with class groups and their behavior patterns will be used.

1. **Pharaoh** acquires the traits of a typical Middle-Eastern king. His intuition, undoubtedly, lessens – as even the most outstanding pharaohs could not foresee the consequences of their actions; Thutmose III was making regular encroaches upon Palestine, while back in Egypt the wise Hatshepsut was aptly squeezing him out of power; Echnaton initiated a religious reform, with the consequences of which his nearest descendants were unable to cope; at the most crucial point of the battle Ramses II was encircled, wherefrom he had to heroically break out. It appears that only Amenhotep II managed – owing to diplomacy – to get a foothold in Palestine and Syria, though his ventures hardly paid off. At the same time, the pharaohs' ego and will were outstanding, like, for example, was the case with many Assyrian kings. Nous and propensity for sensual pleasures matched the image of the kings of that time, as evidenced by the data available on both the excellent education of the pharaohs and on the number of their wives and concubines. Pride, vigor, relentlessness – these are characteristic traits of their behavior.
2. The minds of **priests** are the closest to the mind of a "balanced personality". Practices and asceticism led to strengthened intuition. Servicing gods facilitated the strengthening of the ego. Necromancy and ritual magic led to strengthened reason and will, whereas lengthy education facilitated the development of nous. All of that was happening due to the potential of senses. Plus, a rigorous selection of candidates. A restrained, omniscient and capable of many things priest is, probably, the most likable figure of that epoch, an intellectual and a scholar of the Ancient World.
3. The **aristocracy** of the New Kingdom did not stand out clearly. Yes, that class produced military commanders and administrators, and its representatives were well-educated. But there is no evidence that they thought themselves to be a united force - as did the French nobility or the Polish gentry – to impose their will upon the pharaoh and priests, or that they were idea-driven warriors, as samurai were, with the code of honor of their own. Accepting tribute from farmers, being entertained by dancers and singers, going hunting and having feasts – those were their cravings and lifestyle as shown by numerous drawings. A useless class, relaxed by living relatively peaceful lives, who were in search of something to occupy themselves with.
4. Finally, the **commoners** – farmers, craftsmen, merchants and other representatives of the "third class". Accustomed to obeyance, fearful of independent decisions, they were people, who had rather cunning than reason, and who found delight in sensual pleasures – so vividly depicted in the Turin Erotic Papyrus.

At the same time, except for the periods of wars and catastrophes, they could be quite happy with their lives – as the government system which was based more on persuasion than on direct violence, a good climate, and a well-organized irrigation system created all necessary prerequisites for that. On the other hand, unlike the plebeians and fringe groups of the epochs to come, the Egyptians had not unlearned how to work.

On the whole, the civilization of Ancient Egypt may be characterized as follows: a peak of the Eneolithic and even Neolithic culture that has lost a large portion of potential, intuition, and will, but has gained rudiments of the rational worldview.

## CONCLUSION: ALMOST SCIENCE

In analyzing the representation of space by the Ancient Egyptian artists and in comparing it with the methods of artists from previous epochs, a distinction becomes obvious at once - it is the interest of the Egyptians to reflecting relations of precisely the three-dimensional, geometric space and finding ways of how to do that. Certainly, in a large number of images the "Neolithic", two-dimensional images prevailed, but two methods of reflecting the depth of space had already come to existence, and, even more so, canonical methods for their representation appeared.

A share of the secular painting started to gradually increase, it began to penetrate the sacral area, too, reducing the relations of the God-Pharaoh and his Goddess-Wife to the relations of common people, in an unsuccessful attempt to describe the symbolism of the sacred marriage with the help of geometric and color relations.

This already evidences a new state of mind.

This new state is stipulated by the activation of the channel of nous (the fourth level of self-organization) that perceives the spatiotemporal relations and gradually becomes dominant. It was this process that generated the contemporary civilization and science.

Noting this tendency, answers must be given to the questions which determine the place of the Ancient Egyptian culture in the general evolutionary process: how far have the Egyptians come forth in perceiving the spatiotemporal relations? Can one say that in depicting the space sciences, even if underdeveloped, methods of modeling were used?

To give a response, ethnographic materials should be considered (Winkler, Niebuhr, Schurz, 1903), and two specific features of the visual perception of the contemporary primitive people should be highlighted[4]:

- the tribes that are at the lowest levels of development, extensively use site plans. For example, the Polynesians may represent the location of faraway islands which cannot be seen from one point of view. And the distances between the images correspond to the distances which need to be covered to get from one object to another, rather than to the distances between the visual images of objects. Thus, a plan is the most natural method of representation;
- the representatives of low developed tribes correlate with great difficulty perspective images with the objects depicted on them. It seems that perspective distortions misrepresent the natural form of the object to the extent when they make them unidentifiable by a primitive human.

These data are quite consistent with the specific features of the ancient Egyptian images. Based on the aforementioned data, it should be presumed that a method of representation of space exists which is granted to a human intuitively, as well as a method acquired in the course of the development of reason in the framework of the scientific understanding of reality.

A comparison of the technical methods of the Egyptians with the peculiarities of the primeval painting and children's drawings (chapter 5, *Paleolithic: unity*) evidence that the first method was applied. Hence, the level of the development of nous of the ancient Egyptians has not gone far from that of a contemporary primitive human or a child. Only pre-requisites existed concerning the Egyptians, who had not yet taken the path of scientific cognition and had only slightly advanced in the understanding of spatiotemporal relations.

As far as the legendary wisdom of the Egyptians expressed in religious myths, achievements in magic, astrology and the like, it was necessitated by a much larger than that of a modern human, but, lesser, as compared with the Paleolithic humans, the activity of the intuition, ego, will and sense. Successes in sciences requiring abstract thinking are insignificant (Gillings, 1972).

As for the second – projection – method of space representation, which is characteristic of the understanding of spatiotemporal relations, there are no signs that it existed in Egypt. This path would start with the first experiments in representing the space and three-dimensional objects on the picture plane in the Hellenistic-Roman art, after a deep decline of the ancient civilizations – and ended with the linear perspective discovered by the great artists of the Renaissance.

Therefore, attempts that are being made by contemporary scientists (Kolotov, 1998) to provide a scientific interpretation to the technique of the Ancient Egyptian painting, after describing in it orthogonal projections or transformations of a drawing, are nothing but inappropriate application of scientific categories to the minds of the Ancient Egyptians that have not reached the scientific level. An appropriate framework of categories within the field of the history of art is yet to be developed for such a task.

## FUTURE RESEARCH DIRECTIONS

The following Chapter will be dedicated to spatial constructions in antique art and the changes in the World Tree myth; and, on their basis, states of mind and behavior patterns of the main classes that existed in that epoch will be reconstructed.

## REFERENCES

Adkins, L., & Adkins, R. (2001). *The Little Book of Egyptian Hieroglyphics*. London, UK: Hodder and Stoughton.

Adolf, E. (2008). Life in Ancient Egypt. Moscow, Russia: Centrpoligraf.

Adolf, E. (2008). State, Army, and Society of Ancient Egypt. Moscow, Russia: Centrpoligraf.

Alex-aka-jj. (n.d.). Retrieved from https://alex-aka-jj.livejournal.com/234163.html

Anonymous, & Budge, E. A. (2014). *Wallis (2014)*. The Egyptian Book of the Dead. Published by Createspace Independent Pub.

Art. Biblioclub. Grobnitsa-Sen_n. (n.d.). Retrieved from http://art.biblioclub.ru/picture_32228_grobnitsa_sen_n/

Balagan. Battle of Kadesh. (n.a.). Retrieved from https://balagan.info/the-battle-of-kadesh-in-david-rohls-new-chronology

Bestiary. Bog-Ra. (n.d.). Retrieved from https://www.bestiary.us/images/bog-ra-v-oblike-kota-zajca-ubivaet-zmeja-apopa

Bestiary. Zmienoobraznaya-Boginja. (n.d.). Retrieved from https://www.bestiary.us/images/zmeinoobraznaja-boginja-meritseger-v-korone-drevneegipetska-stella-novoe-carstvo

Breasted, D. A. (1915). *History of Egypt. From the Earliest Times to the Persian Conquest: in 2 Vol. Moscow, Russia:* Publishing House of M. and S. Sabashnikov.

David, R. (2002). *Religion and Magic in Ancient Egypt*. Penguin Books.

Dodson, A., & Hilton, D. (2004). *The Complete Royal Families of Ancient Egypt*. Thames & Hudson.

E-reading Club. Ehrman. (n.d.). Retrieved from https://www.e-reading.club/chapter.php/1005287/4/Erman_-_Zhizn_v_Drevnem_Egipte.html

E-reading_club. Ehrman. (n.d.). Retrieved from https://www.e-reading.club/chapter.php/1005286/13/Erman_Adolf_Gosudarstvo_armiya_i_obschestvo_Drevnego_Egipta.html

Egyptopedia. Arkhitektura-srednego-tsarstva. (n.d.). Retrieved from http://egyptopedia.info/a/1432-arkhitektura-srednego-tsarstva

Egyptopedia. Iskusstvo Drevnego tsarstva. (n.d.). Retrieved from http://egyptopedia.info/i/715-iskusstvo-drevnego-tsarstva

Egyptopedia. Iskusstvo Novigo tsarstva. (n.d.). Retrieved from http://egyptopedia.info/i/1593-iskusstvo-novogo-tsarstva

Egyptopedia. Kultura Srednego tsarstva. (n.d.). Retrieved from http://egyptopedia.info/i/716-kultura-srednego-tsarstva

Egyptopedia. Selskoe-khozyajstvo. (n.d.). Retrieved from http://egyptopedia.info/s/866-selskoe-khozyajstvo

Egyptopedia. Voennaya politika. (n.d.). Retrieved from http://egyptopedia.info/v/857-vneshnyaya-politika

Erikh, T. (2007). Lunar God: Terra.

Fleming, F., & Lothian, A. (1997). *The Way to Eternity: Egyptian Myth*. Amsterdam, The Netherlands: Duncan Baird.

Gardens. Istoriua parkovigo. (n.d.). Retrieved from http://www.best-gardens.ru/landshaftnyiy-dizayn/istoriya-sadovo-parkovogo-iskusstva/sadyi-drevnego-egipta-krita/

Gardiner, A. (1964). *Egypt of the Pharaohs*. Oxford, UK: University Press.

Gillings, R. J. (1972). *Mathematics in the Time of the Pharaohs*. New York: Dover.

Global History of Arts: in 6 Vol. (1956). Academy of Fine Arts of the USSR. Institute of Theory and History of Fine Arts. Moscow, Russia: State Publ. House “Iskusstvo”.

Gombrikh, E. (1988). History of Arts: OOO Izdatelstvo AST.

Grimal, N. (1992). *A History of Ancient Egypt*. Blackwell Books.

Guner. Bibliotec Culture. (n.d.). Retrieved from https://www.gumer.info/bibliotek_Buks/Culture/gombr/02.php

Hill, M. (2007). *Gifts for the gods: images from Egyptian temples*. New York: The Metropolitan Museum of Art.

History. Egipetskie ieroglify. (n.d.). Retrieved from http://history-doc.ru/egipetskie-ieroglify/

History of Mankind. World History: in 9 vol. Vol. 3. Western Asia and Africa (1903). Saint Petersburg, Russia: Publishing partnership "Prosvescheniye".

Kolotov, S. M. (1998). To the history of imaging methods. *Applied Geometry and Engineering Graphics, 63,* 239–255.

Kovalyov, Y. N., & Nitsyn, A. Y. (2010). Space in the Painting of Ancient Egypt. Technical aesthetics and design. 2010, 7. pp. 23–35.

Maat-object. (n.d.). Retrieved from http://maat.org.ru/object/2004/2004- 02.shtml

Mail. Community-catastrof. (2007). Retrieved from https://my.mail.ru/community/catastrof/14C6C1BFB14FA238.html

Matie, M. E. (1956). Ancient Egyptian Myths Moscow, Russia: Publ. House of the Academy of Arts of the USSR.

Matie, M. E. (1966). Art of Ancient Egypt. Moscow, Russia: Iskusstvo.

Megabook-media. Fivy. (n.d.). Retrieved from https://megabook.ru/media/Фивы%20(фрески%20 гробницы%20Нефертари)

Monuments of World Art: in 9 vol. – Vol. 2: Art of Ancient East. (1968). Academy of Arts of the USSR. Research Institute of Theory and History of Fine Arts. Moscow, Russia: Iskusstvo.

Nosov, K. S. (n.d.). *Siege machinery of the Assyrians.* Retrieved from http://xlegio.ru/siege-machinery/poliorketika-and-paraskeuastika/assyrian-siege-machinery/

Pinch, G. (1995). *Magic in Ancient Egypt*. University of Texas Press.

Ragozina, Z. A. (1902). History of Chaldaea. From the oldest times to the rise of Assyria. St. Petersburg, Russia: Publ. House of A.F. Marks.

Ragozina, Z. A. (1902). History of Assyria. From the Rise of the Assyrian State to the Fall of Nineveh. St. Petersburg, Russia: Publ. House of A.F. Marks.

Ratsel, F. (1902). Ethnic Science: in 2 vol. St. Petersburg, Russia: Publishing Partnership "Prosvescheniye"

Raushenbakh, B. V. (1980). Spatial Constructions in Pictorial Art. Outline of Basic Methods. Moscow, Russia: Nauka.

Robins, G. (2008). *The Art of Ancient Egypt* (Rev. ed.). Cambridge, MA: Harvard University Press.

Sacred Texts. (n.d.). Retrieved from https://www.sacred-texts.com/egy/)

Shafer, B. E. (Ed.). (1991). *Religion in Ancient Egypt: Gods, Myths, and Personal Practice*. Cornell University Press.

Silnaya mplitva Bogu. (n.d.). Retrieved from https://сильная-молитва.рф/molitva-k-bogu-totu

Smith, W. S., & Simpson, W. K. (1998). *The Art and Architecture of Ancient Egypt* (3rd ed. 1998). Yale University Press.

Studblue. Notes-midterm. (n.d.). Retrieved from http://www.studyblue.com/notes/note/n/art-midterm/deck/1189362

Symbol. Shumerskaya pismennost. (n.d.). Retrieved from http://the-symbol.ru/shumerskaya_pismenost.htm

Turaev, B. A. (1922). Ancient Egypt. Moscow, Russia: Publ. House "Ogni".

Verman, K. (1904) History of Arts of all Times and Peoples: in 3 vol. / K. Verman. St. Petersburg, Russia: Publishing Partnership "Prosvescheniye"

Vselova. Zodchiy Hesira. (n.d.). Retrieved from http://vseslova.com.ua/word/«Зодчий_Хесира»._Рельеф_из_гробницы_%28Древний_Египет%29_%28изображение%29-38640

Webcitation. (n.d.). Retrieved from https://www.webcitation.org/63ZdUemZU

Wikipedia. Ancient Egypt. (n.d.). Retrieved from https://en.m.wikipedia.org/wiki/File:Ancient_Egypt_map-hiero.svg

Wilkinson, T. (2010). *The Rise and Fall of Ancient Egypt: The History of a Civilisation from 3000BC to Cleopatra*. London, UK: Bloomsbury.

Winkler, G., Niebuhr, K., & Schurz, G. (1903). The History of Mankind. World History: in 9 vol.: Vol. 3. *West Asia and Africa*. St. Petersburg, Russia: Publishing Partnership.

World History: in 13 vol. (1955-1970). Academy of Sciences of the USSR. Moscow, Russia: State Publ. House of Polit. Literature.

Xlegio. Siege machinery. (n.d.). Retrieved from http://xlegio.ru/siege-machinery/poliorketika-and-paraskeuastika/assyrian-siege-machinery/

## ENDNOTES

1. "There's a green oak-tree by the shores of the blue bay; on a gold chain, the cat, learned in the fable stories, walks around the tree in ceaseless strain:" – it can be seen that the words by A.S. Pushkin, familiar to the Russian reader from childhood, have a parallel dating back to Ancient Egypt.
2. A comparison of the Paleolithic understanding of reincarnation with the Egyptian one enables to draw an interesting conclusion. For a Paleolithic human, the "magic of a grave" was natural, when a grave mound, and, thenceforth, a burial mound or a pyramid were associated with the womb of Mother-Earth pregnant with the deceased. And the deceased was restored to life at his time simply "because it could not be the other way". With Ancient Egypt, with the god of the earth and the goddess of the sky, such a mindset is impossible. And this is an argument in favor of a hypothesis that, as a minimum, the Great Pyramids, had a designation different from the traditional understanding

of their being burial sites. Maybe, while ensuring the entry to Duat and a perspective of rising to light, they substituted, by their functions, the World Tree?

3. God Ra reproached God Thoth for having granted writing to the people, saying that their intelligence would weaken after that – and he was right, in a way!

4. Paleolithic humans who developed rock painting and being at the peak of the development should be distinguished from the contemporary primitive people, who have passed the stages of degradation and savagery.

# Chapter 9
# Antiquity:
## The Path to Rationality


## ABSTRACT

*The choice of culture-data sources for studying the evolution of consciousness uses a sequence of four cultures. The most significant features of the depicting space in the reliefs and murals of the Aegean (Krito-Mycenaean) civilization, the classical period of Hellenism, Etruscan civilization, late republican, and imperial Rome are considered. Representations of the World Tree occupied a peripheral place in ancient mythology, but unexpected is the revival of myth - in the form of a legend about the crucifixion and resurrection of Jesus Christ. The adoption of Christianity completes the ancient era. When analyzing markers of evolutionary changes, the most active channels were identified. The results of the reconstruction of behavior patterns are presented in the form of generalized psychological portraits of representatives of the main estates of the Late Antiquity. The features of their collective behavior are described.*


## BACKGROUND

Authors (Kobylina, 1939; Chubova, 1970; Sokolov 1971; Andreae, 1977; Moon, 1983; Maggiani, 1985; Cook, 1986; Spivey, 1997; Dunbabin, 1999; Williams, 1999; Stierlin, 2004; Steingräber, 2006; Betancourt, 2007; Ramage, Nancy & Andrew, 2009; Tuck, 2015; Yatromanolakis, 2016) consider antique pictorial art mainly from historical and art criticism positions. A periodization based on stylistic features is introduced, and features of styles are explored, in particular, plots and techniques. As part of the study of the latter, the methods of transferring expanse are distinguished, sometimes being compared with those characteristics of earlier civilizations. These materials will be used as a database and subjected, as in the previous chapters, to processing from the standpoint of the S-space self-organization theory. B.V. Rauschenbach (Rauschenbach, 2001) formulated the task of space relations reconstruction on the basis of modern geometric concepts. The analysis of space depiction in ancient painting was carried out in (Panofsky, 2004; Kovalyov, Nitsyn, & Matveeva, 2011).

Numerous studies are devoted to ancient mythology (Eliade, 2002). The surviving works of ancient authors, starting with Hesiod have been written about it. Despite the comprehensive nature of the research,

DOI: 10.4018/978-1-7998-1706-2.ch009

many details of the myths about the World Tree, as well as related ideas about the Great Goddess and cult of the dead, remain unclear. One can only say with confidence that the myth of the World Tree was not central in the minds of the ancient human, and such sacraments as the Eleusinian ones associated with the idea of immortality, apparently, will forever keep their secrets. It is also likely that secrecy is associated with the tabooing of “shamanic journeys”, which was noted for the first practicalities of the Eneolithic and was a general tendency (chapter 7, *Eneolith: The Search for Balance*). Therefore, as for the ancient Egyptian civilization, the myth of the World Tree - with the exception of the beginning of Christianization - serves only as a secondary source for reconstructing the evolution of mind.

## JUSTIFICATION OF THE CHOICE OF A SOURCES OF FACTS

Consideration of ancient civilizations as sources of facts about changes in the way of depicting expanse and evolution of the myth of the World Tree is dictated by the desire to have reference content for assessing the changes in mind that took place in ancient Egypt. How regular are these changes? Did they have a local or global character? To answer such questions, you must have the most dissimilar, but the synchronous sample.

In this sense, ancient civilizations are the most appropriate:

- awareness of the differences (despite the diverse connections and contacts) reaches antiquity: even Herodotus emphasized their antagonism with the civilizations of the Middle East, outlining the line of confrontation between Europe and Asia;
- if the ancient Egyptian civilization is continuous (short periods of foreign domination could not significantly change the “cultural code”), then ancient civilizations were interrupted and revived repeatedly, and on a different ethnic and linguistic basis, maintaining their development path;
- in the end, the ancient civilization devoured the Egyptian, and it is interesting to see what contributed to this;
- and still, later they shared one fate, being absorbed by the new civilization - the Christian, although they gave it some of their ideas and forms.

And the last important factor - like the Egyptian, ancient civilizations are well studied.

## THE ORIGIN OF PROJECTIVE METHODS OF SPACE IMAGING

As in the previous chapters the geographical and chronological boundaries of monuments of ancient art were denoted.

Since ancient times the Mediterranean has been the cradle of many civilizations. One can find monuments of original cultures dating back to the Stone Age period even on small islands now, such as Malta. The ancient states (Egypt, state of the Hittites, the Judean Phoenician kingdom, later Assyria and Persia) occupied the West and Asia Minor coasts at different times. Ancient civilizations, such as Etruscan or Aegean, also existed on the Apennine and Balkan Peninsulas, as well as on the islands (Crito-Mycenae, Figure 1). The latter is considered the forerunner of classical Greece, while the Etruscans were in many ways teachers of the Romans.

*Figure 1. Aegean culture: localization and main cities (Isatorius. Livejournal, n/d).*

**Aegean culture.** The first monuments of the Aegean culture appeared about 5,000 years ago; The culmination of the development was the 3,700–3,400 years, which ended in a catastrophe as a result of the explosion of the Santorini volcano, and final death occurred about 3,000 years ago, that is, existence of the Aegean civilization was in sync with the dynastic period of Ancient Egypt. Consider the examples of pictorial art characteristic of the Creto-Mycenaean period, known primarily for the ruins of the palaces of Knossos and Pylos, the monuments of Akrotiri, and ceramics (Figures 2-8), focusing primarily on the ways of representing.

*Figure 2. The partially reconstructed North Bastion at the palace of Knossos Author: Peter Summer (Peter Summers. Archeology), n/d).*

*Figure 3. The Queen's Megaron, with the dolphin fresco (copy) at Knossos palace ≈3 600 years ago. Crete, Greece. Author: Jebulon (Wikimedia. Quenn-megaron, n/d).*

There is a parallel with the ancient Egyptian images - lack of depth, flat figures, and bright colors. The latter is characteristic of both the Paleolithic (chapter 5) and the period of the New Kingdom (chapter 7). The widespread marine scenes presented in the paintings on the walls of palaces, and on vases are a special feature.

Another feature is the predominance of "secular" painting with an almost complete absence of polymorphic images. All images are quite realistic.

Finally, the well-known standardization of images was note - the location of people in profile, attention to hairstyles, and repetitive clothing. Probably the idea of typing contributed to the emergence of writing.

Comparing the methods of transferring expanse with Egyptian, similarities, primarily with sacred pictorial art was found. And only in a few drawings, one can see interest in the depth of expanse, as well as the emergence of a new technique, similar to axonometric projections (fig. 8, on the bottom panel, on the right), which was absent among the Egyptians.

Taking into account the secular nature of the images and craving for realism, concluding that the Aegean mind, like the Egyptians, was in a transitional stage to the fourth level.

Just like the ancient Egyptian civilization, the Aegean also developed its original writing system, or even several such systems[1] (Figure 9-12).

The development of the Aegean civilization was interrupted by the invasion of Dorian tribes (which is referred to in the epic as the "return of Heraklides") and the subsequent "Dark ages", which, however, were not long.

About 3,000 years ago, civilization resume

Further, geometric, archaic, classical, and Hellenistic periods are distinguished; the latter ended about 1900 years ago.

Figure 4. Fresco from Knossos palace/ ≈3 600 years ago. Crete, Greece. Author: cavorite (Flickr-photos-cavorite, n/d).

Figure 5. Bull Leaping Fresco from Knossos palace/ ≈3 600 years ago, Crete, Greece. (National Geographic. Bull-leaping, n/d).

*Figure 6. "Spring" fresco from Thera, ≈ 3 670 years ago. Akrotiri, Santorini (Thera), Greece (Employees. Minoan_Mycenaen, n/d).*

*Figure 7. Octopus from Thera, ≈ 3 650 years ago. Akrotiri, Santorini (Thera), Greece (Employees. Minoan_Mycenaen, n/d).*

*Figure 8. Minoan Miniature Frieze Admirals Flotilla Fresco in Three Panels. Painted sometime before ≈3 630 years ago. West House, Room 5, South Wall. Akrotiri, Santorini (Thera), Greece. Author: W. Sheppard Bairs. (Minoanatlantis. Minoan_Flotilla, n/d).*

The geometric period is named for the predominant nature of the ornaments on ceramics, which are performed, possibly, with the use of geometric tools (fig. 13-14). To a certain extent, such murals are similar to the paintings of Aeneolithic cultures, for example, Tripoli. This suggests that the Dorian tribes were indeed "wild", being in the third stage of self-organization of mind.

Subsequently, figure compositions began to predominate, initially mostly illustrating certain religious or epic subjects. Recorded polymorphic images (centaurs, sirens). In some cases, postures are very dynamic, and the proportions are distorted. This gives vases a certain similarity to neolithic images. There are black-figure (fig. 15) and red-figure (fig. 16) techniques.

*Figure 9. A Linear tablet from the palace of Zakros, Archeological Museum of Sitia. Author: Riley Winters. (Ancient-origins. Artifacts, n/d).*

*Figure 10. Tablet with an inscription in linear B. Heraklion Archeological Museum (Crete). Author: Christian Vandendorpe. (Wikimedia. Tablette-lineare, n/d).*

It can be stated that the techniques used for both Aegean and Egyptian pictorial art were used to depict expanse, namely:

1) flatness: each figure is outlined in a continuous line, of constant thickness, uniformly filled with paint of the same color;
2) lack of perspective: perspective cuts in parts of the object are completely absent due to their different distances from the viewer;
3) the shift of images of objects in the case when one object obscures the other. This is how a monocular sign of depth is realized -overlapping when an image of a nearby object covers images of

*Figure 11. The side A of the disc of Phaistos, as displayed in the Archaeological Museum of Heraklion after the 2014 renovation. Author: C messier (Wikimedia. Phaistos Disk, n/d).*

*Figure 12. The side B of the disc of Phaistos, as displayed in the Archaeological Museum of Heraklion after the 2014 renovation. Author: C messier (Wikimedia. Phaistos Disk, n/d).*

*Figure 13. Late geometric pyxis and lid with three terracotta horses. The pyxis is a container for food offerings at a burial site. 2 740- 2 700 years ago. Author: Yukatan (Wikimedia. Last geometric pyxis, n/d).*

*Figure 14. Attic miniature hydria, ca. 2 770 BC– 2 720 years ago. (Late Attic Geometric), found in Thebes, Boeotia. Department of Greek, Etruscan and Roman Antiquities, Sully, first floor, room 40 Author: Jastrow. (Wikimedia. Miniature hydria. Thebes, n/a).*

*Figure 15. Hercules kills the centaur. (Mif-Meduza, 2016).*

*Figure 16. Vase of sirens. British museum (Mif-Meduza, 2016).*

objects located further away. This feature can create an impression of the rudiments of a linear perspective, but the image is, in fact, devoid of depth, which follows from the uniformity of the sizes of such objects located at different distances from the viewer;

4) However, early antique pictorial art was not familiar with complex depth transfer techniques known from ancient Egyptian pictorial art, such as combining several images with a picture plane.

Thus, the features of the transfer of expanse in the early antique period show that the mind of the inhabitants of Hellas gradually civilized, being at the same stage of self-organization as the inhabitants of the Ancient East. Namely, the domination of intuition, defining the "magical" unity of human and nature in the primitive mind, has long been replaced by the activation of ego and will/reason, manifesting the self-sufficiency of human and opposition of human and god, human and nature, and of course human and human - whether in war or sporting competition; there is also an activation of the mind perceiving the relations of space-time.

**The classical period** is the most important stage in the development of ideas about space.

Classical Greece is the time of the emergence of geometry as a science. Euclid, who about 2300 years ago developed the axioms of expanse, which subsequently received his name, was its father. But one should not forget Aristotle, who created the logic. Let's fix this important moment - from now on, the criterion of truth was not truth (intuition) itself, but certain forms of human thinking (reason).

The highest synthesis of the ancient mind, aesthetic and philosophical, was to look at a human as the "measure of all things" (Protagoras), using the proportions of the human body, to harmonize and provide integrity to temples and sculptural compositions.

The century of the uplift of culture has also become the heyday of pictorial art. It is by this time that the art of chiaroscuro appears and reaches perfection; Apollodorus of Athens, who received the nick-

*Figure 17. The extent of Etruria and the Etruscan civilization. Based on a map from The National Geographic Magazine Vol.173 No.6 June 1988. Author: Norman Einstein (Wikimedia. Etruscan civilization, 1988).*

*Figure 18. Metamorphosis of the Tyrrhenian pirates 2 530 – 2 500 years ago. Toledo Museum of Art, Toledo. Catalog No. 1982.134. (The-history-blog. Kalpis, n/d).*

name Shadow Gist is considered to be its discoverer. Zeuskis, Parrasius, and Timanf were the followers of Apollodorus. Unfortunately, the originals of their works have not been preserved, they can be judged only by later Roman copies. Therefore, let's turn to the analysis of works of Roman fine art, which is much better preserved.

First, let's mention the pictorial art of the Etruscans – "teachers of Rome" (Figure 17). The samples are shown in Figures 18-20 are typical examples of pictorial art.

The methods of depicting expanse are the same as in the pictorial art of the pre-classical periods of Greece: flat figures without light and shade, the simplest methods of transmitting depth, etc. There were attempts to convey the bulk of the figures of people and ornamentation due to chiaroscuro in the pictorial art of late burials (fig.21).

The Etruscan Underworld demons Vance and Charun prepare to conduct the souls of the mortals into Hades. The ghost of Patroclus stays at left. The souls in Hades were no longer made of flesh and blood, they could feel, breathe and speak if they were revived, but only by drinking blood.

Ancient Roman wall pictorial art can be judged by wall paintings and mosaics of Pompeii, Herculaneum, Stabiae, and Rome. Samples of provincial pictorial art - portraits in the oasis of Al-Fayoum (Egypt), paintings in Delos (Greece) - show similarities with the republican, and then the imperial styles (four in total). Here there are already more advanced, in comparison with Etruscan, methods of transmitting spatial forms and relations (Figures 22-26).

Figure 19. Tomb of the Leopards. 2 450 years ago. (Ferrebeekeeper. Etruscan painting, 2014).

Figure 20. Tomb of the Infernal Chariot. Etruscan necropolis of Pianacce. 2 350 – 2 300 years ago. (Benedante. Tomb internal chariot, 2018).

*Figure 21. Achilles' sacrifice of the 12 Trojan prisoners at the funeral of Patroclus. Part of a wall pictorial art in the Francois Tomb, Vulci, 2 350-2 330 years ago. Museo Torlonia, Rome. (Wikimedia. Franqis Tomb painting, n/a).*

*Figure 22. Roman civilization, 1 900 years ago. Fresco depicting the family portrait. From Pompeii, Italy. Naples, Museo Archeologico Nazionale (Archaeological Museum). (Museo-Archeologic-Napoli, n/d).*

*Figure 23. Roman civilization, 1 900 years ago. Fresco depicting the Three Graces. From Pompeii, Italy. Naples, Museo Archeologico Nazionale (Wikimedia The Turee Gracies, n/d).*

As one can see from fig. 22-23, a significant difference from the works of the early antique period is the transfer of the volume of the figures due to light and shade. This gave additional expressiveness, above all, portraits. The construction of shadows does not have a scientific character, the artist conveys what he sees, transferring his impressions also to mythological plots. Note also that the expanse behind the figures has the character of background and is represented conditionally.

In the same place where the transfer of expanse has a compositional or plot value (fig. 24-25), techniques are used that resemble modern perspective and axonometry.

The greatest expressiveness is achieved when combining light and shade and geometric transmission of the depth of expanse (Figure 26), but where it is not necessary, simpler methods are used (Figure 27) - just as it was done in Egypt.

The Roman period added a little to the classical achievements. Systematization, slow improvement, pragmatism, rationality - this is his motto. And it is only natural that it was then that a military engineer and architect Vitruvius (Vitruvius, 1936) - by no means a mathematician - wrote "10 books on architecture" (1,900 years ago) that describe the design of buildings, squares and structures using ichnography (plan on the surface), orthography (frontal plane) and set design (perspective plan in the front setting of the structure). This is in content, and in many technical details, corresponds to modern practice - except for one eloquent detail: in the original text of a fairly lengthy treatise there is not a single drawing!

However, to avoid incorrect modernization, one should point out several significant differences between ichnography, orthography and set design from scientific methods for constructing orthogonal projections, axonometry, and perspectives. They are as follows:

1) mixing of axonometry and perspectives in a single work;
2) no horizon line;

*Figure 24. Fourth Style wall pictorial art with the architectural landscape, a detail from a panel from the portico of the Temple of Isis in Pompeii, Naples National Archaeological Museum Carole Raddato (Wikimedia. Fourth Style, n/d).*

*Figure 25. Roman; Cubiculum (bedroom) from the Villa of P. Fannius Synistor at Boscoreale; Miscellaneous-Paintings. 2 070- 2 060 years ago. This file was donated to Wikimedia Commons by as part of a project by the Metropolitan Museum of Art (Wikimedia. Cubiculum. Villa Fannius, n/d).*

*Figure 26. Iphigenia as a priestess of Artemis in Tauris sets out to greet prisoners, amongst which are her brother Orestes and his friend Pylades. 1 900 years ago. From Pompeii, Italy. Naples, Museo Archeologico Nazionale (Archaeological Museum). Sala LXIX. Inv.111439. Author: Marie-Lan Nguyen (Wikimedia. Iphigenia. Taurus, n/d).*

*Figure 27. Flora. 1 900 years ago. From Pompeii, Italy. Naples, Museo Archeologico Nazionale (Museo-Archeologic-Napoli, n/d).*

3) multiple vanishing points - for each object;
4) inconsistency of the perspective scale.

Now let's compare Antique and Egyptian paintings (Table 1)

And yet it is very tempting to give an interpretation of the technique of transmitting expanse by antique painters from modern concepts

According to E. Panofskiy (2004), "...antiquity... never deviated from the principle that visual values are determined not by distance, but by angle", under which the subject is visible to the observer. E. Panofsky proves this statement by referring to Euclid's eighth postulate (2004): "equal values, not equidistant, are seen disproportionately to the distances between them". With this statement, Euclid summarizes the

*Table 1. Comparison of Ancient Egyptian and antique paintings features*

| Feature | Ancient Egyptian pictorial art | Antique pictorial art (at the end of the period) |
|---|---|---|
| 1. | Sacred pictorial art is characterized by canonicity, a monotony of images, contrast with the background, less dynamic, passive background.<br>Household pictorial art is characterized by violations of the canon, polytonality of images, contrast with the background (the background color has changed), and dynamics | There are no differences between sacred and everyday pictorial art in terms of techniques for conveying the depth of expanse |
| 2. | Sacred pictorial art: the expanse is two-dimensional, there are no elements of perspective, there is no illusion of depth.<br>Structuring expanse is not related to the image of the World Tree and is transmitted by the complex semantics of symbols and characters of writing.<br>Secular pictorial art: along with two-dimensional images, there are ways to convey the depth of expanse and the sweep of the action in time. The action takes place in the world of the living.<br>A very expressive "emotional expanse" is being organized.<br>Many new plots appear. | The expanse is transmitted as three-dimensional, but non-uniform: similarity with perspective or axonometry is purely external. The ways of transmitting the depth - chiaroscuro, and orthography with many vanishing points - are different from Egyptian ones.<br>Structuring expanse is not related to the image of the World Tree.<br>A very expressive "emotional expanse" is being organized.<br>Many new plots appear. |
| 3. | Most often the center is present, but not in a geometrical sense - the figure of Pharaoh or nobles, villa plan, contour of the lake and like stand out (also in size). Cyclicity is mild. | Most often the center is present, but not in a geometric sense, it is determined by the plot. Cyclicity is almost invisible. |
| 4. | The top and bottom, right or left side is distinguishable, as well as the past and present. | The top and bottom, right or left side are distinguishable; As a rule, the action takes place "here and now". |
| 5. | Natural proportions (especially for figures of gods and rulers) and proportions are broken, figures sometimes look grotesque (Amarnian period) | Disturbance of scales and proportions is present but less pronounced |
| 6. | Images of animals and people are canonized, details of behavior and movement are sketchy; scenes of war and violence are relatively few. The exception is the Amarna period and battle scenes, especially the New Kingdom. | Images of people are given realistically, at the time of dramatic experiences or emotional lift, the details of the behavior and movements of animals are expressive; scenes of war and violence are present, but not stand out |
| 7. | Polymorphic gods are presented in large quantities. | Polymorphism is fading |
| 8. | There are complex magical and religious scenes. | Difficult cult scenes are profaned |
| 9. | Pictograms quickly turn into hieroglyphic writing, which, with the gradual loss of similarity of the sign with the designated object, evolves into hieratic, and then demotic writing | Samples of writing are poorly represented in the drawings |

*Figure 28. Reconstruction of spatial constructions in Greco-Roman pictorial art by E. Panofsky*

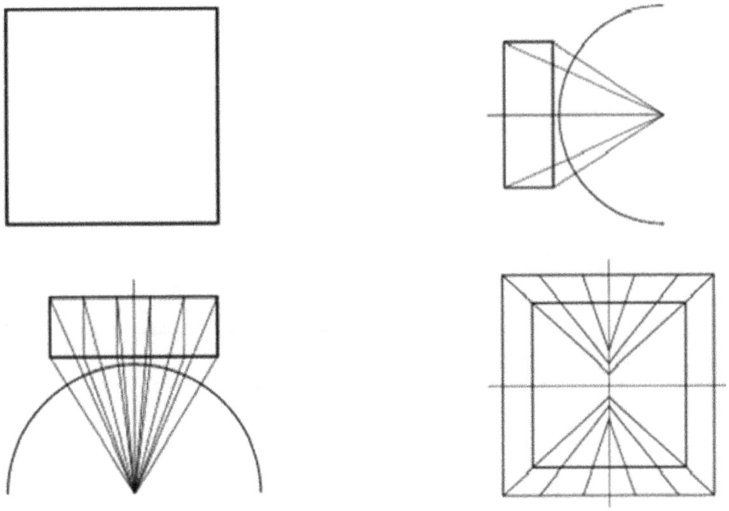

observations made by ancient architects and artists, namely: the visual difference between two equals, but visible from different distances, values are not determined by the ratio of these distances, but by the ratio of visual angles.

E. Panofskiy (2004) proposes the following reconstruction of space cuts in Greco-Roman pictorial art: they "...could be presented based on a spherical projection, that is, a projection circle, both in horizontal and vertical section, but where the arc replaces chord". E. Panofsky places a point of view in the center of the circle and intersects the circle itself (2004): "...the lines of visual rays are the same as a straight line in a modern perspective design, which represents a graphic surface. And if we now construct it using such a "projection circle" (with segments of a circle replaced by corresponding chords), in any case, we get a result that is consistent with the preserved monuments in one decisive moment: the lines extended in depth are never connected at one point, but, hardly converging (since the sectors of a circle, when deployed, several breaks at the ends), connect in pairs at several points lying on the same axis, so that the impression of a fish ridge appears" . Let`s show the picture of the interior from an ancient perspective by E. Panofsky in Figure 28.

Imagine geometric constructions that, in our opinion, lead to an antique perspective in fig.29. Equations describing them:

$$x_1 = z_0 arctg \frac{x}{z + z_0}; \tag{1}$$

$$y_1 = z_0 arctg \frac{y}{z + z_0}. \tag{2}$$

*Figure 29. The first attempt to reconstruct the expanse in Greco-Roman art using the eighth postulate of Euclid*

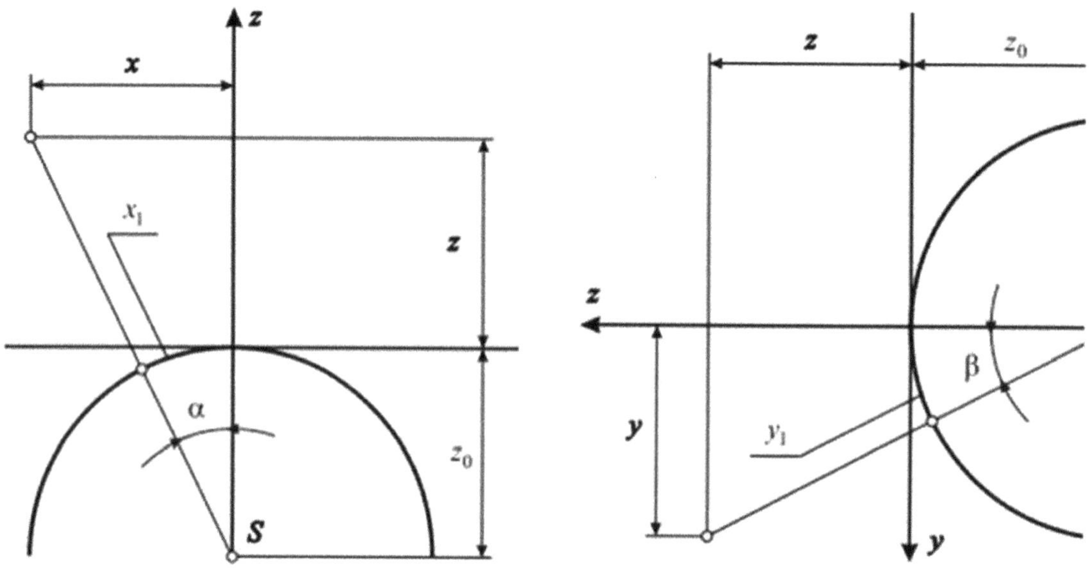


These equations do not correspond to a perspective that has many vanishing points located on the same axis; they rather describe a perspective in which straight lines, perpendicular to the picture plane, converge at one point. Indeed, at an infinite distance of the depicted point from the picture plane, its perspective rushes to the main point of the picture, regardless of the position of the depicted point in space.

However, what should be equations like describing the antique perspective? Suppose that the set of vanishing points located on the same axis corresponds to a set of points of view taken on the same vertical straight line. Fig. 30 demonstrates the construction of a linear perspective of a point in expanse depending on the position of the point of view. Hence the equations describing this dependence:

$$x_2 = x\frac{z_0}{z+z_0}; \tag{3}$$

$$y_2 = (y-h)\frac{z_0}{z+z_0}+h. \tag{4}$$

No one needs to set an equation that describes the dependence of the position of the point of view on the position of the imaged point in space $h=h(x)$. However, the form of this equation and patterns of visual perception reflected by it are completely incomprehensible.

Consequently, contrary to the statements of E. Panofsky, Euclid's postulate on the priority of the visual angle over the apparent size does not lead to the formation of a "fish ridge". In our opinion, a more

*Figure 30. The second attempt to reconstruct the expanse in Greco-Roman art using multiple points of view*

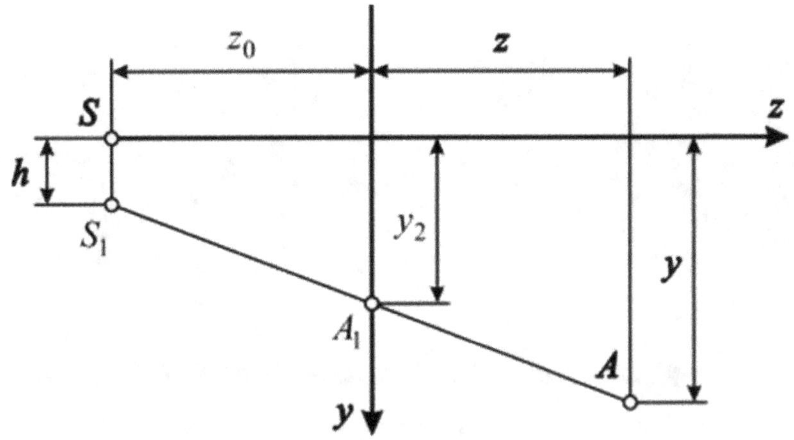


*Figure 31. The third attempt to reconstruct the expanse in Greco-Roman art using four local axonometry*

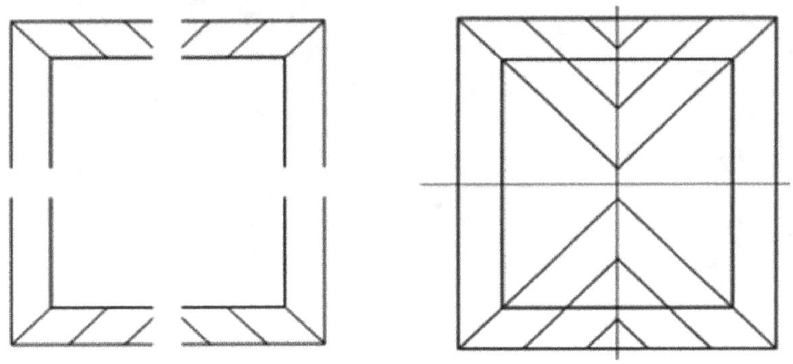

correct explanation of the ancient perspective was given by B.V. Rauschenbach (Rauschenbach, 2001). He suggested that the "fish ridge" effect is a consequence of an artistic device that considers the image as a combination of local axonometry. Fig.ure 31 show the images of the interior, which combines four types of axonometry: from the corner points in the direction of the view of the opposite angle diagonally.

Also, in Figure 31 shows how the intersections of images of parallel straight lines belonging to the ceiling and floor form a "fish ridge."

Hence the conclusion that the artistic technique, which consists of combining four local axonometry in one image, can indeed be considered as the geometric basis of spatial constructions in Greco-Roman pictorial art.

## WORLD TREE IN ETRUSCAN TOMBS

As noted for ancient Egypt, the myth of the World Tree went to the periphery of the human mind. No, he was not forgotten at all; moreover, it is present where it should be expected - on the way to the afterlife, as evidenced by the very interesting paintings of Etruscan tombs - like Figure 19-21 above, so 32-34, located further.

*Figure 32. Tomb of the Leopards in Tarquinia,2 500 2 490 years ago (Ferrebeekeeper. Etruscan painting (2014).*

The tree takes a central, prominent place for all; there is also a "classic" pair of animal guards, in this case, leopards or leopards. However, the feast scene is still the main.

Almost central place, and also accompanied by defenders, is taken by the World Tree in the mural of another grave (Figure 33-34) - but here it is already inscribed in one of the episodes of the Trojan War.

In general, the ancient people did not expect anything good from the underworld. No rush to the eternal life, just a colorless existence in the form of a shadow, always eager to get drunk with blood. Orpheus could not bring Eurydice out of Hades; Odysseus found a second youth, rather than eternal life; Hades only briefly releases Persephone, despite the best efforts of her mother goddess Demeter. Only Hercules, by the grace of the gods, managed to become immortal, hitting Olympus - and maybe he just died, like everyone else. The path is blocked by Cerberus, or such a three-headed dragon, as in fig.20. And neither rich offerings nor human sacrifices will help (fig.21).

And people gradually put up with it. This can be seen from figures 32 and 33: oh you, leaving for Hades, let the tombstone remember the friends with whom you were feasting and the women you loved - so you can understand the plots of their paintings.

*Figure 33. Tomb of the Bulls, Tarquinia. ≈2 5500 years ago. Unknown Etruscan artist. (Realhistory. Etruscan tomb, n/d).*

*Figure 34. The ambush of Troilus by Achilles, Tomb of the Bulls, Tarquinia ≈2 5500 years ago. Unknown Etruscan artist (Wikimedia. Etruscan mural, n/d).*

## CROSS AND ANUBIS

And suddenly, about 2,000 years ago, a completely unexpected turn occurs. A Jew, Jesus comes around - and in Judaism the tabooing of the world tree has reached its maximum: eating the fruit from the tree of the knowledge of good and evil was the cause of exile from paradise; an angel with a fiery sword has since guarded the entrance, depriving a person of the opportunity to get to the tree of life. Moreover, "hanging in a tree, be cursed!". And this Jesus takes the sin of the world upon himself, sacrifices himself, is subject to death through crucifixion on a wooden cross hoisted above the grave of the first man Adam, dies, descends into hell, destroys the gates of hell, conquers death and frees the righteous, sits at the right hand of God-Father and makes the murder weapon a means of finding eternal life, reuniting the three worlds again and bestowing salvation and eternal life on a person!

Let's look at all this with the eyes of a Paleolithic human: from now on, any believer (now neither Judah nor Ellina is, but we are all one in Christ Jesus) can, through the World Tree (namely, it means the Life-giving Cross at the grave of Adam) pass the route of the dead, be reborn becoming immortal.

Let's look through the eyes of a Neolithic human: the great shaman and the great hero defeated the underground lords, and now they have no power over us.

Let's look with the eyes of the Eneolithic human - a great Savior was found, sacrificed himself for all, and thereby saved everyone, and made everyone immortal.

*Figure 35. St. Martyr Christopher. 250 years ago. Icon from the collection of the Museum of the Rostov Kremlin. Author: cheger (Cheger. Livejournal, n/d).*

Let's look with the eyes of an ancient Egyptian - here he is, true Gore, the son of Osiris and Isis, who defeated Seth and killed Apophis, who has in himself the source of eternal life, the pharaoh, who leads his people to the light.

And justifying expectations, so ancient and universal, that they coincide with the very first manifestation of the ego, with the awareness of man himself as such.

Of course, he is the Messiah, he is the Christ, he is the son of the living God!

That was how his loyal followers should have perceived, and that is why the spread of Christianity was so unstoppable.

And the last. There is one mysterious figure in Christianity - St. Christopher (Figure 35), the contemplation of which leaves a stunning impression. Why is the saint depicted with a dog's head (especially considering that the dog was considered an unclean animal)? Not knowing about polymorphic figures from prehistoric pictorial art, not knowing about the Egyptian god Anubis (Figure 36) and not imagining their role as creatures of the other world (polymorphs), and the guides to the world of the dead (Anubis and the dog), it is impossible to understand.

A marvelous conclusion of Paleolithic insights!

It is this, the most unexpected version, that is comparable with the ancient Egyptian version of the myth, summarizing the reasoning in Table 2.

*Figure 36. Anubis is a guide to the world of the dead. Drawing on the inner wall of the tomb of the son of Ramses III. Author: Psyche (Mail. Question 39179570, n/d).*

*Table 2. Comparison of the Ancient Egyptian and Christian versions of the myth of the world tree features.*

| Feature | Ancient Egyptian version | Christian version |
|---|---|---|
| 1. | The structuring of worlds is not associated with the myth of the World Tree. | The structuring of worlds is again symbolized by the World Tree. |
| 2. | The identification of the Goddess with the Tree is missing. | God's use of the Tree to sacrifice himself and move to hell was the place to be. The birth of God in the world of an earthly woman - the Mother of God. |
| 3. | Connection with the cult of the dead: direct - absent; indirect - to access the tree, it is necessary to overcome the resistance of Apis. But the serpent is not a "boundary" creature, it is the personification of chaos and darkness and even gods cannot kill him. | Connection with the cult of the dead - direct |
| 4. | No data, as information of the «Book of the Dead" is associated with movements on the World Tree | Very distant view of the world of the dead in the Book of the Dead and Bible prophecies about the resurrection. |
| 5. | The myth is tabooed, but, going to the periphery of the mind, it is still not forgotten. | The myth is restored in its entirety, the representations of all previous eras; at the same time, Jesus Christ is the central figure, and the cross is only a means |

## MARKERS ANALYSIS

**Image of the expanse.** It is the logic that determines homogeneity, additivity, discreteness, the immutability of the elements of expanse (chapter 2, *Self-organization of complex systems*).

It can be stated that all these characteristics are absent in the paintings of ancient artists who have not yet realized that expanse consists of the same points, and the figures are just combinations of the same points and do not form something qualitatively different from them. From here - a set of vanishing points, etc. (table 1).

On the other hand, based on Euclidean geometry one could create projective geometry and substantiate the method of perspective, but this did not happen.

Taking into account the change in the transfer of expanse in late antique pictorial art as compared with early antique, as well as the level of geometric achievements of the classical period, concluding that they reflect the most important period in the evolution of human mind due to the laws of self-organization complex systems. In the classical and late antique period, the predominance of ego and will is replaced by the activation of the nous, with its characteristic ternary perception, including spatial-temporal relations. This was manifested in the creation of geometry and the general interest in the transmission of the depth of expanse.

At the same time, the methods of transmission were not scientific and represented a sensory perspective based on the immediate impressions of the artist. And here is the strengthening of other channels - sensory. This process is accompanied by the profanation of old myths, sensuality, interest in conveying the beauty of the human body, emotions, objects, the environment, not only in terms of forms but also colors. Such interest is fully consistent with the current state of the human mind. However, the absence of a geometric imaging apparatus shows that the state of the ancient mind was different from the modern one.

**The myth of the World Tree**. At the beginning of an epoch, there is an interest, if not to the myth itself, but the ideas about eternal life associated with it (preservation of intuition at a high level). But by the middle of the epoch, it is fading away (a strong weakening of intuition and will - the potential is spent on strengthening the nous), and the World Tree becomes one of the iconographic images that a person rarely encounters in everyday life. Alexander the Great, who seriously considered himself the son of God, even he sets military-political goals for the scale of Ocumene - but not for the scale of three worlds. The end of an era can be described as general unbelief - the priests remained needed for official ceremonies, for divination, for the fulfillment of demands, but no more. Society has become completely pragmatic and deaf to all matters not connected with ordinary life (ego, nous, sensuality); at the same time, abstract sciences and philosophy developed (reason and nous). Even such an important thing as evocation (luring the enemy's gods to one's side - an echo of the ideas of the tribal war as the earthly expression of the gods' wars), seems to be perceived no more than a curious rite.

And suddenly - an event not following from all this logic - the phenomenon of the Savior, who embodied people's aspirations and restored the myth in its entirety.

Let's dwell on this statement and ask ourselves the question of how this fact was perceived by ancient society; The answer should be sought in behavioral patterns.

## ANCIENT HUMAN MIND

Like Ancient Egypt, ancient society is already quite structured. Before proceeding to the compilation of psychological portraits of estates, discussing whether they will vary from epoch to epoch, or will remain relatively constant.

Let's start at the first emperors of Rome. Octavian, discreet and cautious, able to wait a moment and pick up people; Tiberius, by the end of his life literally opposed to all; Claudius, for studying history, revered as a madman; the unbridled Caligula and Nero, who became real maniacs; weak Otto and Vitellius; Vespasian, sly and stingy like a soldier, who had found the strength to say before his death; it seems that I am becoming a god. What do all these people have in common? Only one thing: a colossal egoism.

If add to this list Alexander and his Diadokhs, or, for example, the heroes of the Iliad, the conclusion will not change: a common feature is only super-ordinary egoism and the related lust for power and claims.

Consequently, when drawing a generalized portrait of a ruler of any epoch, it is the channel of egoism that must be distinguished from all other possibilities of interaction with the environment; and only in individual psychological portraits will difference appear.

Consequently, such a generalized portrait will very much resemble the diagram of the Pharaoh state of mind (fig. 32 in chapter 8, *The Ancient World: Ego, Archaic, Immortality*). Repeating this pattern is not necessary.

Similarly, there is no need to additionally describe the state of mind of a priest or commoner. It can only be stated that the potential and training of the ancient priest will be weaker than that of the Egyptian, but the sensual frenzy of the Roman plebs, corrupted by idleness, on the contrary, will surpass those of the industrious Egyptians. "Bread and circuses" is, after all, not an Egyptian, but a Roman slogan.

As for the nobles, the mobilization of constant wars in the beginning and middle of the era was replaced by the burning of accumulated wealth to its end. The warriors quickly lost their best qualities, becoming either courtiers or provincial sybarites, on whose riches the emperors from time to time were not afraid to lay their hands on.

But what has become new is a significant increase in the intelligentsia, perhaps already conscious of itself as a separate class. Bureaucrats, scribes, engineers, scientists, etc. gradually created their habitat and system of their reproduction (schools, teaching rhetoric, Likey and Academy, Areopagus and Forum,

*Figure 37. The state of mind of the intelligentsia of the ancient era*

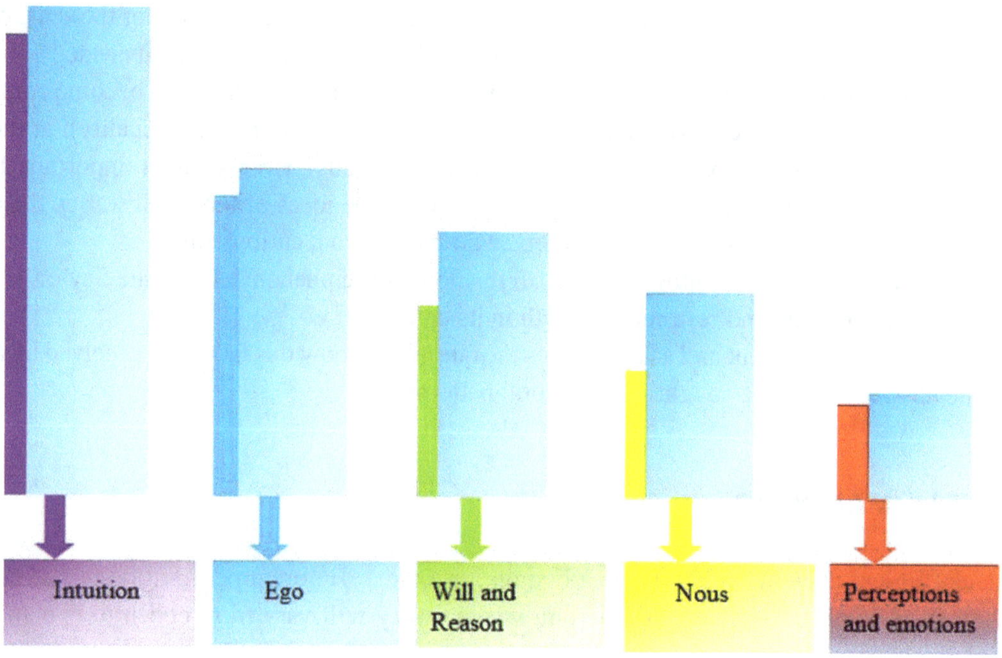


Alexandria and other libraries, etc.), managing to convince the rulers and fellow citizens of the usefulness of all this. This new estate deserves a separate psychological portrait (Figure 37).

Note that this class is heterogeneous and is presented, according to table 5 of chapter 4, *Self-organization: evolution and behavior*, by subtypes of philosophers and scientists, on the one hand, and bureaucrats, on the other. It was the latter who condemned Socrates to death and persecuted Aristotle; they also constituted the backbone of the scribes and Pharisees of all time.

Also note that this is a kind of generalized portrait, whereas the mind of a particular person may differ significantly.

## BEHAVIOR PATTERNS

Let us move away from the scheme established in previous chapters to track the public reaction to the preaching of the new faith - in fact, the restoration of the oldest forgotten myth of humanity.

1. The Jewish **priests** were most hostile towards Jesus and the apostles. And this is natural: after a certain period of debate about the new doctrine, they concluded that it contradicts their paradigm, and Jesus's denunciations are directed personally against them, as scribes and hypocrites, unable to rise from a literal reading of Scripture to understand its symbolic meaning. Having correctly recognized in Jesus not only a competitor but a direct enemy, they achieved both his execution and the death of many apostles - but thereby contributed to the restoration of the paradigm of the old myth.

Other priests showed great tolerance, and the magicians-astrologers, guided by an ancient prophecy and their observations, welcomed his birth.

2. Roman and local **officials and nobles** treated the adherents of Christianity with displeasure as a source of troubles and problems. The meaning and goals of the sermon went far beyond their comprehension and daily worries (you are mad, Pavel! Great scholarship makes you mad - the words of the procurator Fest express this attitude in the best way possible). Depending on the situation, they could both discourage the same Paul from a thirsty crowd of people, and order him to be scourged to please the same crowd in another city. But they also saw no reason to refuse a fair trial to a Roman citizen. And Nero used to solve his political problems - accused of burning Rome.
3. In the same way, the **intelligentsia** distinguished itself by a complete lack of understanding. "We will hear you about this later," said the same to Paul in the Areopagus, when he began to speak about the resurrection of the dead, concluding that "you put something strange in our ears." But out of curiosity - listen.

   Here, the aforementioned weakening of intuition and the strengthening of reason have already fully manifested themselves - not unity with the truth, not its direct perception, but logic, text, authoritative opinion - this was the basis of their conclusions.

   In subsequent centuries, this same circumstance will lead to fruitless disputes of the scholastics.
4. Women and **commoners** reacted most sympathetically to the sermon - that is, those who are distinguished by heightened emotionality and sensuality, not retarded by reason and education, to whom Jesus addressed. Many of them did it for miracles, some for bread, others because they had no other hope.

And no one remembered the myth of the World Tree.

Now let's use this analysis to clarify the psychological portraits of the ancient classes. The **priests** remained the most adequate to the event and showing high intuition; the degree of their activity was determined by how much their interests were affected.

The **rulers** were not up to par, not understanding the situation and not assessing its consequences. The subsequent repression of Christians, and then the adoption of Christianity as an imperial religion - a consequence of the manifested myopia.

The paradigm shift is a very difficult task for the **intelligentsia**, showing its level. He was not high; there were no scientists of Pythagoras or Aristotle caliber, the rest of the phenomenon could not be estimated. This again speaks of the weakening of intuition. It took centuries, the philosophers of the level (and faith) of Basil the Great, Gregory the Theologian, Gregory of Nyssa, to inscribe the problems of synergy and Trinity in the problematics and categories of ancient philosophy, avoiding both numerous heresies and Gnosticism.

Also, the primary motivations of **commoners** are quite understandable. But having taken the path of cultivation and even sacrificing themselves, they received the most — the good sense of mortal life and the hope of eternal life in paradise.

All this speaks of three things.

First, the path of salvation leading to a lost paradise remained open to all. And the key to it is the abandonment of one's ego and sacrificial love, which opens the way to intuitive oneness with God.

Secondly, society is not able to adequately assess extraordinary events - and those who are still able to understand their significance are prone to hostility and aggression, being unable to part with their ideas and privileges. And this is very sad because it is unlikely that modern society, representing all the same estates, can react differently, going beyond the limits of class stereotypes.

Thirdly, total misunderstanding - motivational, linguistic, behavioral, paradigmatic - of one estate to others in antiquity reaches its maximum. This was manifested in the acceptance/rejection of Christianity. Subsequently, rationalism/irrationality will lead to splits within Christianity, to numerous intractable conflicts in society, to external wars. The feeling of the unity of mankind was lost precisely in the Late Antique period.

## CONCLUSION

Based on the model of subjective space, a methodology for the analysis of ancient painting was developed. The connection of the technical features of the works with the corresponding characteristics of previous and synchronous paintings is traced. The tendencies of development of antique painting are established and their connections with the general laws of evolution of human consciousness during the existence of ancient civilization are revealed. As a result, an understanding of ancient art that is free from the misuse of scientific categories that do not take into account the level of consciousness of the ancient person is achieved.

The reconstruction of the myth of the World Tree and related myths has been carried out. The actual restoration of this myth in Christianity is discussed; attention is drawn to the appearance and significance of such a figure as St. Christopher.

The states of consciousness of the main estates were reconstructed; attention is drawn to the fact that they have acquired a typical character, except the new class - the intelligentsia, the strengthening of which is associated with the development of reason in the ancient era. A psychological portrait has been drawn up for him. It is emphasized that the development of the mind occurs due to the weakening of other channels.

Patterns of the behavior of estates concerning the paradigm shift situation are considered. The estates of the priests, rulers, aristocracy, and intelligentsia were too rigid to accept the necessary changes. And only the consciousness of commoners, based on sensory perception of the world, turned out to be quite suggestible and labile.

## FUTURE RESEARCH DIRECTIONS

The next chapter will consider spatial constructions in medieval art, as well as reconstructed states of mind and patterns of behavior of the main classes.

## REFERENCES

Ancient-origins. Artifacts. (n.d.). Retrieved from https://www.ancient-origins.net/artifacts-ancient-writings/reading-between-lines-decrypting-scripts-minoans-and-mycenaeans-008180

Andreae, B. (1977). *The Art of Rome*. New York: H. N. Abrams.

*Art of nations and nations of the world*. (Vol. 1). (1967). Moscow, Russia: Short Art Encyclopedia.

Benedante. Tomb internal chariot. (2018). Retrieved from http://benedante.blogspot.com/2018/10/the-tomb-of-infernal-chariot.html

Betancourt, P. P. (2007). *Introduction to Aegean Art*. Philadelphia: INSTAP Academic Press.

Cheger. Livejournal. (n.d.). Retrieved from https://cheger.livejournal.com/485661.html

Chubova, A. P. (1966). Roman pictorial art. Album reproductions. Leningrad, Russia.

Chubova, A. P. (1970). Art of Europe of 1-3 centuries, Moscow.

Cook, R. M. (1986). Greek Art. Penguin.

Dunbabin, K. M. D. (1999). Mosaics of the Greek and Roman World. Cambridge, UK: Cambridge University Press.

Employees. Minoan_Mycenaen. (n.d.). Retrieved from http://employees.oneonta.edu/farberas/arth/Images/ARTH209images/minoan/octopus_vase.jpg

Ferrebeekeeper. Etruscan painting. (2014). Retrieved from https://ferrebeekeeper.wordpress.com/2014/06/13/etruscan-painting/

Flickr-photos-cavorite. (n.d.). Retrieved from https://www.flickr.com/photos/cavorite/98591365/in/set-1011009/

Isatorius. Livejournal. (n.d.). Retrieved from https://isartorius.livejournal.com/photo/album/1107/?page=28

Kobylina, M. M. (1939). *Art of Old Rome*. Moscow, Russia.

Kolpinskiy, Yu. D. (1970). Monuments of world art: 9 vol. Russian Academy of Arts. Institute of Theory and History of Fine Arts. Moscow, Russia: Art, V. 3: The Art of the Aegean World and Ancient Greece.

Kolpinskiy, Yu. D., & Britova, N. N. (1983). Monuments of world art: 9 vol. Russian Academy of Arts. Institute of Theory and History of Fine Arts. Moscow, Russia: Art, V. 7: The Art of Etruscans and Ancient Rome.

Kovalyov, Y. N., Nitsyn, A. Y., & Matveeva, E. L. (2011). The Development of Space Depictions in Ancient Painting. Technical Aesthetics and Design. 2011, 9. pp. 10–19.

Maggiani, A. (1985). *Artistic crafts: Northern Etruria in Hellenistic Rome*. Italy: Electra.

Mail. Question 39179570. (n/d). Retrieved from https://otvet.mail.ru/question/39179570

Mif-Meduza. (2016). Retrieved from https://mif-medyza.ru/2016/10/28/древнегреческие-вазы-геометрические/ Minoanatlantis.

Minoan_Flotilla. (n.d.). Retrieved from https://www.minoanatlantis.com/pix/Minoan_Flotilla_Fresco_Three_Panels_800px.jpg

Minoan_Mycenaen. (n.d.). Retrieved from http://employees.oneonta.edu/farberas/arth/ARTH209/minoan_mycenaean.html

Mircea, E. (1998). *A myth of the Eternal Return.*

Mircea, E. (2002). History of faith and religious ideas. Vol. 1. From the Stone Age to the Eleusinian Mysteries, Vol. 2. From Gautama Buddha to the triumph of Christianity: Criterion Employees.

Moon, W. G. (1983). *Ancient Greek Art and Iconography*. Madison, WI: University of Wisconsin Press.

Museo-Archeologic-Napoli. (n.d.). Retrieved from https://www.museoarcheologiconapoli.it/en/room-and-sections-of-the-exhibition/frescoes/

National Geographic. Bull-leaping. (n.d.). Retrieved from https://www.nationalgeographic.org/media/bull-leaping/

Panofsky, E. (2004). *Perspective as a "symbolic form". Gothic architecture and scholastics*. Saint Petersburg, Russia: Azbuka-Klassika.

Peter Summers. Archeology. (n.d.). Retrieved from https://www.petersommer.com/blog/archaeology-history/minoan-frescoes

Ramage, N. H., & Ramage, A. (2009). Roman Art: Romulus to Constantine. New Jersey: Pearson Prentice-Hall.

Rauschenbach, B. V. (2001). *The geometry of the picture and visual perception*. Saint Petersburg, Russia: Azbuka-Klassika.

Realhistory. Etruscan tomb. (n.d.). Retrieved from http://realhistoryww.com/world_history/ancient/Misc/Med/Etruscan_tomb_painting1.htm

Sokolov, G. I. (1971). *Art of Old Rome*. Moscow, Russia.

Spivey, N. (1997). *Etruscan Art*. London, UK: Thames and Hudson.

Steingräber, S. (2006). *The abundance of Life: Etruscan Wall Painting*. Los Angeles, CA: J. Paul Getty Museum.

*Stierlin, H.* (2004). Greece: From Mycenae to the Parthenon. TASCHEN.

The-history-blog. Kalpis. (n/d). Retrieved from http://www.thehistoryblog.com/wp-content/uploads/2012/06/kalpis.jpg

Tuck, S. L. A. (2015). *History of Roman Art*. Malden, MA: Wiley Blackwell.

Vitruvius. (1936). *Ten books about architecture*. Moscow, Russia: Publishing House of the Academy of Architecture.

Wikimedia. Cubiculum. Villa Fannius. (n.d.). Retrieved from https://commons.wikimedia.org/wiki/File:Cubiculum_(bedroom)_from_the_Villa_of_P._Fannius_Synistor_at_Boscoreale_MET_DP170950.jpg

Wikimedia. Etruscan civilization. (1988). Retrieved from https://en.wikipedia.org/wiki/File:Etruscan_civilization_map.png

Wikimedia. Etruscan mural. (n.d.). Retrieved from https://commons.wikimedia.org/wiki/File:Etruscan_mural_achilles_Troilus.gif

Wikimedia. Fourth Style. (n.d.). Retrieved from https://commons.wikimedia.org/wiki/File:Fourth_Style_wall_painting_with_architectural_landscape,_a_detail_from_a_panel_from_the_portico_of_the_Temple_of_Isis_in_Pompeii,_Naples_National_Archaeological_Museum_(14584129434).jpg

Wikimedia. Franqis Tomb painting. (n.a.). Retrieved from https://commons.wikimedia.org/wiki/File:François_Tomb_Painting_02.jpg

Wikimedia. Iphigenia. Taurus. (n.d.). Retrieved from https://commons.wikimedia.org/wiki/File:Iphigenia_in_Tauris_MAN_Napoli_Inv111439.jpg

Wikimedia. Last geometric pyxis. (n.d.). Retrieved from https://commons.wikimedia.org/wiki/File:Late_geometric_pyxis_Ancient_Agora_Museum.jpg?uselang=ru

Wikimedia. Miniature hydria. Thebes. (n.a.). Retrieved from https://commons.wikimedia.org/wiki/File:Miniature_hydria_Thebes_Louvre_A566.jpg?uselang=ru

Wikimedia. Phaistos Disk. (n.d.). Retrieved from https://en.wikipedia.org/wiki/Phaistos_Disc#/media/File:Δίσκος_της_Φαιστού_πλευρά_A_6380.JPG

Wikimedia. Phaistos Disk. (n.d.). Retrieved from https://en.wikipedia.org/wiki/File:Δίσκος_της_Φαιστού_πλευρά_B_6381.JPG

Wikimedia. Quenn-megaron. (n.d.). Retrieved from https://commons.wikimedia.org/wiki/File:Queen%27s_Megaron_with_dolphins_Knossos_Palace.jpg

Wikimedia. Tablette-lineare. (n.d.). Retrieved from https://commons.wikimedia.org/wiki/File:Tablette_lineaire-B.jpg?uselang=ru

Wikimedia. The Turee Gracies. (n.d.). Retrieved from https://commons.wikimedia.org/wiki/File:The_Three_Graces,_from_Pompeii_(fresco).jpg

Williams, D. (1999). *Masterpieces of Classical Art, 2009*. British Museum Press.

Yatromanolakis, D. (2016). *Epigraphy of Art: Ancient Greek Vase-Inscriptions and Vase-Paintings*. Oxford, UK: Archaeopress.

## ENDNOTE

1 Some researchers, in particular, Jerome M. Eisenberg, consider the disc of Phaistos a fake.

# Chapter 10
# Middle Ages:
## Failed Return


## ABSTRACT

*The sources to mind evolution study were chosen. The methods of the depicting space in the painting of the European Middle Ages and painting of previous and synchronous cultures are considered. The trends in the development of medieval pictorial art are established and their connections with the general laws of evolution of the human mind are revealed. When analyzing markers of evolutionary changes, the most active channels were established, and the forecast following from the scenario of self-organization of complex systems was checked. The results of the analysis are presented in the form of a psychological portrait of one of the most outstanding women of the Middle Ages - Eleanor of Aquitaine. The behavior patterns of the Middle Ages main estates representatives were described.*


## BACKGROUND

In the period between the fall of the Western Roman Empire and until the collapse of Byzantium (circa 1500-550 years ago) – these particular dates are considered to be provisional timeframes of the Middle Ages – Europe constituted a formation that was extremely heterogeneous from the political, religious and cultural perspective.

The heroic age of Justinian striving to restore the Roman Empire in the east, the ruins of Italy, Gaul, Iberia in the south and west, lands devastated by the Huns in the center of Europe… In a matter of just two hundred years, that picture was replaced by the Mauritanian conquest of the Pyrenean Peninsula, the Carolingian Renaissance in France, Germany, and Italy and by the devastation of Byzantium exhausted in its fights against the Avars, Slavs, Persians, and Arabs. 200-300 years more – and the newly organized kingdoms began the Reconquest in the south, converted to Christianity in the east and north, sent their subjects on the Crusades, assailed Constantinople and led endless wars with each other. And – the rise of Italy and Northern Europe that paved the way for the Renaissance; the conquest of Grenada as a prelude to the conquest of America; an almost accomplished liberation of Rus that had become a Third Rome

DOI: 10.4018/978-1-7998-1706-2.ch010

after the final downfall of the Second one, with which that boisterous epoch that reflected, as it were, the entire history of mankind, ended.

The Middle Ages are sometimes called the Dark Ages hinting at the cultural and technological regress as opposed to the Antiquity.

Was it the case? The Byzantine sources brought to us the information about disputes sparkling at marketplaces solely because of theological controversies; about convocations discussing articles of faith at the time when hostile battering rams where ruining the walls of their cities, and – sometimes – about miraculous rescues from foes. The focus of the society's interests had shifted in the direction of the Christian mysticism – and the mind, so to speak, rushed back to the first and second levels, gaining there a new unity – the unity with God... And that, indeed, resulted in the abandonment of temporal affairs. The same very Byzantium that had inherited the Rome's achievements untouched lost its technological advantages in a matter of just 600 years. Thereafter, the interest in temporal affairs was aroused again, and the mind returned to the fourth level.

The Medieval understanding of space immediately precedes the creation of the perspective methods of representation of space forms and relations that emerged in the epoch of the Renaissance. Does the Medieval understanding of space have an independent value? Or is it only important as a preparatory stage? How do the changes in that understanding correlate with the patterns of changes in space representation methods during the previous historical epochs? These questions need to be answered after tracing space representation methods from the Early Middle Ages to its "Autumn" (Huizinga, 1988), and by comparing the resulting patterns with the synchronic monuments of other cultures.

The sources (Beckwith, 1964; Mango& Hawkins 1964; Rice, 1968; Henderson, 1977; Tiazhelov, 1981; Weitzmann & Chatzidakis, 1982; Cormack, 1985; Dodwell, 1993; Campbell, 1998; Nesselstrauss, 2000; Adams, 2001; Benton, 2002) consider Medieval pictorial art from the perspective of art history. A periodization is introduced and peculiar features of styles are researched, including storylines and techniques. The chronology is based on historical events – the establishment and the fall of the feudal formation. A distinction is made between the differences characteristic of certain regions of Western Europe, in particular, of the Byzantine, Italian, French, Spanish, and German. In the framework of researching Western Europe, space representation methods are singled out and are occasionally compared with the methods characteristic of the antiquity period.

The purpose of this research is to classify Medieval images by the methods of representing space forms and relations (a similar task was set in Raushenbakh, 1986, 2001; Kovalyov & Nitsyn, 2012); to juxtapose the space representation methods with the achievements of the synchronic cultures; to pinpoint the role of the Medieval pictorial art in the establishment of the contemporary understanding of space by comparing it with the synchronic schools of the Islamic and Chinese regions.

The other task is to study the mind evolution, and specialists put attention on this problem (Grant, 2001; Huizinga, 1988; Whitton, 1998).

And the last task is reconstructions of behavior patterns, and for these publications (Huizinga, 1988; Stuard, 1987; Giles, 1995; Whitton, 1998; Bernhard & Jussen, ed. 2001) were devoted. The reconstruction of a psychological portrait of Eleanor of Aquitane is planned; many publications are about this outstanding person (Chambers, 1941; Meade, 1977; Turner, 2009).

## HELL, PURGATORY, PARADISE

Three periods of the development of the Medieval pictorial art are singled out (which can be notionally named as the Early, the Middle and the Late[1]), each of which has visible peculiar space representation features. Each of them will be considered according to the following scheme: an overview will be given, a visual set of the works of painting will be presented, peculiar features of space representation will be singled out, a comparison will be undertaken with synchronic works of other cultures and an interpretation will be provided from the perspective of the complex systems self-organization theory.

### Early Middle Ages

Early Middle Ages (circa 1500-1200 years ago) left a large number of monuments in the Balkans and Italy. This period may be generally characterized as that of disturbances and fierce wars, the Dark Ages indeed, except Byzantium and Italy which had managed to fight off the barbarians and to retain, to a certain extent, the Roman heritage. The main monument of the Early Byzantine period is Hagia Sophia Cathedral in Constantinople which was built under the reign of Emperor Justinian (Figures 1 & 2).

However, the early Byzantine paintings are very badly preserved there, due to their destruction by iconoclasts in the latest period.

That is why two lines of images will be given. Byzantium will be represented by the icons from St Catherine's Monastery in the Sinai Peninsula (Figs. 3 & 4) and Ravenna in Italy ((Figs. 5 & 6)[2], as well as by several miniatures characterizing the state of the pictorial art in other territories of Europe (Figures 7 & 8).

*Figure 1. Hagia Sophia Cathedral (the present state), Istanbul, Turkey, circa 1500 years ago Author: tigra991 (Kemclub. Photo 3674067, n/d).*

*Figure 2. Interior of Holy Wisdom (Hagia Sophia) temple. Author: Dean Strelau. (Wikimedia. Interior of Hagia Sophia, n/d).*

*Figure 3. Virgin and Child with angels and Sts. George and Theodore. the encaustic icon on the panel, ≈1400 years ago. Saint Catherine's Monastery (Wikipedia Encaustic Virgin, n/d).*

*Figure 4. Saint Peter, the encaustic icon from Saint Catherine's Monastery, Mount Sinai. ≈1400 years ago. (Marysrosaries. Saint Peter, n/d).*

*Figure 5. The mosaic of Emperor Justinian and his retinue, Basilica of San Vitale, Ravenna, Italy. ≈1450 years ago. Author: Carole Raddato (Wikipedia. Mosaic Empereor Justinian, n/d).*

*Figure 6. The mosaic of Empress Theodora, Basilica of San Vitale, Ravenna, Italy. ≈1450 years ago. Author: José Luiz Bernardes Ribeiro (Wikimedia. Mosaic of Theodora, n/d).*

*Figure 7. Evangelist Matthew. Miniature from the Canterbury Codex Aureus, circa 1250 years ago. Royal Library, Stockholm (Biblioclub. Zolotoy Kodeks, n/d).*

*Figure 8. Book of Kells, Folio 7v, Madonna and Child. Scanned from: Meehan, Bernard; The Book of Kells': an illustrated introduction to the manuscript in Trinity College Dublin. London: Thames and Hudson, 1994, p. 12. circa.1200 years ago. (Wikimedia. Madonna. Child, 1994).*

*Figure 9 The Temptation of Jesus, the Book of Kells, circa 1200 years ago (Studfiles, prewiew-5254880, n/d).*

As can be seen, the Early Middle Ages (circa 1500-1200 years ago) are characterized by a departure from the level achieved in the late antique period (the perceptive perspective with a multitude of vanishing points or axonometry, Chapter 8, *The Ancient World: Ego, Archaic, Immortality*), to a more primitive one (the absence of the depth of space or its representation via a conventional background). This is seen in comparing the paintings of a villa in Boscoreale (Figure 25, Chapter 8) first with the icons of the St Catherine's Monastery and then with the book miniatures of the latest period. Perhaps, only the overlapping of the remote objects by the objects positioned closer to the viewer is a constantly utilized indication of depth. Some other details (representation of a circle as an ellipse, play of light and shadows in the folds of clothing, etc.) are rather spontaneous and reflect certain local relations of different figures rather than spatial relations on the whole. In comparing the Byzantine samples with the monuments from those parts of Europe that fell off of the Empire, a decline in their level of artistry can be evidenced.

Attention should be paid that in contrast with the primitive painting, this one is characterized by the lack of expression, of dynamics, of polymorphy of images and by the presence of a distinct orientation of the sky and the earth.

As compared with the Egyptian sacral painting and the early Antique paintings with their fundamental planeness, attention should be drawn to attempts to depict, nevertheless, the depth of space. However, the tools used to reflect cannot but be characterized as most primitive as opposed to the late Antique paintings.

It is these differences that make it possible to speak about degradation: on the one hand, the intuitive-magical directness of primeval painters was lost (1-3 levels of the self-organization of subjective space), on the other hand, the late Antique understanding of the space is lacking (a weakened 4-the level). Therefore, the development of the understanding of space in the Middle Ages begins from a sufficiently low level – albeit, not from scratch – and takes place within the boundaries of the fourth level.

*Figure 10. Paintings in Qusayr Amra, Jordan (circa 1200 years ago). Author: Igor Diksa. (Igordiksa, n/d).*

*Figure 11. Painting from the Umayyad Mosque, Damascus, Syria (circa 1250 years ago). Author: Ondatr. (Arimoya. Iskusstvo Islamskogo Mira, n/d).*

Comparisons will be made with a synchronic Islamic painting under the Umayyad reign (Figure 10-11), and with a painting from the early Chinese – the Northern Qi, the early T'ang (Figure 12-13) cultures.

It can be seen that both the Islamic and the Chinese pictorial art employs similar devices: a local overlapping by the nearer objects of the farther objects, a system of plane plans, axonometry. The play of light and shadow is visible. A distinctive feature is a use by the Chinese artists of the aerial perspective and a restrained palette.

If allowances are made for the level of artistry, it may be said that the devices of depicting the depth of space (and the interest to three-dimensional space on the whole), were global.

This is where the following verses of Shen Yue (circa 1550-1500 years ago) may spring to mind:

*Inaccessible rocks*
*tower in an endless chain afar,*
*with the twilight gathering*
*in a vague blue distance,*
*On the endless expanses,*
*where the wind blows, gusty and fierce,*
*Herbs grow by the roadside,*
*In the autumnal time.*

*Figure 12. Gu Kaizhi. "The Fairy of the Luo River". Fragment: The Fairy. Gugong, Beijing, circa 1650 years ago. (Wikimedia. Gu Kaychdji. Fea, n/d).*

*Figure 13. Fresco in the Lou Rui tomb. The Northern Qi, circa 1500 years ago. Author: Natalia (Truisheva. Kitayskaya zhivopis, n/d).*

## The Middle Period

The Middle period (circa 1200-600 years ago) included such heterogeneous events as the Carolingian renaissance and the Ottonian Renaissance, the Crusades, the emergence of universities and the disputes about universals, a fancy to alchemistry and the enhancement of inquisition, wars between the Guelphs and the Gibelins, and a larger portion of the Hundred Year's war.

The visual sets for Byzantium (Figures 14-15), the Ancient Rus (Figures 15-16) and Western Europe (Figures 17-20) will be presented separately.

*Figure 14. The Virgin and Child mosaic, in the apse of Hagia Sophia. 1150 years ago. Istanbul, Turkey. (Pallasweb. Virgin and Child, n/d).*

*Figure 15. The Empress Zoe mosaics in Hagia Sophia (right). 950 years ago. Istanbul, Turkey. Photographer: Myrabella. (Wikipedia. Empress Zoe, n/d).*

*Figure 16. Ancient Rus. The Virgin Orans in St. Sophia temple. ≈970 years ago. Kyiv. Ukraine. Source: ZwHmjIymnSeDvQ at Google Cultural Institute (Wikimedia. Oranta. Kyiv, n/d).*

*Figure 17. Ancient Rus. Andrei Rublev. The Trinity. ≈600 years ago. State Tretyakov Gallery. Moscow, Russia (Wikimedia. Trinity. Rublev, n/d).*

*Figure 18. Charles the Bald welcomes monks from Tours who bring the Vivian Bible which includes this miniature. ≈1000 years ago. Bibliothèque Nationale, Paris Author: Count Vivien (Wikimedia. Karl III_monks, n/d).*

*Figure 19. The history of Adam and Eve. A miniature from Moutier Grandval Bible, circa 1 080 years ago. London, the National Museum Library. Author: Enott (Iskusstvoved. Kniznaya miniatyra, n/d)*

*Figure 20. The Adoration of the Magi. The Missal of Robert of Jumièges.1000 years ago, Rouen, the City Library. (Forum. Gallery-original-view, n/d).*

*Figure 21. Christ holding a globe of the world. The "Westminster Retable", Westminster Abbey, London. ≈1000 years ago. (Westminster-Abbey. About Abbey, n/d).*

*Figure 22. Medieval University. Miniature, circa 670 years ago. Author: ultima_cruzado. (Ultima Crusado. Livrjournal, n/d).*

The Middle Period is characterized by the rapid emergence of quite complex spatial compositions and specific medieval devices of spatial relations representation, namely, of the inverted perspective for nearer objects and axonometry for the farther ones. The same devices were then used in the Byzantine and the Russian icon painting. The latter is evaluated by contemporary researchers (Raushenbakh, 2001) as such which are more appropriate for the peculiarities of human perception than the linear perspective. A considerable improvement of spatial representation methods on complex surfaces should be noted, as well as of the light and shadow depiction methods.

Now the space depth representation technique used by the European artists will be compared with that used by the medieval Muslim (fig. 23-24) and Chinese artists (Figure 25-27).

Plane figures attempt to depict the space by placing objects on different plans, simplified artistic language, lack of life-like portrayals and fine details that may be observed. The images look more like ancient Egyptian (or early Mesopotamian) than like the paintings of the Umayyad epoch.

A considerable progress can be evidenced in representing the space of a large depth with the help of aerial perspective devices, in sizes becoming reduced (being non-linear) as the distances from the viewer grows, in the absence of a distinct skyline and vanishing points, as well as in the delicate representation of mood, and a very high level of artistic skill.

*Figure 23. Yahya ibn Mahmud al-Wasiti. Illustration to the manuscript al-Hariri "Macams". Abu Said standing in front of governor Rahba (the tenth Macam). (circa 780 years ago). Paris. The National Library. Author: ngasanova. (Ngasanova. Livejournal, n/d).*

*Figure 24. Yahya ibn Mahmud al-Wasiti. Illustration to the manuscript al-Hariri "Macams". Conversation on the outskirts of an aul (the forty-third Macam). Paris. The National Library. Author: ngasanova (Ngasanova. Livejournal, n/d).*

*Figure 25. Li Cheng. A Solitary Temple Amid Clearing Peaks. ≈1070 years ago. Nelson-Atkins Museum of Art. Kansas City, the USA. N 47-71(Wikimedia. Li Cheng. A Solitary Temple, n/d).*

*Figure 26. Dong Yuan. WintryGroves and Layered Banks. »1070 years ago. Kurokawa Inst. Hyogo (Wikimedia. Dong Yan. Wintry Groves, n/d)*

*Figure 27. Dong Yuan. Xiao and Xiang Rivers. ≈1080 years ago. Palace Museum, Beijing. (Wikimedia. Xiao and Xiang rivers, n/d).*

## Late Middle Ages

Late Middle Ages (circa 600-500 years ago) is characterized by a cultural life that was quite rich in events – against the backdrop of the failure of the Crusades (whereas the Reconquest was completed) and of the Turkish take-over of Constantinople, as well as by numerous wars, a classic example of which is the Hundred Year's War. Italy, in spite of its political disunity and foreign conquests, remained the culturally richest part of Europe. After Constantinople had fallen and many Byzantine scholars emigrated to Italy, that status of it was only reinforced.

Several samples of paintings of the period in question will be examined further. Two trends are visible in that the "Medieval classics" persisted, with its simplest methods of representing the depth of space, (fig. 28-30), and novations related with the development of the perceptive perspective and the light-and-shade technique (Figures 31-34).

Formally, the methods of representing space through a perspective are similar to the late antique ones (multiple vanishing points, an absence of a distinct skyline), however, the coordination of scales on different plans, the depiction of light and shadow look more perfect; the aerial perspective is present. Portrait painting in the period of the late Middle Ages acquires the traits of psychological veracity. All of that indicates a growing role in sensory perception.

Now a comparison with the works of the Shiite artists will be examined (Figures 35-37).

The following traits can be observed: the depiction of humans (allowed by the Shiites), planeness of figures and simple methods of representing the depth of space (perceptive axonometry, the nearer objects

*Figure 28. Conquest of Constantinople By the Crusaders In 1204. ≈550 years ago. Miniature from the Chronicle of the Emperors by David Aubert (Wikimedia. Conqest of Constantinople, n/d).*

*Figure 29. Battle of Crécy between the English and French in the Hundred Years' War. From an illuminated manuscript of Jean Froissart's Chronicles. ≈610 years ago. Bibliotheque Nationale de France, Paris, France. (Wikimedia. Battle of Crecy,n/d).*

*Figure 30. Joan of Arc. Miniature. ≈550 years ago. Archives nationales de France. (Wikimedia. Joan of Arc, n/d).*

*Figure 31. Robert Campin. ≈580 years ago. Portrait of a Man in a Red Turban, the London National Gallery (Wikipedia. Kompen Rober, n/d).*

*Figure 32. Jan van Eyck. The Virgin of cancellor Rolin. ≈585 years ago. Department of Paintings of the Louvre. Paris, France. INV 1271 (Wikimedia. Jan van Eyck, n/d).*

*Figure 33. Jan van Eyck. The Madonna in the Church. ≈580 years ago. Gemäldegalerie der Staatlichen Museen zu Berlin. 525C. (Wikimedia. Jan van Eyck. Madonna, n/d).*

*Figure 34. Jan van Eyck. The Crucifixion; The Last Judgment. ≈580 years ago. Metropolitan Museum of Art. 33.92ab. New York, USA (Wikimedia. Jan van Eyck. Diptych, n/d).*

*Figure 35. Junaid. Humayun peeping at Humay who is standing at the gates. Miniature. "Three Poems" by Khaju Kermani. ≈620 years ago. The British Library, London. (Wikipedia. Persidskaya Zhivopis, n/d).*

*Figure 36. Behzad. The Seduction of Yusuf. "Bustan" by Saadi. Herat, ≈530 years ago. Cairo, the National Library. (Wikipedia. Persidskaya Zhivopis, n/d).*

*Figure 37. Iskander on the Throne. Miniature. "Demott Shahname". ≈685 years ago. Louvre, Paris. Photographer: Evgeniy Ardaev. (Wikipedia. Persidskaya Zhivopis, n/d).*

being overlapped by the figures of the farther ones, the breakdown into plans). Faces are plane, or their relief is depicted by different colors. On the whole, the space representation technique had not only failed to develop, but it had also become even more archaic. The minds of the Islamic artists fixed the relations of the 2-3 levels of the subjective space by representing the images symbolically, not realistically.

Pictures made by the Chinese artists will now be considered (Figures 39-41)

Here, suspended development can also be stated, but due to another reason – mutinies and foreign invasions did not contribute to the preservation of culture. The methods of space representation are precisely the same as in the previous period, - albeit, the Tang epoch artists could have hardly been surpassed in any case.

Thus, by the eve of Renaissance, with the rational substantiation of the linear perspective inherent in that period, the perceptive perspective of the Middle Ages had reached a peak of its development, after driving the perceptive perspective to the maximum level of perfection and after creating all the bases for a scientific understanding of the perspective in the times to follow.

At the same time, the pictorial art of the eastern artists demonstrates the following trends:

- in the early period – the space representation technique of the Islamic pictorial art surpasses the West European one and is somewhat behind that of Byzantium; the Chinese artists elaborate the first methods to represent the three-dimensionality of figures and the depth of space;
- in the middle period – the Islamic pictorial art demonstrates the signs of lagging and even of regress (they once again begin to regard images as symbols); the Chinese technique of the aerial perspective reaches its peak, whilst the change in the sizes with the growth of distance from the viewer is approximately traceable;

*Figure 38. Huang Gongwang, Clearing After Sudden Snow, ≈680 years ago. Palace Museum, Beijing. (Wikipedia. Huan Gunvan, n/d).*

*Figure 39. Qin Xuan Wang Xizhi Watching Geese. ≈680 years ago. Metropolitan Museum of Art, N-Y. (Wikipedia. Tsan Syan, n/d).*

*Figure 40. Ni Zan. Water and Bamboo Dwelling. ≈680 years ago. Museum of Chinese History, Beijing (Wikipedia. Ni Tsan, n/d).*

- in the late period – a regress in the interest to representing the depth of space in the Islamic pictorial art continues and is reinforced after the take-over of Baghdad by the troops of a Mongolese khan Hulagy accompanied not only by the destruction of libraries but of the cultural heritage as a whole – after that the Islam became significantly more conservative. The same processes take place in China: a fall of the Tang dynasty, Manchurian conquests, successful assaults of the Mongols led by Genghis Khan and Hubilie Khan undermined the countries' forces and contributed to the stagnation of the pictorial art.

From then on, the development of the perspective becomes "Europe-centered" (as is seen in the subsequent comparisons), however, both in the Dark Ages in Europe, and after the Mongolese assaults in Asia, the idea of depicting a three-dimensional space was never forgotten, thus demonstrating the global nature of human mind evolution. Now the Medieval and the Antique pictorial art will be compared (Table 1).

*Table 1. Comparison of the Medieval and Antique Pictorial Art Features*

| Feature | Anti que Pictorial Art (at the end of the period) | Medieval Pictorial Art of Europe (late period) |
|---|---|---|
| 1. | No differences between the sacral and the domestic painting from the techniques of depicting the depth of space. | No differences between the sacral and the domestic painting from the techniques of depicting the depth of space. |
| 2. | Space is depicted as three-dimensional, yet inhomogeneous: any similarity with the perspective and axonometry is purely external. The methods of representing the depth are light-and-shadow and orthography with numerous vanishing points.<br>Space structurization is not connected with the image of the World Tree.<br>A highly expressive "emotional space" is organized.<br>Many new plots appear. | Space is depicted as three-dimensional; the homogeneity is not perceived; inverted and perceptive perspective are used, which are also similar to the axonometry of construction.<br>Aerial perspective and light-and-shadow are used.<br>Space structurization is not connected with the image of the World Tree.<br>A psychologically reliable "emotional space" is organized.<br>The majority of plots deal with religion; by the end of the period, a lot of secular plots have appeared. |
| 3. | Most often, a center exists, but not in the geometric sense, it is determined by the plot. Cyclicity is almost invisible. | Most often, a center exists, but not in the geometric sense, it is determined by the plot. Cyclicity is almost invisible. |
| 4. | Top and bottom, right and left sides are discernible; the action normally takes place "here and now". | Top and bottom, right and left sides are discernible; the action normally takes place "here and now". |
| 5. | There is a distortion of scales and proportions, but it is less pronounced. | There is a distortion of scales and proportions. |
| 6. | Images of people are expressed realistically, at the time of dramatic experiences or emotional uprise, details of conduct and movements of animals are expressive; scenes of war and violence are present but are not emphasized. | Images of people are characterized by psychological veracity and well-pronounced individuality; images of animals are schematic; scenes of war and violence are present. |
| 7. | The polymorphy is disappearing. | The polymorphy is episodic. |
| 8. | Complex cult scenes are profaned. | Complex cult scenes are significantly represented. |
| 9. | Samples of writing are poorly represented in drawings. | Samples of writing are poorly represented in drawings (except for miniatures). |

*Figure 41. Icelandic manuscript AM 738 4to. Arni Magnusson Institute. Iceland. 17th century. Photographer: Cornischong (Wikipedia. Mirovoe derevo, n/d).*

## WORLD TREE IN THE MIDDLE AGES

The World Tree is frequent in the pagan mythology of the Scandinavian peoples and is at the heart of it. A lot of images have remained since then, for example, Figure41 (which itself dates back to the 17th century) depicts all details of the canonic images of the Tree. The World Tree symbols are preserved as an element of a pattern in the Slavic embroidery, in the ancient petroglyphs, etc. After the distribution of Christianity, by the end of the Middle Ages, its initial meaning had been forgotten.

On the other hand, the representatives of the Abrahamic religions had never forgotten the meaning of the World Tree in the manner as it is represented in the Bible. A fragment of the painting of the Umayyad mosque in Damascus may serve as an example of this (Figure 42).

However, of most interest are the images related to Hermetic communities of various kinds: magic, alchemic, Kabbalistic.

Here, the dating is, of course, the most complex problem. The Kabbalistic books feature the image of the Sephiroth Tree (Figure 43), as well as of Adam Kadmon, who is identified with it (fig. 44). The first editions of the Kabbalistic books appeared in the 16 century; traditionally, the appearance of Kabbalah is dated back to the times of Abraham; and Joseph Flavius in the "Judean War" – already in the 1st century – mentioned one of the Kabbalistic teachings – about the evolution of a soul during its transmigration from less perfect bodies into more perfect ones. It is hard to get rid of a perception about the influence of Gnosticism, the natural philosophy of five elements, reminiscences of the crucification of Jesus Christ, etc. Probably, the formation of this teaching can be dated back, on the whole, to the Middle Ages or the Late Antiquity.

*Figure 42. Painting of the Umayyad mosque, Damascus, Syria (circa 1250 years ago). Author: Ondatr. (Arimoya. Iskusstva Islamskogo mira, n/d).*

*Figure 43. The Kabbalistic Tree of Life with the names of the Sephiroth and paths in Hebrew. Based on Figure 10, page 155, of The Bahir: An ancient Kabbalistic text attributed to Rabbi Nehuniah ben HaKana, first century, C. E., Aryeh Kaplan trans., First edition, 1979, Samuel Weiser, New York. User: AnonMoos. (Wikipedia. Tree of Life, n/d).*

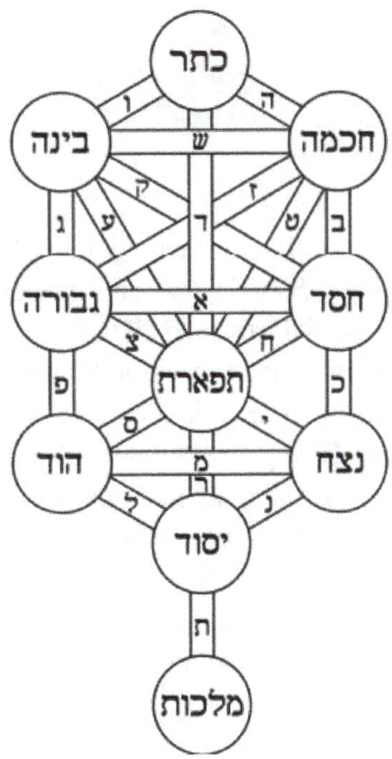

*Figure 44. Adam Kadmon. 19-th century. From José Luis Moreno blog. (Silenciollama-blogpost. Adam Kadmon, n/d).*

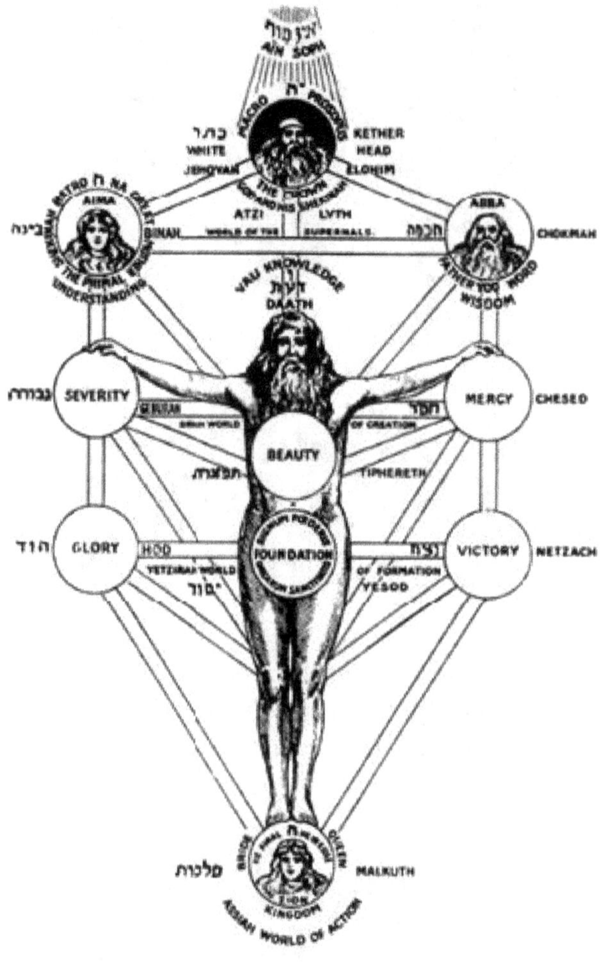


Of interest is a Chinese parallel, that of Taoist teaching about San-Cai – the Heaven-Human-Earth triad. According to it, the world is a trinity (San-Cai) which is represented in the images of Heaven, Human, and Earth. Father-Heaven is the celestial vault and all that is subtle and incognizable and that exists in the spheres of the worlds of spirits, deities, and ancestors. Mother-Earth gives life to everything that is born on it. The Earth is a living organism; all material objects surrounding human beings have merged into the image of the Earth. Within this Trinity, the Human is a point of contact of the Heaven and the Earth, where the laws of the former and the latter are manifested. It is not difficult to see that the image of the Human, like the image of Adam Kadmon, is an embodiment of the World Tree.

One more image is a version of alchemists and magicians (Figure45) from a book published in 1613. The central role of the World Tree is unquestioned – but solely for cognition.

Now, the Hermetic version will be compared with the Christian one (Table 2).

*Figure 45. Title page of "Occulta Philosophia ver den verborgenen Philosophischen Geheimnussen...". Frankfurt, 1613. (Openlibrary. Occulta philosiphia, n/d).*

*Table 2. Comparison of the Christian and Hermetic Versions of the World Tree Myth Features*

| Feature | Christian Version | Hermetic Version |
|---|---|---|
| 1. | The structuring of the world is symbolized by the World Tree. | The World Tree is a symbol of creation and cognition. |
| 2. | Use of the Tree by God for sacrificing himself and for transmigration to hell was the case. Delivery of God into the world by an earthly woman – Mother of God. | Use by a human being of spiritual development, acquisition of power, magical practices, cognition. |
| 3. | Connection with the cult of the dead – direct. | Is absent or shrouded. |
| 4. | A very distant perception of the underworld in the "Book of the Dead" and the biblical prophecies about the resurrection. | Absent. |
| 5. | The myth is restored in all entirety of perceptions of the previous epochs, with Jesus Christ being the central figure and the cross – a mere tool. | A grossly curtailed version of the myth with or without preservation of the identification of a human being with the Tree. |

## MYSTICISM AGAINST LIFE

"God is Love" – Jesus bequeathed while defining Christianity as a religion of love – for God and neighbor. Though it would appear that the Middle Ages, an epoch, when religion was so powerful as, probably, it had never been in any other century of the human history, ought to have become a time of peace and love, hardly any other supposition could have been that far from reality. The Middle Ages is the period of total and continuous confrontations – religious, ideological, political, cultural, economic which were repeatedly transformed into wars and massive reprisals.

Such a state of things was occasioned by many reasons; however, our goal is to determine psychological prerequisites, such which would demonstrate the state of mind of a human being of that epoch.

The church will be the starting point. "I am the Path, the Truth and Life" – Jesus Christ once said while determining what would appear an indisputable and eternal criterium in any possible controversies. It would seem that monks had found, over many centuries, a reliable path to cognize God and Truth, that of struggle with fleshly lusts, non-sophistication, meekness (implementation of the folding up of space, Chapter 1), the enhanced intuition, with prayers and ceremonies used to steer it to the required direction, and, finally, Love which means God and Truth. In other words, such a path means to practice and direct perception, not disputes and reasoning. But, instead, theological provisions were defended by, for example, Thomas Aquinas, with the help of a sophisticated logic dating back to the "deific Aristotle"; whereas one of the most famous judicial proceedings of the Middle Ages – that against Jeanne d'Arc – could not have gone without an opinion of the Paris University which was also logic-based. And that was not an exception, but a customary practice of the time.

Limitless egos and expansion of major and minor rulers only be contained by equal opponents. Yes, the church often intervened in such conflicts - but they only changed shape, moving from open military clashes to endless accusations that their kings should judge.

Knights, ideally impeccable warriors, transformed their sublime aspiration to mud, blood, carnage, and robbery of the Crusades.

Artistic personalities, like an eternal student Francois Villon, wrote excellent poems describing the world and the senses in the forms of images that oppose each other.

*I die of thirst beside the fountain*
*I'm hot as fire, I'm shaking tooth on tooth*
*In my own country, I'm in a distant land*
*Beside the blaze, I'm shivering in flames*

this is how a ballade from the poetry completion in Blois begins (Villon, 1965). However, a substantial portion of Villon's poems was written in the cant of the Parisian thieves with whom he was on a very friendly footing, and who was even given a death sentence for being an accomplice to robberies.

The commoners of neighboring towns hated each other so much that even today those senses reverberate in the opposition of, for example, Milan and Atalanta, or of football fan groups.

Finally, the intellectual elite had been bogged down in century-old disputes. Very illustrative is a dispute about universals which dates back to as early as Plato and Aristotle. Do common concepts exist in reality?

Universalia sunt realia, universalia sunt ante rem – Johannes Scottus Eriugena and Anselm of Canterbury would affirm.

Universalia sunt nomine, universalia sunt post rem – Johannes Roscelin and Occam would object.

The dispute as such, being of most importance from the viewpoint of the fundamentals of science[3], had been conducted through scholastic methods in the form of "logical games of sense", evidencing thus the involvement of mind into a discourse characteristic of the third level of self-organization. It, certainly, could not have been settled in the framework of such an approach.

Alchemistry, with its ideas of a philosopher's stone by which metals can be transformed into gold, and of an elixir of immortality that gives eternal life to the body, was likewise largely based on oppositions. This is how the relations between the elements were described in the "Aristotle Square" (Figure 46).

A focus was made not on a reciprocal transformation of qualitatively different elements like it was the case with the ancient natural philosophers, the Indian adepts of Sankhya or Chinese Taoists, but exactly on their setting off and counter-balancing.

On the whole, the alchemistry demonstrated a significant "earthliness" of the dreams of rulers and intelligentsia of that period. It should be noted, however, for the sake of fairness, that a magical interpretation always existed in the Hermetic societies – a spiritual transmutation of an alchemist himself takes place in the course of searching for a philosopher's stone. The same very doctrine exists in an "internal transmutation" of Taoists.

*Figure 46. Aristotle Square. Image from an alchemic treatise. S Levchenkov. (Wikipedia. Kvadrat Aristoteles, n/d).*

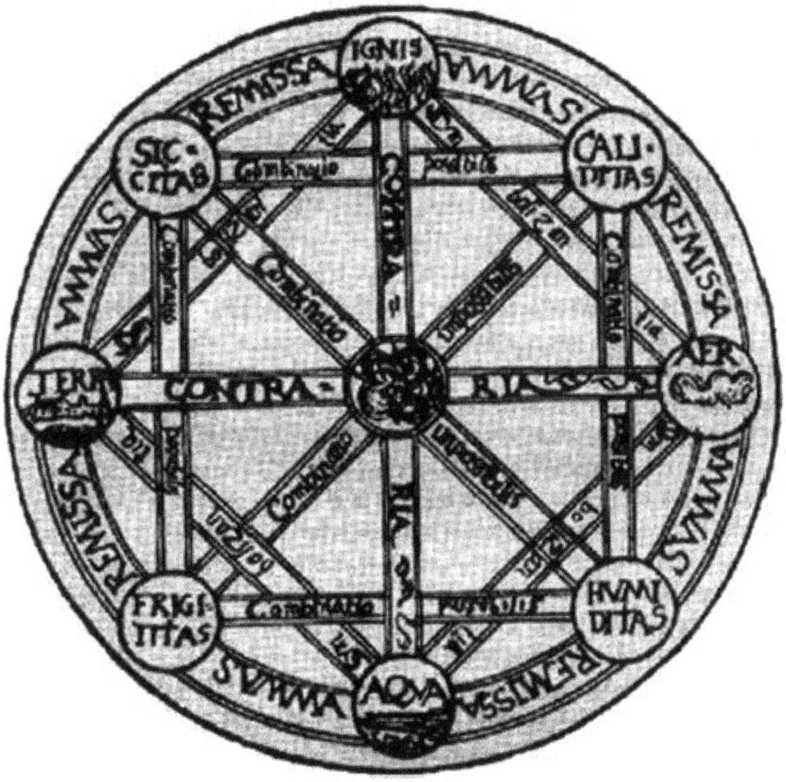

Thus, it was already at that time that the human mind had wandered so far, an intuitive perception, that even century-long efforts of the church had only managed to push it a little back – albeit, not to the Paleolithic, but, maybe, to the Neolithic state? But, no, not to the Neolithic – seeing that the social stratification had not been overcome and that the interest to the depiction of either a three-dimensional space or of shades of senses was still there.

## MARKERS ANALYSIS

### Space Representation

It may be stated that despite the attempts of exerting religious pressure, on the whole, no shifts had taken place in the methods of space representation either to the Neolithic or to Paleolithic methods.

This means that the process of the public conscience evolution had already become irreversible – though a path for the development of intuition had been offered which was accessible by everybody.

Homogeneity, additivity, distinctiveness, invariability of the elements of space are lacking in the pictures of medieval artists both in Europe, in the Islamic world and China – but everywhere, even if contained by a religious doctrine, the activity of reason can be observed, with a characteristic ternarity of perception, including spatial and temporal relationships. This was manifested in a general interest in depicting the depth of space.

The representation methods did not become scientific; they constituted a perceptive perspective based on immediate impressions of an artist.

This suggests that the nous had not yet become a dominating channel.

Interest in the reflection of human senses in Europe and of beauties of nature in China demonstrated an enhancement of sensory channels. And the norms of the religious moral failed to suppress that interest.

Therefore, in the period under consideration, the space representation technique was developing in the framework of the 4th stage of self-organization (with setbacks to the 3rd stage observable in the Islamic painting), that process is global. However, a presence of interest to senses and beauties of nature illustrates a vivid presence of the 5th stage (sensory perception).

### The World Tree Myth

The appearance of the Savior and the reconstruction of the World Tree Myth in all its entirety have turned out not to be inter-related in the minds of the greater majority of people. Perhaps, it was only amidst the Kabbalists – the people who thought Jesus to be a false Messiah – that these events had received at least an associative connection in the image of Adam Kadmon and Sephiroth Tree (fig. 44). And even with the Chinese, who were situated very far from the place setting, the image of the World Tree turned into a human being serving as a link between the Heaven and the Earth.

An advancing weakening of intuition, strong opposition characteristic of a nous used for understanding harmony and unity through geometric images of triangles can be evidenced. All of that is illustrated in fig. 45.

## MEDIEVAL HUMAN MIND

The factor of the church and the fact of its comprehensive influence – informational, ceremonial, semantic, historical, abusive, etc. – upon the state of the human mind has become a novelty and deserve a separate conclusion. What was the goal, what was the ideal image for the formation of which so much effort had been taken? It was undoubtedly an image of a monk; his psychological portrait is shown in Figure 47.

For the sake of comparison, an attempt will be made to make a psychological portrait of a woman – of, perhaps, the brightest personality of the High Middle Ages – that of Eleanor of Aquitaine (French: Aliénor d'Aquitaine*)*. Since in the minds of contemporary men the image of Eleanor is eclipsed by the figure of Jeanne d'Arc, certain facts of her biography should be reminded, which formed the basis of the created portrait. According to (Chambers, 1941; Meade, 1977; Turner, 2009), Eleanor of Aquitane (circa 1124-1204), the Duchess of Aquitaine and Gascony, the Countess of Poitiers, the Queen of France and the Queen of England, a mother of two English kings, was born in a family of William X and Anor de Chastellerault, who was a daughter of a years-long lover of her grandfather William IX, a talented troubadour and poet, and not so successful a governor, who was twice excommunicated by the church because of his excesses. Hence, an atmosphere in which she grew at the court is understandable.

After the early death of her father, she, as a rich heiress, at the age of 13 or 15 years, was given in marriage to a son of the French king. Pretty soon she got entangled in various intrigues, primarily, by deserving the "right to love" of her 15-year-old younger sister and of the 50-year-old Raul de Vermandois, which resulted in several wars, many years of ex-communications and interdictions.

Subsequently, she took part (possibly, even in the role of a commander of a female squad) in a Crusade organized by her husband Louis VII, who had already been by that time the king of France. Understandably, that adventure was taking place in the conditions far from the contemporary understanding of comfort.

*Figure 47. State of mind of a medieval monk. A strongly pronounced intuition of a contemplator – but, at the same time, of a man who is active and who often demonstrates an "unreasonable zeal" – it's a fanatic.*

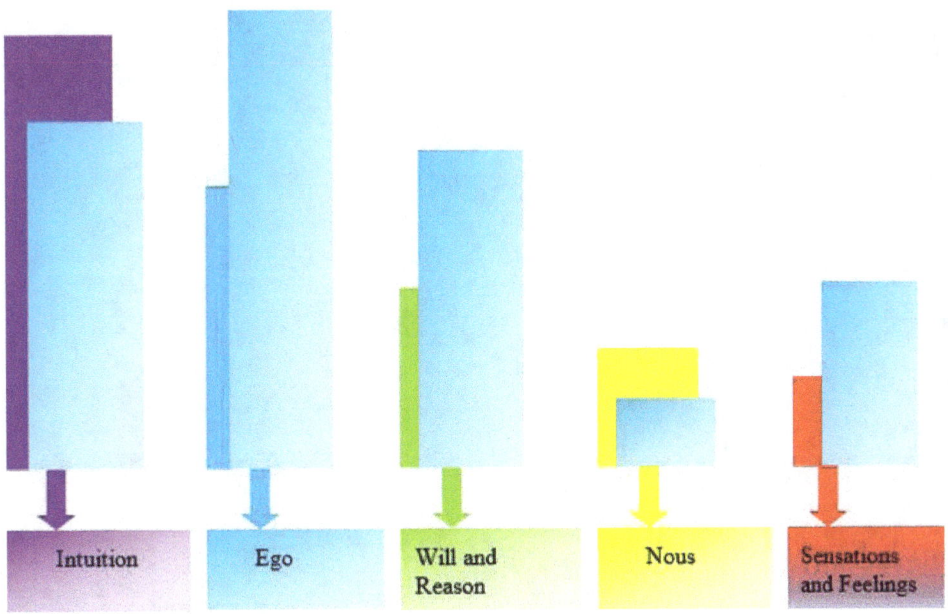

Upon bearing a child after the crusade, and on becoming disappointed in her husband (possibly, under the influence of the military failures as well) she got divorced, and, several months later married the future English king Henry II, who had obtained a heritage which was larger than his native England.

Both the first and the second events gave rise to scandals, with which the Pope had to interfere; and it is, probably, this marriage that might be the prime cause of the would-be Hundred Years' War.

A romance with Henry II had also been developing boisterously, including the episodes of complete confidence, when the king provided his wife with the right to dispose of the treasury, and up to the confinement of Eleanor, after she had induced the king's children to excite a riot against him.

Upon liberation only after the demise of her husband by her favorite son, Richard the Lionheart, she, during the time of his numerous, and, ultimately, unsuccessful wars, had sovereignly governed England, while being most active in arranging marriages of her numerous children and minor relatives. After the death of Richard and accession of the unloved son John the Landless, she returned to Aquitaine and before retiring from her affairs to the Abbey of Fontevraud, where she died at the age of 80, she had successfully withstood a siege by the troops of her grandson.

Let the assurances of exceptional intelligence and beauty of that woman remain on the conscience of the courtiers (no in-life portraits have remained), but out of the bare facts, there arises an image of an outstanding personality with vast potential.

Personality, whose intuition was not strong enough for her to be able to foresee the implications of her deeds. Personality, who was extremely selfish, high-handed, proud, volatile, and, at a younger age – romantic in nature. Who was always ready to protect what was hers, but who was, actually, far from being a "Great Monarchess".

Who had a strong will, but a deal lesser reason.

Who was not prone to orderliness and who viewed the world through the prism of matrimonial interests.

Who was sensual and who valued feelings highly, at least, when she was young (Figure 48).

*Figure 48. A probable psychological portrait of Eleanor of Aquitaine.*

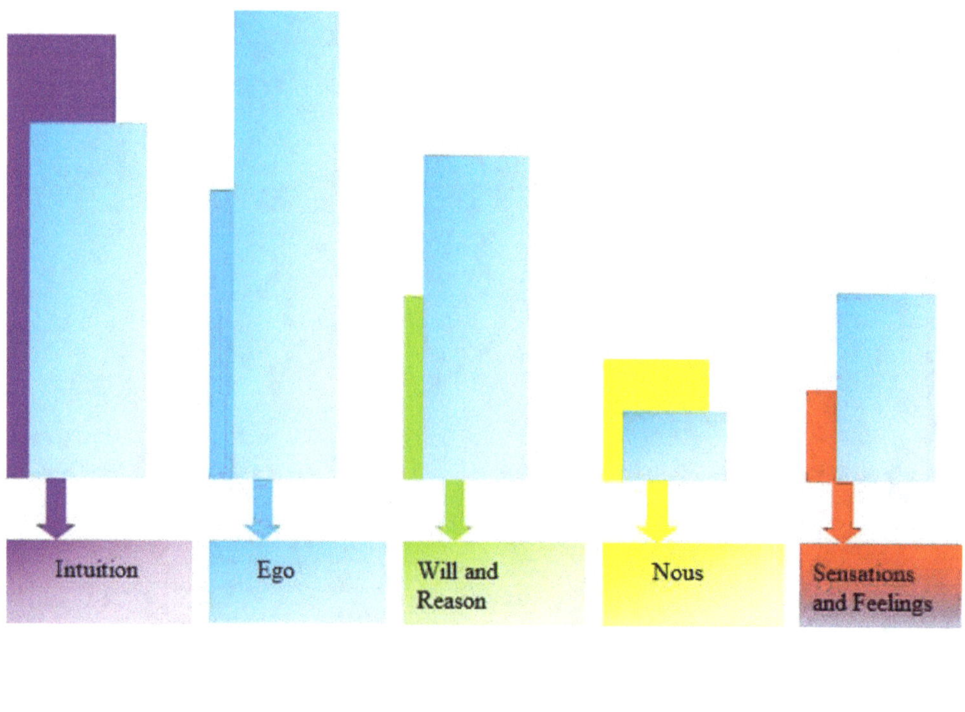

This is the diametric opposite – with the exception for the intransigent will – to the psychological portrait of a monk. Incompatibility of these types and impossibility to make Eleanor's type (which was, to a certain extent, a typical image of a human being of the Medieval period) fit a monk's mind without massive reprisals is quite evident. Ultimately, it was this contrast that resulted in extensive witch-hunts, though even the church itself initially denied mere existence of them.

## BEHAVIOR PATTERNS

Tables 1-6 of Chapter 4, Self-organization: evolution and behavior, will be used, and also publications (Huizinga, 1988; Stuard, 1987; Giles, 1995; Whitton, 1998; Bernhard, Jussen, ed. 2001).

**Priests and monks**. The phenomenons of saints were extremely rare – as has been the case in all the times, starting from the Ancient World. The type of a fanatic was most common, the one with "unreasonable zeal", striving to destroy foes instead of loving them (fig. 47). There was a type of a man who is educated, but who is, at the same time, a good master (an entrepreneurial researcher, for example, Sugerus Abbot). There were a number of those striving for power or of hypocrites anxious for merely "lucrative jobs".

**Rulers**. Selfish people. Behaved proudly and haughtily. Subsequent conduct depended on the potential and the development of other channels.

**Knights** – largely, a "warrior" type; many of them were invaders, less frequently - strugglers.

**Intelligentsia** – the type of scholar (weak will, strong reason, and nous). A great number of artistic personalities, with balanced personalities being very rare.

**Commoners** – mainly, a subtype of a philistine. Those, whose feelings reached a level of exaltation, had very much chance to perish in the period of the Autumn of the Middle Ages, by falling victims to the Inquisition after the slander by neighbors. The conduct was mean, almost slavish in villages, and a bit more independent in cities.

## CONCLUSION

The connection of the features of painting in Europe, the Islamic world, China with the corresponding characteristics of previous monuments is traced. The trends in the development of medieval painting are established and their connections with the general laws of evolution of the human mind are revealed. It is confirmed that evolution has a universal character.

Reconstruction of the myth of the World Tree. It is noted that, with the exception of the pagan cultures of Scandinavia in Europe, China in the East, it was preserved and rethought in various sealed communities.

Reconstructed, on the example of a psychological portrait of Eleanor of Aquitaine, the typical state of the human mind of the Middle Ages; its main feature is a rigidly binary worldview, intransigence, slow progress of the nous, the emergence of interest in feelings. The psychological portrait of the monk confirms this picture, supplementing it with the features of fanaticism. In marginal circles, however, sensory behavior is fundamental.

The patterns of behavior of the main classes are considered. The Middle Ages can be said to be the standard of behavior of the estates; it was then that their typical features developed to an extreme degree.

## FUTURE RESEARCH DIRECTIONS

The following chapter will be dedicated to reconstruction of space depicting in the art of the Italian Renaissance and in the art of the Northern Renaissance, as will be reconstruction of the states of mind and behavior patterns of the main estates of that period.

## REFERENCES

Adams, L. S. (2001). *A History of Western Art* (3rd ed.). Boston, MA: McGraw Hill.

Arimoya. Iskusstvo Islamskogo Mira. (n.d.). Retrieved from http://forum.arimoya.info/threads/Искусство-исламского-мира.4021

Beckwith, J. (1964). *Early Medieval Art: Carolingian*. Ottonian, Romanesque, Thames, & Hudson.

Benton, J. R. (2002). *Art of the Middle Ages. World of Art*. London, UK: Thames & Hudson.

Bernhard, J. (Ed.). (2001). *Ordering Medieval Society: Perspectives on Intellectual and Practical Modes of Shaping Social Relations*P. Selwyn, Trans.). Philadelphia, PA: University of Pennsylvania Press.

Biblioclub. Zolotoy Kodeks. (n.d.). Retrieved from http://art.biblioclub.ru/picture_12453_zolotoy_kodeks_iz_kenterberi_evangelist_matfey/

Binski, P., & Messing, A. (2009). The Westminster Abbey Retable. History, technique, conservation.

Campbell, L. (1998). *The Fifteenth Century Netherlandish Paintings, National Gallery Catalogues (new series)*. New Haven, CT: Yale University Press.

Chambers, F. M. (1941). Some Legends Concerning Eleanor of Aquitaine. *Speculum: A Journal of Medieval Studies, 16*(4), 459–468. University of Chicago Press. .

Constable, G. (1995). The Orders of Society, chap. 3 of Three Studies in Medieval Religious and Social Thought. Cambridge, UK: Cambridge University Press, 1995, pp. 249–360.

Cormack, R. (1985). *Writing in Gold, Byzantine Society and its Icons*. London, UK: George Philip.

Crusado, U. Livrjournal. (n.d.). Retrieved from https://ultima-cruzado.livejournal.com/photo/album/417/?page=6

Dodwell, C. R. (1993). The Pictorial Arts of the West: 800–1200. Pelican History of Art. New Haven, CT: Yale University Press.

Forum. Gallery-original-view. (n.d.). Retrieved from http://www.vvv.ru/forum_gallery/original_view.php?id=92007&tid=1

François, V. (1965). *The Poems* (G. Kinnell, Trans.). Houghton Mifflin.

Grant, E. (2001). *God and Reason in the Middle Ages*. Cambridge, UK: Cambridge University Press. doi:10.1017/CBO9780511512155

Henderson, G. (1977). *Early Medieval Art* (Rev. ed.). Harmondsworth, UK: Penguin.

Huizinga, J. (1988). The Autumn of the Middle Ages. A study of the forms of life and thought in the XIV and XV centuries in France and the Netherlands. Moscow, Russia: Science.

Igordiksa. (n.d.). Retrieved from http://igordiksa.com/?p=7871&lang=ru

Iskusstvoved. Kniznaya miniatyra. (n.d.). Retrieved from http://iskusstvoed.ru/2018/09/01/книжная-миниатюра-эпохи-каролингов-к

Kemclub. Photo 3674067. (n.d.). Retrieved from http://kemclub.ru/photo/3674067/

Kovalyov, Y. N., & Nitsyn, A. Y. (2012). Development of Space Representations in Medieval Painting. *Proceedings of the Tauride State Agrotechnological University. 14 International Scientific-Practical Conference "Contemporary Problems of Geometric Modeling of SPGM 14"*. Melitopol, Ukraine: TDATU. 2012. pp. 63-69.

Mango, C., & Hawkins, E. J. W. (1964). The Apse Mosaics of St. Sophia in Istanbul. *Report on Work Carried out.*

Marysrosaries. Saint Peter. (n.d.). Retrieved from http://www.marysrosaries.com/collaboration/index.php?title=File:Saint_Peter_Icon_from_Saint_Catherine%27s_Monastery,_Mount_Sinai_001.jpg

Meade, M. (1977). *Eleanor of Aquitaine: A Biography*. Hawthorn.

Nesselstrauss, C. G. (2000). The art of the Early Middle Ages. St. Petersburg, Moscow: Azbuka-Klassika.

Ngasanova. Livejournal. (n.d.). Retrieved from https://ngasanova.livejournal.com/627043.html

Openlibrary. Occulta philosiphia. (n.d.). Retrieved from https://openlibrary.org/books/OL24342362M/Occvlta_philosophia

Pallasweb. Virgin and Child. (n.d.). Retrieved from https://www.pallasweb.com/deesis/virgin-and-child-hagia-sophia.html

Panofsky, E. (2004). Perspective as a "Symbolic Form". Gothic Architecture and Scholasticism. St. Petersburg, Russia: Azbuka-Klassika.

Raushenbakh, B. V. (1986). Systems of Perspective in the Pictorial Art. General Theory of Perspective. Moscow, Russia: Nauka.

Raushenbakh, B. V. (2001). The geometry of a Picture and Visual Perception. St. Petersburg, Russia: Azbuka-Klassika.

Rice, D. T. (1968). *Byzantine Art* (3rd ed.). Penguin Books.

Silenciollama-blogpost. Adam Kadmon. (n.d.). Retrieved from https://silenciollama.blogspot.com/2010/11/adam-kadmon.html

Stuard, S. M. (1987). *Women in Medieval History and Historiography*. Philadelphia, PA: University of Pennsylvania Press. doi:10.9783/9781512807295

Studfiles, preview-5254880. (n.d.). Retrieved from https://studfiles.net/preview/5254880/

Tiazhelov, V. N. (1981). The Art of the Middle Ages in Western and Central Europe. Moscow, Russia: Iskusstvo.

Truisheva. Kitayskaya zhivopis. (n.d.). Retrieved from http://turisheva.ru/2015/05/08/kitayskaya-zhivopis-iv-vi-veka-n-e/

Turner, R. V. (2009). *Eleanor of Aquitaine: Queen of France, Queen of England*. Yale University Press.

Weitzmann, K., & Chatzidakis, M. (1982). *The Icon*. London, UK: Evans Brothers.

Westminster-Abbey. About Abbey. (n.d.). Retrieved from https://www.westminster-abbey.org/ru/about-the-abbey/history/retable

Whitton, D. (1998). The Society of Northern Europe in the High Middle Ages, 900–1200. In G. Holmes (Ed.), The Oxford Illustrated History of Medieval Europe. Oxford, UK: Oxford University Press. pp. 115–174.

Wikimedia. Battle of Crecy. (n.d.). Retrieved from https://commons.wikimedia.org/wiki/File:Battle_of_crecy_froissart.jpg?uselang=ru

Wikimedia. Conquest of Constantinople. (n.d.). Retrieved from https://commons.wikimedia.org/wiki/File:ConquestOfConstantinopleByTheCrusadersIn1204.jpg

Wikimedia. Dong Yan. Wintry Groves. (n.d.). Retrieved from https://commons.wikimedia.org/wiki/File:Dong_Yuan._Wintry_Groves_and_Layered_Banks_ca._950_(181,5x116,5cm)_Kurokava_Inst._Hyogo.jpg?uselang=ru

Wikimedia. Gu Kaychdji. Fea. (n.d.). Retrieved from https://commons.wikimedia.org/wiki/File:Гу_Кайчжи._Фея_реки_Ло._Деталь_-_Фея._Гугун,_Пекин.jpg?uselang=ru

Wikimedia. Interior of Hagia Sophia. (n.d.). Retrieved from https://commons.wikimedia.org/wiki/File:Interior_of_Hagia_Sophia.jpg

Wikimedia. Jan van Eyck. Diptych. (n.d.). Retrieved from https://commons.wikimedia.org/wiki/File:Jan_van_Eyck_-_Diptych_-_WGA07587.jpg?uselang=ru

Wikimedia. Jan van Eyck. Madonna. (n.d.). Retrieved from https://commons.wikimedia.org/wiki/File:Jan_van_Eyck_-_The_Madonna_in_the_Church_-_Google_Art_Project.jpg

Wikimedia. Jan van Eyck. (n.d.). Retrieved from https://commons.wikimedia.org/wiki/File:Jan_van_Eyck_070.jpg

Wikimedia. Joan of Arc. (n.d.). Retrieved from https://commons.wikimedia.org/wiki/File:Joan_of_Arc_miniature_graded.jpg

Wikimedia. Karl III_monks/ (n.d.). Retrieved from https://commons.wikimedia.org/wiki/File:KarlII_monks.jpg

Wikimedia. Li Cheng. A Solitary Temple. (n.d.). Retrieved from https://commons.wikimedia.org/wiki/File:Li_Cheng_-_A_Solitary_Temple_Amid_Clearing_Peaks_-_Google_Art_Project.jpg?uselang=ru

Wikimedia. Madonna. Child. (1994). Retrieved from https://commons.wikimedia.org/wiki/File:KellsFol007vMadonnaChild.jpg

Wikimedia. Mosaic of Theodora. (n.d.). Retrieved from https://commons.wikimedia.org/wiki/File:Mosaic_of_Theodora_-_San_Vitale_-_Ravenna_2016_(4).jpg

Wikimedia. Oranta. Kyiv. (n.d.) Retrieved from https://commons.wikimedia.org/wiki/File:Oranta-Kyiv.jpg?uselang=ru

Wikimedia. Trinity. Rublev. (n.d.). Retrieved from https://commons.wikimedia.org/wiki/File:Angelsatmamre-trinity-rublev-1410.jpg

Wikimedia. Xiao and Xiang rivers. (n.d.). Retrieved from https://commons.wikimedia.org/wiki/File:Xiao_and_Xiang_rivers.jpg?uselang=ru

Wikipedia. Empress Zoe. (n.d.). Retrieved from https://en.wikipedia.org/wiki/File:Empress_Zoe_mosaic_Hagia_Sophia.jpg

Wikipedia. Huan Gunvan. (n.d.). Retrieved from https://ru.wikipedia.org/wiki/Хуан_Гунван#/media/File:Huang_Gongwang._Clearing_After_Sudden_Snow._ca._1340._104,6_cm_long.Palace_Museum,_Beijing.jpg

Wikipedia. Kompen Rober. (n.d.). Retrieved from https://ru.wikipedia.org/wiki/Кампен,_Робер#/media/File:RobertCampinWorkshop-Man-1314.jpg

Wikipedia. Kvadrat Aristoeles. (n.d.). Retrieved from https://ru.wikipedia.org/wiki/Файл:Quadrat_Aristoteles.jpg

Wikipedia. Mirovoe derevo. (n.d.). Retrieved from https://ru.wikipedia.org/wiki/Мировое_древо#/media/File:Yggdrasil_AM_738_4to.jpg

Wikipedia. Mosaic Empereor Justinian. (n.d.). Retrieved from https://en.wikipedia.org/wiki/File:The_mosaic_of_Emperor_Justinian_and_his_retinue.jpg

Wikipedia. Ni Tsan. (n.d.). Retrieved from https://ru.wikipedia.org/wiki/Файл:Ни_Цзань.Воды_и_бамбуковый_домик._1343_Музей_кит._истории,Пекин.jpg#/media/File:Ни_Цзань.Воды_и_бамбуковый_домик._1343_Музей_кит._истории,_Пекин.jpg

Wikipedia. Persidskaya Zhivopis. (n.d.). Retrieved from https://ru.wikipedia.org/wiki/Персидская_живопись#/media/File:Demotte_Shahname_001.jpg

Wikipedia. Persidskaya Zhivopis. (n.d.). Retrieved from https://ru.wikipedia.org/wiki/Персидская_живопись#/media/File:Junaid_001.jpg

Wikipedia. Persidskaya Zhivopis. (n.d.). Retrieved from https://ru.wikipedia.org/wiki/Персидская_живопись#/media/File:Yusef_Zuleykha.jpg

Wikipedia. Tree of Life. (n.d.). Retrieved from https://en.wikipedia.org/wiki/File:Tree_of_life_bahir_Hebrew.svg

Wikipedia. Tsan Syan. (n.d.). Retrieved from https://ru.wikipedia.org/wiki/Цянь_Сюань#/media/File:2_Qian_Xuan_Wang_Xizhi_Watching_Geese._(23,2x92,7cm)_Metropolitan_Museum_of_Art,_N-Y.jpg

Wikipedia Encaustic Virgin. (n.d.). Retrieved from https://en.wikipedia.org/wiki/File:Encaustic_Virgin.jpg

## ENDNOTES

1 The author is aware of the historical and art historical periodizations of the Middle Ages, both geography - and time-based ones. However, this is not a book in art history and periodization is chosen here to best characterize the changes in the human mind.

2. The town of Ravenna used to be a long-time capital of the Gothic kings in Italy; after the defeat in wars with Byzantium under the reign of Justinian, the Ghots withdrew to the Iberian Peninsula, whilst the Byzantines positioned themselves in Ravenna.

3 if universalias are only names, then sets are determined only by their elements – this is a fundamental premise of a classical set theory; if universalias exist, sets are not reduced to their elements – this is a no less fundamental premise of the complex systems theory.

# Chapter 11
# Renaissance:
## Legalization of Feelings


## ABSTRACT

*The influence of the Greco-Roman art and the art of the Middle Ages on depicting space in the pictorial art of the Italian Renaissance is considered. It is shown that the linear perspective, opened by the masters of the Italian Renaissance, not only reproduces the image of a homogeneous and continuous space, but also due to the presence in the picture of a single vanishing point of straight lines perpendicular to the picture plane, creates a feeling of unlimited space. Methods of the depicting space by the Northern Renaissance masters are reconstructed; the differences from the linear perspective are shown and the advantages of using a curved perspective are analyzed. In analyzing the markers of evolutionary changes, the most active channels were obtained. The results are shown in the form of generalized psychological portraits. The behavior patterns of the main estates representatives were reconstructed.*


## BACKGROUND

The Renaissance, bright, close to our time and the shortest of all considered earlier, is very well documented.

There are a large number of memoirs, political and historical works, theological and scientific works written in hot pursuit of various events, often extremely polemical, which allows us to feel the atmosphere of this most interesting period. And this atmosphere encompassed the souls of artists, giving rise to inspiration and desire to create in them, not like their medieval predecessors. There are also treatises that reveal the essence of space depicting, and a considerable part of them is written by the artists themselves. Of course, our contemporaries – art historians, philosophers, historians, psychologists – could not leave the Renaissance without their attention. In their writings, various assessments of these or other events are given, as well as the characteristics of:

- creativity and art (Taylor, 1981; Prokofiev, 1985; Nikulin, 1999; Dirk De Vos, 2002; Eco 2004; Paoletti, Gary, 2011; Sinisgalli, 2011; Summit, 2011);

DOI: 10.4018/978-1-7998-1706-2.ch011

- society (King, 1991; Shlain, 1991; Hause & Maltby, 2001; Patrick, ed. 2007; Ruggiero, 2015; Monfasani, 2016);
- mind and science (Butterfield, 1965; Debus, 1978; Grant, 1994; Field, 1997; Lahr, 1997; Kemp, 1989; 2000; Stephens, 1990; Cattaneo, 2016);
- persons (Lvov, 1971; Znamerovskaya, 1975; Velchinskaya, 1977; Gershenzon-Chegodaeva, 1983; Wilde, 1994; Criminisi, van Mander, 2007);
- and the epoch as a whole (Ferguson, 1951; Losev, 1978; Burke, 1990).

There are also numerous works affecting not only art history, but also the geometric issues of the depicting of space by Renaissance masters (Benesh, 1973; 1983; Wright, 1983; Prokofiev, 1985; Edgerton, 1977; Velchinskaya, 1977; Field, 1997; Grashchenkov, 1988; Kubovy, 1988; Kemp, 1989; 2000; 1999; 2006; Hockney, 2001; Hockney, Falco, 2003; Talbot, 2003; Panofsky, 2004; 2006; Kovalyev & Nitsyn, 2009).

All this makes it possible to better understand which direction the vector of evolutionary changes points to - after all, the tasks of studying the evolution of the mind, as well as behavioral patterns continue to be central even in the magnificence of the Renaissance!

## GENERAL CHARACTERISTICS OF THE EPOCH

Depending on the worldview, the same phenomenon can be perceived by the human mind in completely different ways. Renaissance (Quattrocento) in the liberal-humanistic paradigm is progress, unconditional benefit, the eve of unprecedented prosperity; in religious – apostasy, new paganism, Luciferian riot, the eve of the Apocalypse. The conflict of these assessments, the sources of which can be traced back to the Middle Ages, testified another swing of history – now the ancient deification of human was perceived as the ideal. For a proper understanding of the content of this stage, it is necessary to give it a comprehensive description.

### Renaissance in Italy

Approximately 600-550 years ago, a movement that proclaimed human as the highest value – humanism – was born in Northern Italy (Monfasani, 2016). Later it reached other parts of Europe. The human right to the all-round development of abilities, immediate and unlimited satisfaction of needs, mainly material ones, was now recognized as the integral.

This worldview was the result of a long process of apostasy that covered the Christian world. Let us single out the following symptoms of the spiritual ailment of the Church:

- dogmatic disputes about the filioque, which eventually split the Christian world approx. 950 years ago, and which led to the loss of correct understanding of the hierarchy of the Holy Trinity, created dissatisfaction of spiritual needs and shifted the vector of interests to scholastic exercises;
- already mentioned (Chapter 10) disputes of scholastics about universals, which led, in modern terms, to the set-theoretic description of systems as an additive, led to the loss of understanding of the emergence of complex objects and caused a redirection of scientific search from *systems* to *elements*;

- the struggle of the Palamites and the Barlaamites around the Divine energies: the Barlaamites, who defended the intrinsic value of sciences and arts, were expelled from Byzantium and found shelter in Italy – which testified the final immersion of the local intellectuals into life actualities and made the appearance of humanism almost inevitable.

This was also contributed by other circumstances as well. From the time of the Crusades, spices from India, silk from China, gold ornaments from Damascus, knowledge, and wisdom of the East flowed through the ports of Genoa and Venice to Europe. Consumerism and secret wisdom lured not only the unfortunate Knights Templar, but also the rulers, and many prelates; the aristocracy lost its lands in an attempt to buy up luxury goods; usurers from all parts of Syria and Palestine came in on the shoulders of the crusaders in the towns of Northern Italy.

Humanists have deeply questioned the teaching of Aristotle on the structure of the Universe. It is not entirely clear why it was the teaching of Aristotle about the geocentric structure of the world that fell under the attacks of the humanists. Indeed, if the Earth occupies the central position, and the entire Universe is limited to concentric spheres, human acquires a special significance in the Universe, and all celestial phenomena are evaluated in terms of their usefulness for human activity. Apparently, the fact is that the teaching of Aristotle had been the basis of the cosmogonic views of the Church since the time of St. Augustine (Cattaneo, 2016), and Christianity had been the main subject of criticism of humanists. In fact, the Catholic Church is the last stronghold of medieval ideas about human nature as the source of temptation and the cause of eternal death. The Church members continue to be hypocritical, and who sincerely, like Savonarola, appealed to the parishioners with the preaching of an ascetic way of life, restraint from meeting the needs resulting from the animal nature of human, and glorification of

*Figure 1. Andrea del Verrocchio and Orsino Benintendi. Bust of the ruler of Florence Lorenzo de Medici. ≈540 (?) years ago. National Gallery of Art. Washington. D.C., United States of National Gallery of Art. Washington, D.C., United States of America. No. 1943.4.92. (Academic. Verocchio_Lorenzo de Mrdici, n/d).*

*Figure 2. The preacher Girolamo Savonarola. Monument in Ferrara, Italy. Photographer: Vitold Muratov. (Wikimedia. Savanorola. Rerrara, n/d).*

*Figure 3. Lucas Cranach the Elder. The Judgment of Paris. ≈505 years ago. Kimbell Art Museum. Fort Worth, Texas, United States of America. (Wikimedia. Lucas Cranah. Judment, n/d).*

the divine principle as an earnest of his future salvation (Figures 1-3). These theories contradicted the progressive views that required a human to take everything from life that can be bought for money. If such preaching prevents good people from honestly making money and spending it nicely, then the hell with all such preachers! Savonarola must be executed!

After the break with Christianity, humanists find solace in ancient art, being interested, though, not so much by ancient myths, but by the morals and the cult of beauty of the human body[1].

And here is how the philosopher A.F. Losev described the morals of the Renaissance era (Losev, 1978):

"In the Renaissance, they told fortunes by corpses, conjured public women, made love drinks, summoned demons, performed magical operations when laying buildings, practiced physiognomy and palmistry, threw crucifixes with terrible blasphemy into the sea and buried donkeys in the ground to induce rain during a drought. They believed in large numbers in ghosts, in the evil eye and in general in any kind of bad spell, they believed in black horsemen who allegedly intended to destroy Florence for its sins, but removed by the spells of the righteous; bewitched children, animals, and field fruits. They believed that women copulated with demons and were sorceresses, although sometimes good ones. Prostitutes to attract men used different drugs, which included hair, skulls, ribs, teeth, and eyes of the dead, human skin, baby navel, soles of shoes and pieces of clothing extracted from graves, and even corpse meat from the cemetery, which they could sneakily give their lovers to eat. They pierced figures of wax and ash by famous choruses to influence those who those figures impersonated; they took revenge on the prophets for their predictions. ... When a famous person died, rumors immediately spread that he had been poisoned, and very often these rumors were fully justified. … Executions, murders, exiles, bashing, tortures, conspiracies, arsons, robberies continuously follow each other. ... The clergymen run butcher shops, taverns, gambling houses, and brothels, so there was a repeated need to issue decrees prohibiting priests from "becoming panders of prostitutes for the sake of money", but all in vain. Monks read the Decameron and indulge in orgies, and in the dirty drains, children's skeletons are found as consequences of these orgies. Writers of the time compared monasteries either with bandit dens or with whorehouses. Thousands of monks and nuns live outside the monastic walls. In Como, due to strife, real battles take place between Franciscan monks and nuns, the latter bravely resisting the attacks of armed monks. People get drunk and feast in churches; in front of miraculous icons images of genitals are hung healed by these icons". That was when that the first time the Catholic Church recognized the existence of witches. The Bull of Innocent VIII Summis desiderantes (1484) consecrated new times with fire of inquisitorial bonfires; the search for witches, sorcerers, heretics became the duty of the clergy, and in their help, Malleus Maleficarum – methodic guidelines for identifying, arresting and interrogating witches and sorcerers was published. It is estimated that since then over 150 years, 30,000 witches have been burned in Spain, Italy, and Germany.

And, of course, numerous wars complement the picture, sometimes giving rise to very colorful characters, such as condottiero Bartolomeo Colleleone (Figure 4) or Sigismondo Pandolfo Malatesta, condottiero and lord of Rimini, Fano, and Cesena.

And the dance of the dead – dance macabre – a reminder of the end of life, which many would like to forget (Figure 5)…

So, the Church, meant to ensure the salvation of the human soul and its life in eternity, inevitably came into conflict with the spirit of the times, found itself in an extremely weakened state of spiritual crisis and disarmament, organizational confusion (struggle against Albigensians, Avignon Captivity of the Papacy), moral degradation (disgraceful behavior of the clergy, trade-in indulgences) and was unable to respond to the historical challenge. In fact, it could either go with the flow, preserving the structure,

*Figure 4. Andrea del Verrocchio. The equestrian statue of the Condottiero Bartolomeo Colleleone (c. 530 years ago). Photo by: @annaart. (Golos. Andrea del Verocchio, n/d).*

*Figure 5. Illustrations from the Nuremberg Chronicle, by Hartmann Schedel (1440-1514) (Wikipedia. Nurenberg. Dance of Death, n/d).*

hypocritically condemning hedonistic aspirations, but accepting them internally, or be reconstructed from the outside, with breaking the structure under the slogan of reviving the original purity of faith. The first option became predominant in Italy, the second – in Northern Europe (the Reformation). Accordingly, the keynote of public sentiment in the south was the desire to *take everything from life* covered with reminiscences on the themes of Antiquity, while in the north the state of crisis was much more acute, and the struggle was much more fundamental and cruel. It's time to consider the situation in the North.

## Renaissance in Northern Europe

In the struggle with the degrading Catholic Church, thinkers in Northern Europe sought support in what was closer to them, namely, in the teachings of the early Christians about active goodness, moderation in satisfying carnal needs, and humility of pride (Patrick, ed. 2007). An example of such aspirations was the Modern Devotion movement founded by Geert Groote (c. 670-630 years ago), a native of the Dutch city of Deventer. It was popularized by Thomas à Kempis, the content of whose book *The Imitation of Christ* was precisely composed by the preaching of Geert Groote.

A special place in the system of spiritual values of the Modern Devotion was occupied by the works of St. Augustine, whose teaching on the attitude of man to God's grace penetrated all the Protestant movements in Northern Europe, for example, the movements led by Martin Luther and Jean Calvin. Therefore, the view of St. Augustine that the earthly space is enclosed within the sphere of fixed stars, in the middle of which the Earth is located, was the immutable truth for most of the learned people of Northern Europe, i.e. Aristotle's philosophy and his geocentric system of the world remained the ideological prerequisite for the formation of the picture space of the painters of the Northern Renaissance.

A devotee of Geert Groote was Nicholas of Cusa, who in his youth had the lot to be a graduate of the Modern Devotion school in Deventer. Like the teachings of Geert Groote, the philosophy of Nicholas of Cusa grows on the basis of pantheism and expresses a worldview imbued with reverent worship of beauty and the infinite diversity of the earthly world. "Every creation of God is perfect, even if compared to others, it seems less perfect. ...God, who is one, resides in a single Universe. The Universe is in a universal aggregate of things, defining in it. In this way, we understand that God, existing in the single Universe, dwells in all things" writes Nicholas of Cusa in his essay On Learned Ignorance. By the way, he also gave rise to return to the philosophical turn of the famous thesis "all in all" – famous for the fact that many people know these words and love to show off them, but few know what they mean...

The philosophy of Nicholas of Cusa was largely consonant with the mindset that reigned in the Dutch schools of painting, thus having a mediated effect on the visual art of Northern Europe. Despite the fact that pictorial art of the Netherlands bore the imprint of a deep religious feeling, ideas about the divine world disclosed with the help of pictures of routine life, determined the sincere immediacy, novelty and heartfelt sharpness of perception of the earthly world. Artists praised its beauty, seeing in all phenomena of reality a manifestation of the Divine and divine order. Whatever was depicted in the paintings of the masters of the Northern Renaissance: whether it is a person, a household item, or part of the plant and animal world, everything was done with the utmost care, scrupulousness and love for the smallest details. At the same time, each written detail occupied the only possible place in the picture, and their totality did not in the least destroy the overall composition and only contributed to disclosure of the artist's intention, as if turning the paintings into illustrations of one of the provisions of the treatise On Learned Ignorance: "Unity of things or the Universe in the set and back: the set in the Unity".

Works of the masters of the Northern Renaissance coincided with the era of great geographical discoveries, the most prominent events of which include the discovery of the American continent by Christopher Columbus, the discovery of the sea route to India by Vasco da Gama, the first world tour of Fernando Magellan 520-490 years ago. These discoveries for the first time gave incontestable evidence of the teachings of ancient Greek thinkers Pythagoras and Aristotle that the Earth has a spherical shape.

Great geographical discoveries set for the cartographers of Western Europe the task of describing the surface of the globe, and approx. 500 years ago, Abraham Ortelius compiled the first atlas Theatrum Orbis Terrarum (van den Broecke, 1996), in which an attempt was made to give a true picture of the unity of the globe. This means that if the earth's surface is continuous, then its image on the plane should be just as continuous, and the inevitable distortions should be made with uniform changes.

However, the era in which the artists of the Northern Renaissance worked was marked not only by geographical discoveries: while the first European explorers of the Earth discovered new countries, Nicolaus Copernicus accomplished a feat that consisted in the scientific substantiation of the teachings of the ancient Greek astronomer Aristarchus of Samos that it was not the Sun with the planets that moved around the Earth, but the Earth, like other planets, revolves around the Sun. This expanded the understanding of European scientists about the structure of the Universe, pushing its boundaries beyond the celestial firmament, which, according to the teachings of Ptolemy, are lined with stars and the Sun.

Such was the scientific achievements, which made a revolution in the artistic worldview of the masters of the Northern Renaissance. The image of spheroidal space no longer reminds of the optical laws of reflection in the convex surface of a spherical mirror. Now the image of the sphere of the master of the Northern Renaissance is associated with the idea of the cosmic abode of humanity, the great bowl of earthly life, a repository of various human passions.

The influence of the teachings of Nicholas of Cusa also manifests itself in the form of the picture space chosen by the masters of the Northern Renaissance. Indeed, if the will of God is such that a person perceives earthly space as the inner space of the sphere of fixed stars, then the pious feeling that the artist experienced to every creation of the Lord God attracted him to show that in the picture the earthly space was curvilinear and closed inside the heavenly sphere.

This representation most deeply penetrated the landscape pictorial art of the masters of the Northern Renaissance. Pictures of wide spatial distances, which open from the top of a high hill, inspired the artists of the Northern Renaissance to develop images of curvilinear circular spaces, while urban architectural environment became for the masters of the Italian Renaissance the area of application of strict linear perspective rules that reproduce the image of an infinite straight space.

Thus, the spiritual basis of space depicting in the pictorial art of the Italian Renaissance was the teaching of Aristarchus of Samos on the heliocentric device – the Universe, which is closely linked with the idea of the infinity of world space.

On the contrary, the ideological prerequisite of shaping the picture space used by the painters of the Northern Renaissance was the philosophy of Aristotle and his geocentric system of the world, which contains a statement about the infinity of world space (Cattaneo, 2016). If the will of God is such that a person perceives earthly space as the inner space of the sphere of fixed stars, then the reverent feeling that the artist experienced to every creation of the Lord God attracted him to show the earthly space curved and closed inside the celestial sphere in the picture. It is known from psychology that direct visual impressions can be recognized and interpreted only in a certain context, whether it be inborn gestalt or acquired ideological views.

The Renaissance ended about 420 years ago.

## ANALYSIS OF SPACE IMAGES: ITALY

Let`s note that the discovery of the linear perspective coincided with the publication of the work by Aristarchus of Samos "On the Sizes and Distances of the Sun and Moon "where, among other things, trigonometry issues are discussed. In addition, it got widespread "Perspective" by Vitelo, which is the translation of Ibn al-Haytham's work "The Book of Optics " (Hockney Falco, 2003).

It can be assumed that one of the prerequisites for creating a linear perspective is the idea of the infinity of space put forward by Aristarch of Samos.

The second prerequisite is the statement of Ibn al-Haytham that each point of space emits light rays, which after reflection and refraction on the surfaces of objects fall into the eyes of man.

Probably, the combination of these ideas prompted Filippo Brunelleschi to the ingenious conjecture that the image of the object in the picture can be represented as the result of its intersection with the visual cone.

So, the replacement of the humanitarian paradigm led to complete rejection of the medieval canon of space depiction. This was also manifested in the shift of interest from God to a human in all its manifestations, for which space was just a diluted background, for the depiction of which the simple linear perspective was exactly useful, despite all its distortions.

The Renaissance gave many prominent personalities, whose works and ideas were included in the golden fund of the theory of images. Let's name only a few:

- Filippo Brunelleschi, architect, and sculptor discovered a linear perspective;
- Pietro Della Francesco del Borgo, an artist, determined the apparatus and method of constructing the perspective;
- Leonardo da Vinci, an artist, and scientist improved the technique of drawing, coloring, composition, light and shadow effects, building an image on cylindrical surfaces;
- Albrecht Dürer, an artist, described perspective images using a plan and a facade, used the "perspective grid" method;
- Daniele Barbaro, a mathematician, and philanthropist created a compendium of methods for the perspective representation of three-dimensional bodies and their shadows;
- Guido Ubaldo del Monte, an aristocrat, mathematician, described the "way of architects", the perspective on curvilinear surfaces, the reconstruction of figures from perspective images.

An indicative sign of consciousness of people of the Renaissance is the emergence of dozens of treatises on imaging. These treatises are very unlike modern books on descriptive geometry; they do not have graphic illustrations. Here is a quote from the book of Leon Batista Alberti (1404-1472) *On Architecture* which describes the construction of perspective:

"First, where I have to draw a picture, I draw a quadrangle with right angles and take it as an open window. Inside this quadrangle, where it is convenient for me, I set a point that occupies the place where the central ray hits, and therefore I call this point the central one." And so on.

Let us consider the paintings of Italian artists (Figures 6-12).

Areas of space that the linear perspective recreates with distortion are hidden. The desire to hide flaws leads to a contradiction: the picture space looks limited and does not go beyond the limits delineated by the window frame, while the linear perspective is based on the assumption of the unlimited length of the depicted space.

*Figure 6. Raphael Santi. School of Athens. ≈510 years ago. Rome, Vatican (Wikimedia. Raffael-058, n/d).*

*Figure 7. Leonardo da Vinci. The Last Supper (approx. 515 years ago). Santa Maria Delle Grazie, dining room, Milan, Italy (Google. Leonardo. Tajnaya-vecera, n/d).*

*Figure 8. Leonardo da Vinci. Madonna Litta. ≈530 years ago. The Hermitage, ГЭ-249 St. Petersburg, Russia. (Wikimedia. Leonardo. Madonna Litta, n/d).*

*Figure 9. Sandro Botticelli. Portrait of a young woman (Simonetta Vespucci?) ≈530 years ago. Berlin Art Gallery. Germany. (Wikimedia. Botticelli. Simonetta, n/d).*

Nature is needed to create theatrical scenery in which human drama is played out. Depth of space is often transmitted by aerial perspective.

The landscape is depicted as emotionally diluted and conventional. It is the image of an ideal, artificial nature, it is created under no vivid impression and is not warmed by a sense of belonging to the world around it; it plays the role of a background on which the most subtle shades of spiritual experiences are shown.

Now let's carry out the reconstruction of the main elements of the construction of perspective, taking as an example the fresco The Trinity by Masaccio (Znamerovskaya, 1975).

Tommaso di Giovanni di Simone Cassai, nicknamed Masaccio (612-585 years ago), decorated the image of a gravestone not with a figure of a nobleman or a maiden with a mournful face, but an ugly human skeleton. However, the image of human remains is not the only thing that can surprise the viewer of the fresco The Trinity (Florence, Church of Santa Maria Novella. Approx. 590 years ago): not a single work on art history devoted to pictorial art of Quattrocento, can do without mentioning this amazing phenomenon in the world of art. Indeed, the fresco is the first picture work in which the correct construction of perspective with a single vanishing point of images of straight lines rushing into the depth of picture space in the same direction is given. It is possible to add to this that the plane of the slab on which the figures of the donators are leaned against is located at the level of the artist's eyes and is depicted as a straight line coinciding with the horizon line.

Let`s note that the technique consisting in combining the point of view with the object plane was not used by the Masaccio predecessors and is almost never found in the works of his contemporaries. Indeed, the monuments of the Quattrocento, in which the horizon line coincides with the image of the object plane, are few: the fresco The Family of Marquis Ludovico Gonzaga by Andrea Mantegna, the fresco Founding the Vatican Library by Melozzo da Forli and that's all. Why did Masaccio present the object plane as a straight line passing through the main point of the picture and coinciding with the horizon line? Maybe to hide from the viewer the distance from the figures of the crucified Christ, God the Father, Our Lady and John the Apostle to the artist? Indeed, despite the fact that the perspective of the chapel creates the impression of the depth of space, the figures appear to belong to the plane aligned with its facade.

Usually, in order to find the distances from the points of the object plane to the base of the picture, the depth scale is used, however, it cannot be applied for The Trinity, because the object plane on it degenerates into a straight line. Is there any other way by which it can be found out what position in space is occupied by the figures depicted in the depths of the chapel? Before answering this question, it is necessary to find out which central projection apparatus Masaccio used. At the same time, the task of reconstructing the apparatus is adjacent to one of the most important tasks in realistic visual art – checking the correctness of images made in sketches, drawings by presentation and pictures. Thus, the development of methods for checking perspective patterns on canvases, in which the height of the eye view is zero and the object plane coincides with the plane passing through the eye view, is an urgent task in art history and the artist's practice.

For this reason, the spatial constructions of frescoes have long attracted the attention of art historians. For example, V. N. Grashchenkov argues that in the fresco by Masaccio, "... solved as a monumental altar image, it is not difficult to notice one structural contradiction. The architecture is built according to all the rules of perspective reduction of orthogonal lines. ...It gives the full visual effect of depths of the depicted space. The perspective of the fresco is constructed in such a way that the horizon line, located at the level of our view, coincides with the plane of the slab on which the kneeling donators are located.

*Figure 10. Piero Della Francesca. Portrait of Battista Sforza and Federico da Montefeltro. Diptych. ≈550 years ago. Galleria degli Uffizi. Florence, Italy The Yorck Project (2002) (Wikimedia. Pierro della Francesko, n/d).*

*Figure 11. Leonardo da Vinci. Mona Lisa. ≈515 years ago. Department of Paintings of the Louvre. INV. 779. Author: C2RMF: Galerie de tableaux en très haute définition. (Wikimedia. Leonardo. Mona Lisa, n/d).*

*Figure 12. Masaccio. Holy Trinity. Fresco. ≈590 years ago. Santa Maria Novella, Florence, Italy. From the book: John T. Spike, Masaccio, Rizzoli Libri illustrati, Milano 2002 (Wikimedia. Massccio. Trinitia, 2002).*

...This simple technique makes a difficult illusory understanding of the distance between the figures. They seem close together. ...Ultimately, the compositional unity of the figures forms a semblance of a spatial pyramid, which tends to the front plane of the image" (Grashchenkov, 1988). Moreover, there are works in which schemes of building a perspective on a fresco are given, but, unfortunately, definition of its main elements is limited to establishing obvious facts, for example, that "the arch lines solved in the Renaissance perspective system, ...converge at the base of the cross on which Christ is crucified." (Zamerovskaya, 1975). Let us assume that the central projection apparatus is set if, according to the sketch, drawing, and picture, it is possible to determine the main elements necessary for solving the problem of building a perspective of a space point. These elements include the main point of the picture, the horizon line, the distance from the center of projection to the picture plane and the bottom of the picture.

Let us do the restoration of the main elements of the construction of the perspective on the fresco The Trinity. Figure 12 shows the fresco, and Figure13 – some constructions. Let us construct the main point of the picture with the help of images of straight lines, which are located at right angles to the picture plane in the picture space. For example, let us draw straight lines through the capitals of the columns limiting the space of the Capella, and at the point of their intersection is the position of the main point of the picture *P*. Let us draw a straight line through the main point of the picture *P* parallel to any horizontal line belonging to one of the architectural forms *h*. Let us combine the basis of the picture *k* with the lower border of the fresco, despite the fact that the object plane is defined as the plane on which the depicted objects are located. This was done because the lower boundary of the fresco *The Trinity* concerns the floor of the church of Santa Maria Novella, and according to the same definition, the object plane is also the plane on which the viewer stands. Let us show on Figure 13 constructions with the help of which the main point of the painting *P*, the horizon line and the base of the painting *k* on the fresco were obtained.

*Figure 13. Obtaining the main point of the picture, the horizon line and the bottom of the picture*

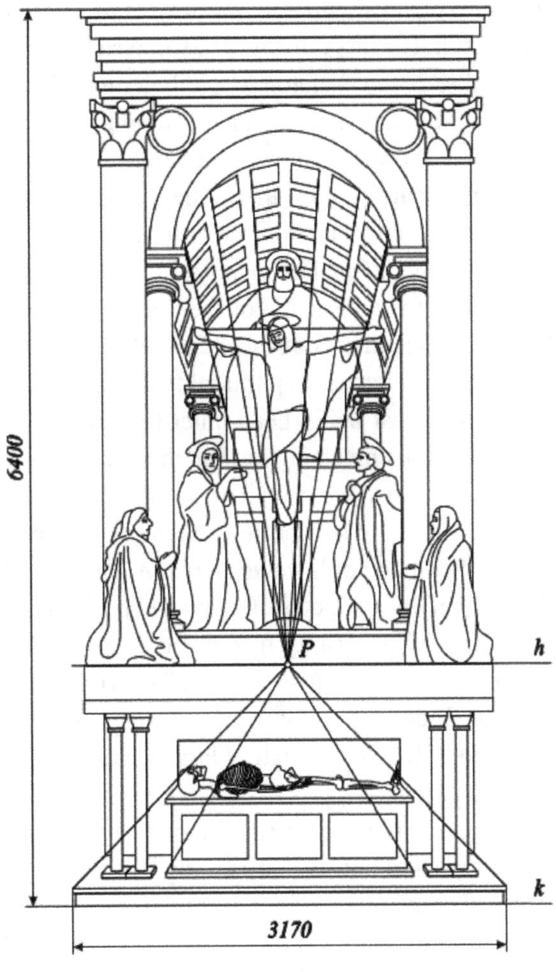

*Figure 14. Scaling the image*

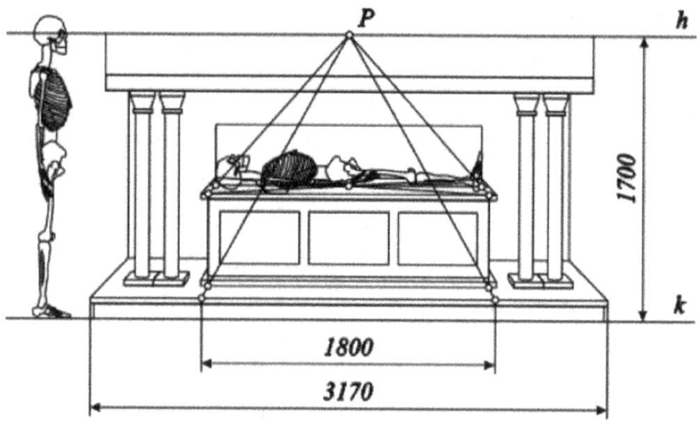


Let us define the scale of the fresco (Figure 14). According to (Nitsyn, 2006), its width is 3.17 m, and its height is 6.4 m. Let us calculate the height of a person who could own a skeleton lying on a sarcophagus lid. Let us take out on the picture plane a segment of a straight line with the length corresponding to the height of the image of the skeleton. Let us perform the geometric constructions shown in Figure 14, and if the fresco was painted in full size, the skeleton would belong to a man whose height is 180 cm.

Let`s position the skeleton vertically and find that the horizon line passes through its eye pits. This indicates that the height of the skeleton corresponds to the height of the artist who painted the fresco. However, the results of anthropological studies show that at the end of 600 years ago, the average height of a person was approximately 165 cm. What explanation can be given to such a discrepancy? First, suppose that the dimensions of the images of objects in The Trinity fresco make up 109% of their actual sizes. Secondly, if you carefully examine the fresco Raising of the Son of Theophilus and St Peter Enthroned (Florence, Church of Santa Maria del Carmine, Brancacci Chapel, about 580 years ago), then in its right corner you can see a young man who fell into a deep thoughtfulness and gazing through the viewer.

Let`s note that the young man is almost half a head taller than the fresco characters standing with him in the background, and his head looks much larger than the heads of the characters standing in the foreground. This suggests that the young man was a man of strong build and tall. Many art critics suggest that the image of the young man is Masaccio self-portrait. It can be assumed that on the Trinity fresco Masaccio also depicted himself in the image of his own skeleton. Perhaps Masaccio had a premonition that he would soon be killed? Note that the attributes of death can be found in many works by Masaccio.

What kind of explanation is true: 1) The Trinity fresco was painted on a slightly larger scale, or 2) Masaccio painted his own skeleton in full size? It seems that The Trinity was written to create a visual illusion of the chapel, i.e. real size. Moreover, why when creating works of monumental pictorial art to choose the scale at which the size of images is slightly larger than the size of objects? Therefore, the second assumption can be assumed as true.

Let`s find a distance point (a vanishing point of images of straight lines lying in horizontal planes and inclined to the picture plane at an angle of 45°), assuming that the cylindrical arch of the chapel is lined with square-shaped caissons. Indeed, the Corinthian capitals and pilasters framing the portal, and composite orders of supports supporting the arch of the chapel indicate that Masaccio used the traditions of classical architecture in the development of architectural forms on the Trinity fresco. It is believed that the architectural background in The Trinity fresco was designed by Filippo Brunelleschi, who revived classical architecture or was inspired by him (Zamerovskaya, 1975). This is evidenced by the fact that Masaccio in The Trinity fresco reflected the technique first introduced into the architecture by Filippo Brunelleschi: the caissons of the cylindrical arch form an even number of rows, while in the classical cylindrical arch the number of rows formed by caissons is always odd.

Let`s draw straight lines through the perspectives of points that correspond to the vertices of the caissons located in the same row, and we find that the constructed straight lines converge at a point taken on the horizon line. Since the diagonals of the caissons are inclined to the base of the picture at an angle of 45°, the point of intersection of their perspectives can be viewed as a distance point *D*. Let us show the construction of the distance point *D* in The Trinity fresco.

When constructing, it was taken into account that the plane to which the selected row of caissons belongs is parallel to the main line of sight and tilted to the object plane. Therefore, the perspectives of straight lines having the same direction and lying in different parallel planes converge to a single point lying on the picture plane trace passing through the eye view and parallel to the planes in which the depicted straight lines are. In this case, the angle of slope of the picture plane trace parallel to the plane in which the caissons lie and passing through the main point of the picture *P* is determined by the angle of slope of the chord, which tightens the arc of the section of the cylindrical surface parallel to the picture of the plane. Let us rotate the picture trace plane parallel to the plane in which the caissons lie and passing through *P* to align with the horizon line *h* and obtain the distance point *D*. Since the images of objects on the fresco are made in full size, the length of the segment connecting point *P* and point *D*, i.e. the distance from the eye point to the picture plane is approximately 6 m 6 (Figure 15).

*Figure 15. Constructing the distance point D in The Trinity fresco*

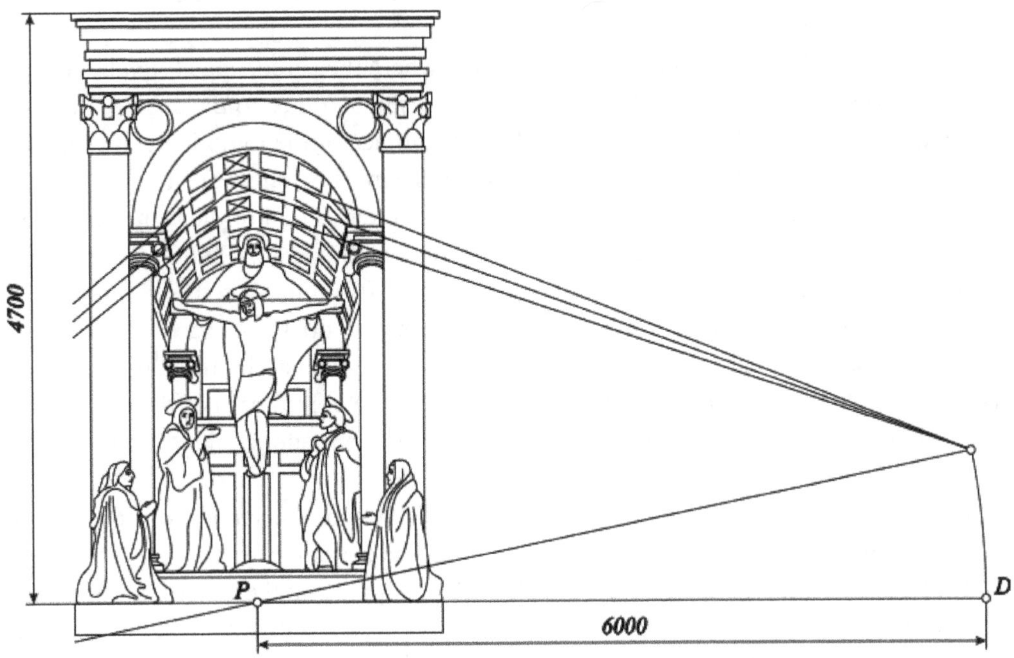

*Figure 16. Reconstructing the chapel in The Trinity fresco*

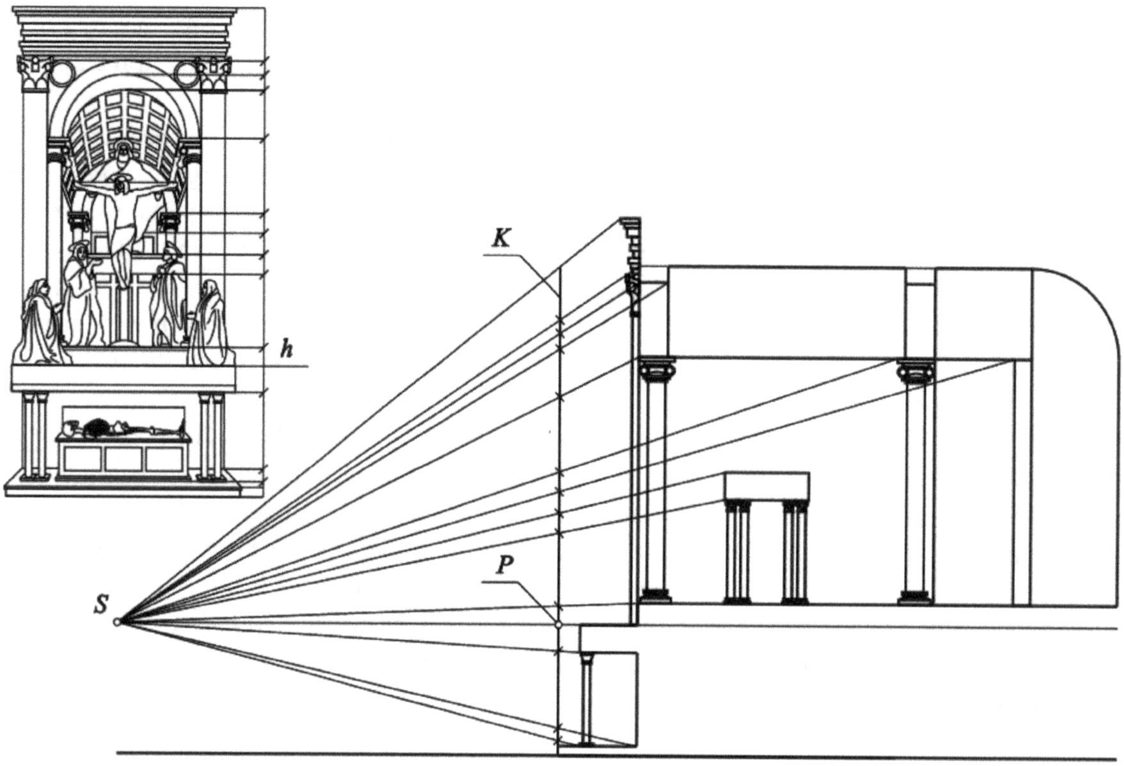


Let`s perform a reconstruction of the chapel depicted on the fresco using the found key elements of the perspective (Figure 16).

Suppose that the plane of the back wall of the niche, where the sarcophagus stands, coincides with the plane of the front wall of the portal erected in front of the entrance to the chapel. Let us combine the picture plane *K* with the plane passing through the front face of the tomb foot, and find the natural size of the tomb foot height equal to the height of its image in The Trinity fresco. Let us perform the geometric constructions shown in Figure 16, and determine the position of the front wall of the chapel portal, as well as the height and dimensions of the slab, on which donators, full of dignity, knee.

Let`s note that if the distance from the eye view *S* to point *P* is 6 m, then the width of the slab with kneeling donators corresponds to 0.77 m. A person whose height does not exceed 170 cm has a width between shoulders equal to approximately 0.4 m, and, therefore, he can be fitted on the slab almost 2 times wider the width between his shoulders. This confirms the correctness of the conjecture that the plane in which the back wall of the tomb lies coincides with the plane passing through the front wall of the portal.

Then, complete the reconstruction of the chapel with the help of the same geometric constructions, by means of which the restoration of the structure of the tomb was performed. At the same time, the height of the pulpit, on which the figure of God the Father supporting the crucified Christ stands, was chosen equal to the height of the eye view and is 1.7 m. This choice provides enough space to place the figures of Our Lady and John the Apostle between the pulpit and the columns installed at the entrance in the chapel and does not contradict the fresco, on which the height of the pulpit is commensurate with the height of the figure of John the Apostle. As Figure 16 shows, the result of the reconstruction of the chapel in The Trinity fresco has the appearance of the church chapel built in classical traditions, namely: the ratio of the chapel height measured by the distance from the floor to the essay of the cylindrical arch to its depth measured by the distance from the front wall of the portal to the back wall of the apse is expressed by the number 0.6816. Since this proportion is in sufficient proximity to a number called the "golden section", it can be concluded that point *D* is correctly defined in The Trinity fresco.

Let`s note that the way of reconstruction of the chapel presented in Figure 10.19 exactly repeats the method of constructing the perspective discovered by Filippo Brunelleschi and presented by Leon Battista Alberti in his treatise "On Painting" (c. 580 years ago) (Sinisgalli, 2011).

Now, the problem of restoring the main elements of the construction of the perspective and reconstructing the architecture depicted in The Trinity fresco has been solved. It was determined that the distance from point *P* to point *D* in The Trinity fresco is 6 m. Despite the fact that during the experiment to determine the dependence of the visible size of the object on its distance from the observer, it was found that in a linear perspective reflecting visual impression from contemplation of nature, the distance from the eye view to the picture plane should be equal to 8 m, the result is quite satisfactory.

Now consider the choice of the eye view and the picture plane in the central projection apparatus.

The legend says that Filippo Brunelleschi was the first to express the brilliant idea that a perspective or a view through a window can be thought of as a section of a "visual pyramid" consisting of straight lines drawn from the eye view to the points of the object being depicted with plane perpendicular to the direction of the look. It follows that the human eye is nothing else but the eye point or the top of the "visual pyramid", and the picture plane is a transparent canvas through which the artist views the objects to be depicted. In this case, the statement that the human eye is the center of projection is given without any evidence and is implied as a matter of course. Furthermore, the model of human vision proposed by Filippo Brunelleschi does not specify how far away from the human eye the picture plane should be removed and at what distance from the picture plane the object to be depicted should be placed. Apparently, Filippo Brunelleschi believed that the sections of the "visual pyramid" by planes separated from the human eye at different distances differ from each other only in the scale of the image. However, the statement that if the "visual pyramid" is cut by the plane, regardless of its position in relation to the human eye, there will always be a drawing close to visual perception in section, is also given without any evidence and is implied as a matter of course. Thus, despite the fact that the discovery of Filippo Brunelleschi received universal recognition among Renaissance artists and for more than 500 years has been the theoretical basis for creating drawings close to visual perception, the choice of the eye view and position of the picture plane remains not fully solved tasks in perspective theory.

After Filippo Brunelleschi, many generations of scholars worked on generalizing experience in building perspectives accumulated by artists and architects, and improving the central projection apparatus. Among the most famous works on perspective is the treatises by Leon Battista Alberti "On Painting" (c. 580 years ago), Pietro Della Francesco del Borgo "On the Perspective in Painting" (c. 530 years ago) and Albrecht Dürer "Instruction on Measurement" (about. 490 years ago). Thanks to them and the work of their followers, recommendations were developed on choosing the distance from the projection center to the object to be depicted, according to which the position of the projection center is chosen so that the object is visible from the projection center at an angle of no more than 40°, and the main point of the picture deviates from the middle images of an object no more than 1/3 of its width. These restrictions are introduced in order to avoid perspective distortions that are observed in the peripheral areas of the image.

Note that the central projection apparatus proposed by Brunelleschi reflects the Euclidean representation known since ancient times, according to which the visual rays come from the human eye, and their points of intersection with the surface of the object are made visible. This explains why in the model of visual perception proposed by Brunelleschi; the projection center is combined with the human eye and is called the eye view. Nevertheless, is it possible to create a drawing close to visual perception using the central projection apparatus, in which the eye view coincides with the center of projection? In our opinion, if the perspective is an image on the picture plane which can be likened to the image formed on the retina, it is natural to assume that in the central projection apparatus the eye view should be chosen at the point belonging to the picture plane.

As mentioned above, in the human vision model proposed by Filippo Brunelleschi, the choice of the position of the picture plane in relation to the projection center can be arbitrary. It can be assumed that if construct the central projections of the object located at different distances from the picture plane, then the obtained perspectives will differ from each other only by the scale of the image. However, in our opinion, this is an erroneous opinion. First, if construct the perspectives of the object for different distances from it to the picture plane, then their perspectives, that is, the perspective reductions of the object due to different removal of its parts, will differ from each other. Second, when contemplating perspectives built for different distances from the center of projection to the picture plane, the viewer will get the impression that the object being depicted is at different distances from it. This is explained by the fact that the perspective of the straight line parallel to the picture plane and remote from it at a distance equal to the distance from the projection center to the picture plane is located in the middle of the height of the horizon line above the trace of the picture plane. Thus, it is necessary to answer the following questions: where is the eye point – at the center of projection or at the point belonging to the picture plane, and at what distance from the picture plane should the projection center be chosen so that the perspective of the object is close to its visual perception.

The solution to the problem of choosing the distance from the eye point to the picture plane was found during an experiment to determine the dependence of the visible size of the object on its distance from the observer. The experiment was that the experimenter asked the assistant with a ruler in his hands to move away from him at a specified distance and compared the visible size of the ruler with his measuring tool. So, the experimenter noticed that when the assistant had moved from the experimenter at a distance of approximately 8 m, he had the impression that the assistant was standing in the middle of the space stretching from a straight line passing through the traces of his feet to the horizon line. Recall that through the perspective of the object plane point, a horizontal straight line can be drawn dividing the height of the horizon above the picture plane trace into two equal parts, if the distance from the subject plane point to the picture plane is equal to the distance from the eye view to the picture plane. Consequently, in the central projection apparatus, with which you can create a perspective close to the visual impression of contemplation of nature, the distance from the projection center to the picture plane should be approximately 8 m. At the same time, the experimental results show that the selected value of the distance from the projection center to the picture plane is in sufficient proximity to the confidence interval of 8.05...12.05 m for its mathematical expectation calculated with a confidence probability of 0.95 (Nitsyn, 2009). Imagine on Figure 17 the results of the experiment to determine the distance from the center of projection to the picture plane, in the form of a histogram of distances to the object plane point, which, according to the assistant, is located in the middle of the height of the horizon line above the straight line passing through the traces of his legs.

*Figure 17. The histogram of distances to the object plane point, which seems to be located in the middle of the height of the horizon line*

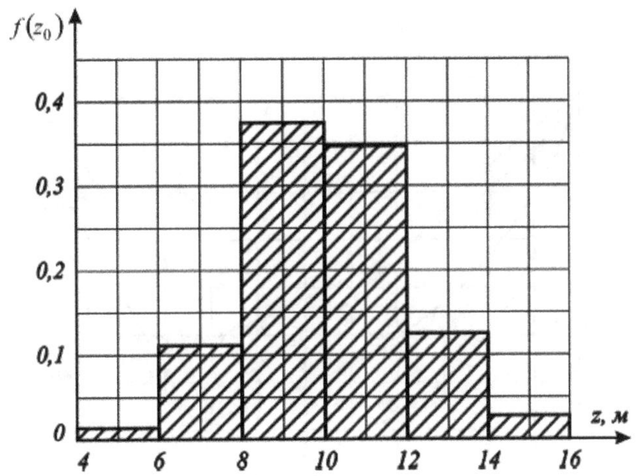


Let`s check the correctness of the choice of the distance from the center of the projection to the picture plane. Imagine in Figure 18 the results of the experiment to measure the visible size of the object. The essence of the experiment was to determine the visible size of the object by measuring the angle of divergence between visual images of lines, which in reality are parallel (Nitsyn, 1999).

Now compare the results of the experiment with the graphs of the dependence of the central projection of the object on the distance between the projection center and the picture plane: a) a graph obtained for the distance from the projection center to the picture plane equal to 4 m; b) a graph obtained for the distance from the projection center to the picture plane equal to 8 m; c) a graph obtained for the distance from the center of projection to the picture plane equal to 16 m. The comparison shows that the graph of the central projection of the object from its distance to the observer corresponding to the distance from the center of the projection to the picture plane equal to 8 m, is sufficiently close to the curve line that approximates experimental data.

Consider Figure 19, which presents experimental data on the speed of change in the visible size of the object, which was obtained by processing the experimental results (Nitsyn, 1999). Let us compare the experimental data with graphs of the dependence of the speed of change of the central projection of the object on the distance between the projection center and the picture plane: a) a graph obtained for the distance from the projection center to the picture plane equal to 4 m; b) a graph obtained for the distance from the projection center to the picture plane, equal to 8 m; c) a graph obtained for the distance from the center of projection to the picture plane equal to 16 m. The comparison shows that the graph of the speed with which the central projection of the object changes from its distance to the observer, for the distance from the center of projection to the picture plane equal to 8 m is sufficiently close to the curve approximating the experimental data.

Therefore, in order to create the perspective close to visual perception from the contemplation of nature, it is necessary to remove in the central projection apparatus the center of projection from the picture plane at a distance equal to 8 m.

*Figure 18. The results of the experiment to measure the visible size of the object and the graphs of the dependence of the projection of the object on the distance between the center of projection and the picture plane: a) equal to 4 m; b) equal to 8 m; c) equal to 16 m*

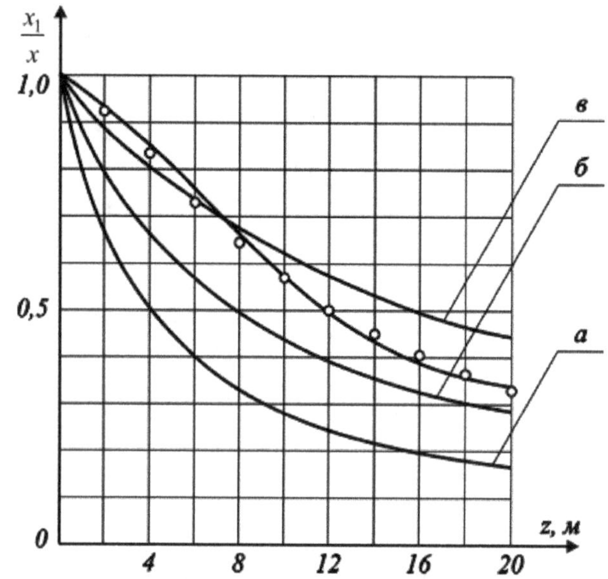


*Figure 19. The results of the experiment to assess the speed of change in the visible size of the object and the graphs of the speed of the perspective contractions on the distance between the projection center and the picture plane: a) equal to 4 m; b) equal to 8 m; c) equal to 16 m*

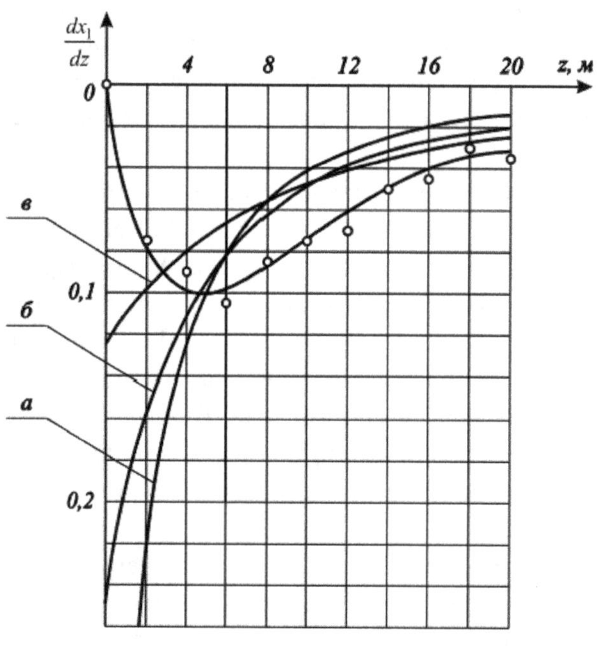

*Figure 20. The perspective of the object when the eye view is located in the center of the projection S*

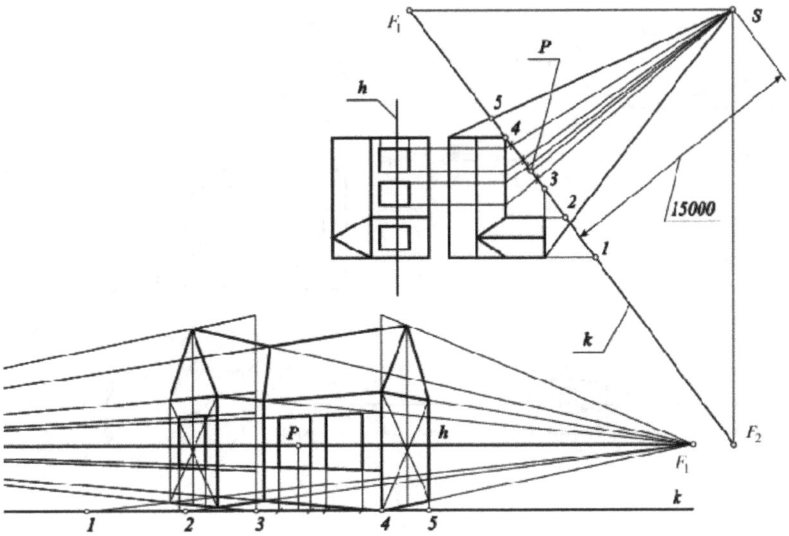


Let us solve the problem of choosing the eye view. Let us construct the perspective of the geometric object, the orthogonal projections of which are shown in Figure 20. Combine the eye of the observer with the projection center *S* and consider it as the eye view. Arrange the geometric object so that the sides of the angle of the eye view with the top in the center of the projection *S* and equal to 28°4`, are tangent to the outline of its horizontal projection. Perform the central projection of the geometric object by the way of vanishing points of perspectives of straight lines parallel to the main directions, or "by the method of architects". Note that the picture plane is drawn through the edge of the geometric object that is closest to the eye view, and the distance from the eye view to the geometric object is approximately 15 m. Let us show it on Figure 20 the perspective of the geometric object built at the location of the eye view in the center of projection *S*.

Then construct the perspective of the same geometrical object under the assumption that the eye of the observer is at the main point of the picture *P* and consider it as the eye view. Let us arrange the geometric object so that the sides of the view angle, with the top at the main point of the picture *P* and equal to 28°4`, are tangent to the sketch of its horizontal projection. Note that the distance from the eye view to the edge of the geometric object that is closest to the eye view is also equal to approximately 15 m. In order to create a correct idea of the depth of space when contemplating the perspective, let us locate the center of projection at a point 8 meters from the picture plane. Perform the central projection of the geometric object using scales of heights and depths. Let us show on Figure 21 perspective of the geometric object built at the location of the eye view at the main point of the picture.

Let us imagine on Figure 22 perspectives of the same geometric object: a) when the eye view is located in the center of projection *S*; b) at the location of the eye view at the main point of the picture *P*. What perspective creates more correct idea of the depth of space, if both perspectives show geometrical objects that are 15 meters from the eye view at the same distance? In our opinion, the perspective built at the location of the eye view at the main point of the picture *P*, and the center of projection *S* – at a point 8 meters from the picture plane corresponds to visual impression from contemplation of the geo-

*Figure 21. The perspective of the object when the eye view is located at the main point of the picture P*

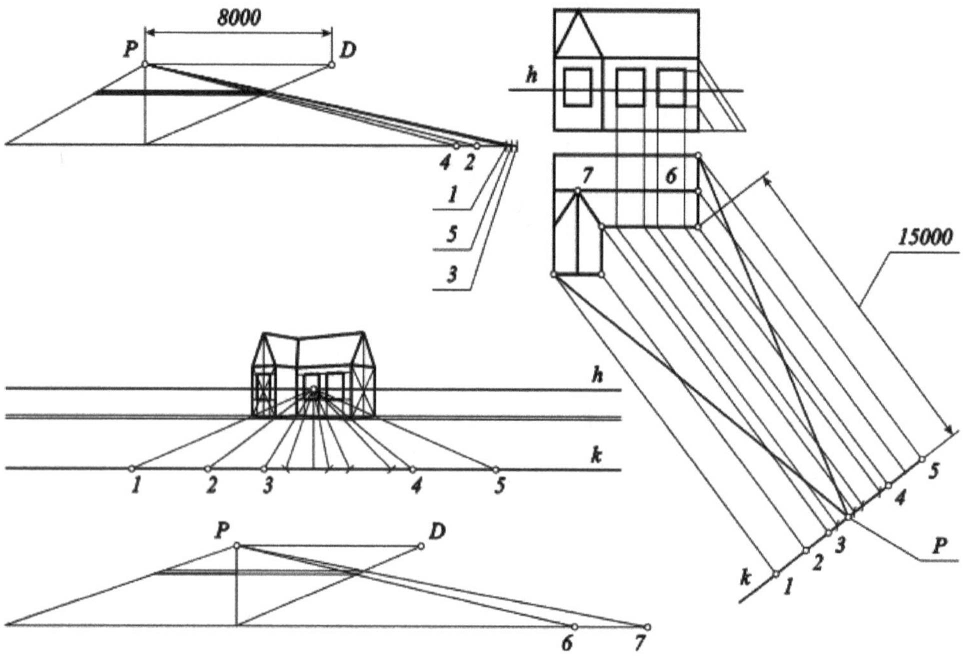


*Figure 22. The perspectives of the object remote from the eye view for the same distance with the eye view located: a) in the center of the projection S; b) at the main point of the picture P*

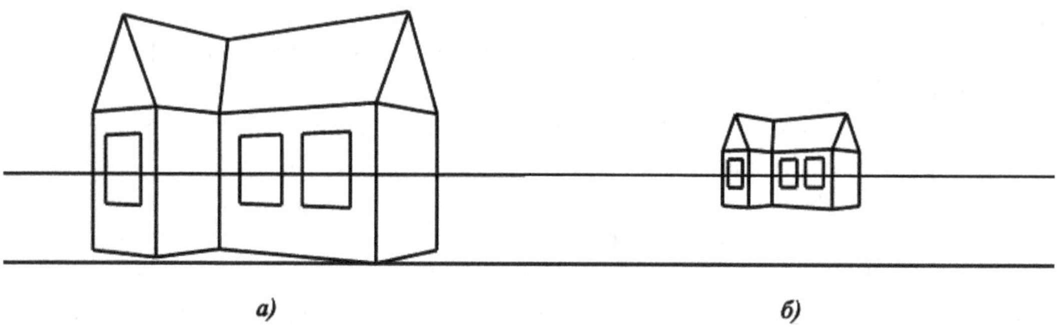


metrical object remote from the viewer for a distance equal to 15 m. At the same time, the perspective, built at the location of the eye view in the center of projection *S* does not create a correct idea of the distance to the geometric object that the viewer observes from a distance of 15 m.

Thus, the problem of choosing the eye view and choosing the distance from the center of projection to the picture plane is solved. It was proved that in order for the perspective of the object to be close to its visual perception, the eye view should be at the main point of the picture, and the distance from the projection center to the picture plane should be 8 m.

The proof of the statement about the choice of the eye view was based on the comparison of the perspectives of the same geometric object built at the location of the eye view in the center of the projection, and in the main point of the picture. The comparison of perspectives showed that in order to correctly convey the depth of space on the picture plane, the eye view should be chosen at the main point of the picture or at the point belonging to the picture plane. To prove the statement about the choice of the distance from the projection center to the picture plane, the results of an experiment to study the dependence of the visible size of the object on its distance from the observer were presented. Moreover, this statement was confirmed by the results of a study of the central projection apparatus in the paintings of the Renaissance artists. For example, in the scene of The Annunciation of the polyptych of Saint Anthony by Piero Della Francesca, spatial constructions were made under the condition that the distance from the projection center to the picture plane is 8 m, and on The Trinity fresco by Masaccio – provided that the distance from the projection center to the picture plane is 6 m.

In our opinion, the development of the method for constructing the perspective that corresponds to the visual impression of contemplation of nature cannot be considered complete. Therefore, further research is aimed at improving the central projection apparatus in accordance with the current level of knowledge about the psychology of visual perception.

## ANALYSIS OF SPACE IMAGES: NORTHERN EUROPE

Now let's take a look at the North of Europe and present a number of works by local artists.

Space and time spread into the depths, and upwards, and in breadth far beyond the limits outlined by the frame of the picture. Open spaces are presented widely – not only by the length but also by the time of the plot development (Figure 23):

*Figure 23. Hans Memling. The Passion of the Christ (c. 550 years ago). Gallery Sabauda, Turin, Italy Web Gallery of Art (Wikimedia. Gans Memling, n/d).*

Space consists of heaven, earthly world and hell (reincarnation of the World Tree) and is actively involved in the universal apocalyptic drama (Fomin, 1974), (Figures 24-26).

*Figure 24. Hieronymus Bosch. The Last Judgment. ≈520 years ago. Academy of Fine Art, GG-579-581. Vienna, Austria Author: Arezzo88 (Wikimedia. Bosh. Judment, n/d).*

*Figure 25. Hieronymus Bosch. Visions of the Hereafter. ≈520 years ago. Doges Palace, Venice, Italy (Wikimedia. Bosh.Vision of Hereafter, n/d).*

*Figure 26. Hieronymus Bosch. Exterior (shutters) of The Garden of Earthly Delights ≈520 years ago. Museo Nacional del Prado. P02823 Madrid, Spain. Originally uploaded to the English Wikipedia by: User: Blankfaze (Wikimedia. Bosh. The Garden, n/d).*

The curved perspective is used (Lvov, 1971), (Figure 27):

The use of light and shadow has been mastered and improved (Figures 28-29):

Nature has an independent value, and this is another reason for giving depth to an open space (Figure 30):

Note that the gallows were perceived as part of the landscape which to no extent interfere with outdoor recreation – a characteristic feature of morals of the times when Duke Alba actively eradicated heresy in the Netherlands.

The landscape is often not just a background but involved in dramatic events. On the contrary, the characters could be typical but not individual (Figure 31):

The artist's idea of whether space is finite and bounded by a sphere of fixed stars or an infinite medium containing an infinite number of celestial bodies is crucial for the formation of visual perception of space.

Let's consider the case of the infinity of world space. Imagine how points are stretching to infinity. Since experience suggests that all objects stretching into the distance shrink in size and, eventually, tighten to a point, and it seems to the observer that points of space move closer to each other until they merge. This point is usually called the vanishing point of straight lines, which in objective space are perpendicular to the picture plane, although it might be correct to call it a vanishing point of straight lines stretching to

*Figure 27. Pieter Bruegel the Elder. The Tower of Babel. ≈470 years ago. GG_1026 Kunsthistorieshes Museum Vienna, Austria. from Google Art Project. (Wikimedia. Bruegel the Elder. Tower, n/d).*

*Figure 28. Petrus Christus. Portrait of a young girl. ≈550 years ago. Gemäldegalerie der Staatlichen Museen zu Berlin (Wikimedia. Petus Cristus, n/d).*

*Figure 29. Pieter Bruegel the Elder Portrait of an old woman ≈457 years ago The Alte Pinakothek, Munich. Germany (Wikipedia. Bruegel the Elder-001, n/d).*

*Figure 30. Pieter Bruegel the Elder. A Magpie on the Gallows (c. 450 years ago). Hessisches Landesmuseum Darmstadt, Germany (Wikipedia. Bruegel the Elder. Soroka, n/d).*

*Figure 31. Pieter Bruegel the Elder. The Triumph of Death (c. 460 years ago). Prado, No. P001393, Madrid, Spain. (Wikipedia. Bruegel the Elder. Triumph of Death, n/d).*

infinity. It is noticed that if in the picture you specify a point, into which stretch straight lines parallel to the direction of sight, it would seem that the picture space is infinite. Therefore, the vanishing point can be considered as a visual image of the infinity of space. The masters of the Italian Renaissance knew or guessed about the ability of the vanishing point to create a visual illusion of the infinity of the picture space. This explains why in the paintings of Italian painters there is always a vanishing point, usually in the geometric center of the picture. If several vanishing points are introduced into the picture, this is usually not related to the construction of space but is driven by compositional considerations.

Let`s consider the geometric properties of the picture, if in the artist's view the world space is finite and closed. Imagine an observer who stands in the center of a spherical dome covering the Earth. Since all the sight directions coming from the center of the sphere are equivalent, any point taken on the horizon line can be regarded as a vanishing point of straight lines parallel to the sight direction. It follows that for an observer, who is inside the sphere and located at its center, there is an infinite number of vanishing points that correspond to the directions of sight lying in the plane of the horizon. However, the Dutch and German artists could not ignore their visual sensations, which said that the vanishing point does exist. To reconcile the teaching of Aristotle on the geocentric structure of the world with his visual perception, the masters of the Northern Renaissance developed some clever tricks.

First, to hide the vanishing point, they introduced not one but not less than two vanishing points. The composition of the picture includes several families of lines, which are visual images of straight lines parallel to different directions of sight, and vanishing points are points from which the data of the families of diagonal straight lines diverge (Figures 32-33).

*Figure 32. Pieter Bruegel the Elder. The Fall of Icarus (c. 460 years ago). Royal Museum of Fine Arts No. 4030, Brussels, Belgium (Wikipedia. Bruegel the Elder. The Fall of Icarus, n/d).*

*Figure 33. Drawing lines of sight and vanishing points*

Second, the artists acted on the principle: no vanishing point – no problem of the infinity of the picture space. In order not to show the vanishing point, it was either taken outside the picture space or combined with the left or right edge of the picture, which made it possible to speak about the composition, not with the central but the diagonal vanishing point. This technique was used by Pieter Bruegel the Elder (Figures 34-.35) and Hieronymus Bosch.

Third, to divert the vanishing point from the viewer's attention, the masters of the Northern Renaissance shifted it to the upper edge of the picture, as if the artist had painted it from a high eye view (Figures 36-37).

Now let us answer the question: does the perspective – linear or curvilinear – better correspond to the visual perception of the physical space (psychophysical space)?

First, let`s consider the geometric properties of the psychophysical space, which arise from its spheroidal form. To emphasize the peculiarities of the psychophysical space geometry, let's compare it with the space that is reproduced on the picture plane with the help of a linear perspective.

Imagine the perspective as the central projection of the points of three-dimensional space to the points of the two-dimensional picture plane. At the same time, the space picture can be represented as a section of projecting straight lines connecting the center of projection with points of space, a plane perpendicular to a straight line drawn from the eye view at a right angle to the picture plane. For a structure from the picture plane, an eye view and a family of projecting lines could be a model of visual perception, it is necessary to make some essential assumptions:

- human vision is monocular, that is, a person sees with one eye. Indeed, the geometric apparatus of the linear perspective includes only one picture plane, which conditionally replaces the spherical surface of the retina, and only one projection center, considered as an approximate model of the eye lens;
- point of view is fixed, that is, the person looks in one direction. In fact, in a linear perspective, the construction of a two-dimensional image of a point in space is made with the same main direction of projection, which is perpendicular to the picture plane;
- section of projecting straight lines connecting the eye view with points of space, the picture plane is an adequate reflection of the psychophysical image of space.

Unfortunately, these assumptions are in clear contradiction with nature and mean complete abstraction from the psychophysical features of visual perception (Rauschenbach, 2001): first, human vision has binocularity and mobility of the eye view, and second, central projection completely ignores basic psychophysical conditions of visual perception, namely: the mechanism of constancy of size and the mechanism of constancy of form.

Nevertheless, a central projection is in some way approaching visual perception. This is explained by the fact that central projection conveys the main monocular sign of the depth of space, namely, a reduction in the size of objects with increasing distance to the observer. The problem is that a structure from the picture plane, eye view, and family of projecting lines reproduces psychophysical features of visual perception if the visual space is Euclidean, that is, a space that can be described as infinite, static, isotropic and rectilinear (Rauschenbach, 2001).

Imagine infinity as unrestricted motion in any of six directions which are given by axes of the Cartesian coordinate system introduced in three-dimensional Euclidean space. Define static as a fixation of the space-time continuum at some point in time, that is, the assignment of the time coordinate of the

Figure 34. Pieter Bruegel the Elder. Hunters in the Snow (c. 455 years ago) Kunsthistorisches Museum, No. GG_1838 Vienna, Austria (Wikipedia, Bruegel the Elder. Hunters in the Snow, n/d).

Figure 35. Drawing lines of sight and vanishing points

*Figure 36. Hieronymus Bosch. The Cart of Hay (middle part of the triptych, c. 520 years ago) Prado/ Escorial, No. P002052 Madrid, Spain (Wikipedia. Bosch. The Cart of Hay, n/d)*

*Figure 37. Drawing lines of sight and vanishing points*

orthogonal coordinate system introduced in four-dimensional space-time of a specific numerical value. Isotropy means the possibility of performing equal constructions at any point in space and in any direction. This means that if at an arbitrary point in space construct a cube and repeat this construction at some other point in space, then the constructed cubes will have equal sizes and the same shape. Also, homogeneity can be represented as follows: if at an arbitrary point in space construct a cube and rotate it around the axes of the Cartesian coordinate system at arbitrary angles, then the cube obtained as a result of this affine transformation will have the same dimensions and shape as the prototype. Hence it appears that if a cube is constructed at an arbitrary point in space, then at any other point in space a cube with the same size and shape can be obtained by successively performing affine transformations of motion and rotations around the Cartesian coordinate axes.

However, observations of nature show that the geometry of Euclidean space is opposite to the geometric properties of the psychophysical space.

First, the psychophysical space is finite and closed. Indeed, visual sensations of the observer clearly indicate that above the observer's head a spherical surface rises taking on the image of the celestial vault, the observer's feet rest on a cup-shaped earth surface, and the boundary separating the earth firmament from the celestial sphere is perceived by the observer as a concave arc of a circle when viewed from the eye view of the observer placed in the center of the celestial sphere. Imagine that the observer's eyes emanate visual rays, which can extend to an arbitrarily long distance. However, unlike the Euclidean space, visual rays in the psychophysical space cannot go to infinity: no matter how long visual rays are and in what direction they spread, the observer's eyes will necessarily come across either a spherical celestial dome or a cup-shaped earth firmament. Hence it appears that the celestial sphere and the earth firmament are perceived by the observer as a two-dimensional boundary within which the psychophysical space is enclosed. In this case, the observer is prevented from knowing what is outside the psychophysical space – to do this, he needs to go beyond the boundaries of the world in which earthly life occurs. Moreover, an attempt to find out what is going on outside of the earth space is equivalent to asking the question: in which direction is the Universe bent? As is well known, for the observer placed inside the Riemannian space, an answer to this question can only be given from the standpoint of internal geometry, and this point of view, the Riemannian space cannot be either convex or concave. Hence it appears that the Riemannian space acquires curvature from the eye view of the observer who is in the Euclidean space embracing a higher dimension.

Second, the psychophysical space is dynamic. This means that in the mind of the observer the state of the psychophysical space is constantly changing with the continuous change of the time coordinate of the orthogonal coordinate system specified in the space-time continuum. Moreover, the psychophysical space is a combination of pictures that are perceived by the observer with a constant and continuous change in the eye view and direction of view, – precisely because of the mobility of the eye view, the celestial vault is perceived as a spherical surface, the earth firmament looks cup-shaped, and the horizon line seems to be a concave arc of a circle.

Third, the psychophysical space is anisotropic or heterogeneous. In this case, anisotropy means the dependence of the geometrical properties of the psychophysical space on the direction in which the studied point of space moves. This means that if, at an arbitrary point in space, build a cube and transfer it to some other point in space, then as a result of this transformation, both the dimensions and the shape of the built cube will change. In this case, the size and shape of the cube will depend on the choice of a point in the space in which the cube moves. Also, if a cube is built at an arbitrary point in space and rotated around an axis of the Cartesian coordinate system at an arbitrary angle, then the cube obtained as a result of this affine transformation will have dimensions and shapes different from those of the prototype. Hence it appears that the psychophysical space is not invariant for affine transformations of motion and rotations around the Cartesian axes of coordinates.

Fourth, the psychophysical space is curvilinear. This follows from the fact that the psychophysical space is a closed space enclosed within the spheroidal shell. At the same time, the curvature of the psychophysical space is not a constant value and changes as the position of the point at which the curvature of space changes. Indeed, the dimensions of the object are perceived in full size in the area of space immediately adjacent to the observer and are reduced to zero values at points on the horizon line, along which the spherical celestial dome adjoins the cup-shaped earth surface. Furthermore, parallel straight lines seem to parallel in the region of space immediately adjacent to the observer and converge at a single point located on the horizon line. This means that the curvature of the psychophysical space is equal to zero in the region adjacent to the observer, and increases monotonically as it approaches the peripheral regions located on the border of the spheroidal celestial dome.

Thus, the space that exists in the person's visual sensations and is to be reproduced on the plane of the picture is a closed, dynamic, anisotropic and curvilinear space the curvature of which is variable.

Let us consider the peculiarities of visual perception arising from the geometry of the psychophysical space. First of all, thanks to the spheroidal form of the psychophysical space, the observer perceives lines that are straight lines in objective space, like curves, and lines that are curves in objective space, as straight lines. Because the psychophysical space has a spherical shape, the straight lines, which are located in the objective space perpendicular to the picture plane, look parallel in the region of space immediately adjacent to the observer.

For the first time, the idea of the curvature of the psychophysical space was expressed in the works of ancient Greek philosophers and mathematicians Euclid, Aristotle, Diogenes, Plutarch, as well as the thinkers of ancient Rome Lucretius, Petronius, Sextus Empiricus, Tertullian and others (Rauschenbach, 2001).

The views of ancient scholars on the nature of visual perception have found practical application in the architecture and sculpture of ancient Greece and Rome. Take, for example, the columns of ancient Greek temples, which were always made with entasis, that is, with a certain bulge, so that the cylindrical columns do not look distorted due to the apparent narrowing in the middle. Also, ancient architects invented the courtyards, with which the epistyle and stylobate of ancient Greek temples were made. The meaning of the curvature is that the horizontal architectural elements are made with some thickening in the middle to avoid the impression of a deflection.

Ancient architects invented another trick, with which it was possible to give straightness to the visual images of cylindrical columns, namely: the columns were made in the form of truncated cones with a slight deviation of the straight lines forming from the vertical direction. By the way, this is exactly what explains the funny fact that the columns encircling the Parthenon outside lean inward into the structure and form a pyramid of approximately 2500 meters in height. If the ancient architects did not apply this technique, the columns of the temple would seem to be leaning forward or falling towards the viewer.

Nevertheless, the ancient Greek philosophers and mathematicians could not explain why there are distortions of visual images of architectural elements that are stretched in both horizontal and vertical directions. Despite numerous attempts, neither the masters of the Italian Renaissance, nor their numerous followers managed to do this (for example, Guido Gauc's works on perspective theory are widely known, in which he tried to explain such promising phenomena as curvature by the conflict of the lateral triglyph, i.e. between the outermost columns).

The visual effects associated with the curvature of the psychophysical space were first explained in the classic works by H. Helmholtz, E. B. Titchener, E. Hering, E. Emmert, G.T. Fechner: cylindrical columns seem to the observer to narrow upwards, but his consciousness resists these promising distortions, and the mechanisms of constancy of size and shape come into play, which compensates for perspective changes. The consciousness of the observer restores the true size and shape, and the more the perspective image deviates from the size and shape of the prototype, the stronger the effect of the mechanisms of constancy. Therefore, cylindrical columns, which from the point of view of perspective should seem to narrow upward, are perceived as really expanding. Expansion of the columns is manifested by the more distinctly, the closer to the top of it the observer's view moves. Consequently, the action of the mechanisms of constancy leads to the fact that the visual image of cylindrical columns has some narrowing in the middle. To make the cylindrical columns look straight, the builders of ancient Greek temples made columns with some expansion in the middle, that is, gave the forming columns the shape of convexly curved lines.

Compensation of perspective cuts is performed in order to bring the sizes of visual images of objects closer to their actual sizes. Experiments made it possible to establish that information about distances of objects to the observer enters the human consciousness as a result of the combined action of binocular signs of depth, namely accommodation, convergence, and binocular disparity. At the same time, monocular signs of depth do not contain information, with the help of which one can determine distances to observed objects, and only give an idea of distribution of objects over the depth of space, that is, they allow determining which objects are closer to the observer and which objects are farther from him. Consequently, the mechanism of the constancy of quantity is a model that reflects the action of binocular signs of depth. This means that modern science does not know how human consciousness determines the distances of objects to the observer based on binocular signs of depth, however, thanks to works of eminent psychologists of the last century, modern science knows what the effects of binocular signs of depth lead to.

Since the curvature of straight lines which are objectively parallel is explained by the action of the mechanism of value constancy, it can be argued that the curvature of the psychophysical space is a consequence of the binocular signs of depth.

Thus, in giving the psychophysical space such geometric properties as the spheroidal shape of the celestial dome, the shape of the earth surface and the concavity of the horizon, the influence of binocularity of view also affects. Furthermore, binocular vision can be considered as one of the reasons why the psychophysical space is anisotropic and its geometric properties depend on the direction of sight under which objects are observed.

Now back to the geometry of the paintings of the masters of the Northern Renaissance.

At a time when the masters of the visual art of the Italian Renaissance created the perspective as a section of the "visual pyramid" by the picture plane, in the north of Europe: in Germany, the Netherlands, Austria created artists who had a slightly different view on the system of depicting space (Rauschenbach, 2001; Shlain, 19914; Talbot, 2003; Wright, 1983).

Of course, artists of Northern Europe knew about works of Italian masters and could not fail to note the merits of the way they open up a representation of space on the picture plane. The perspective system of the Italian Renaissance aroused the desire for pictorial development of distant regions of space and for the first time made it possible to overcome discontinuity preserved in ancient times in the image of space. At the same time, artists of Northern Europe could not agree that the visible space to be depicted on the plane should be infinite since this condition completely contradicted their visual perception of space, which was represented by them as a sphere bounded by a cup-shaped earth surface.

Thus, the artists of Northern Europe were faced with a task, the solution of which must satisfy conflicting, mutually exclusive conditions: on the one hand, the picture space must be continuous, that is, it must not contain gaps between the pictorial plane in which the divine being flows, and stretching far away, on the background, where earth life unfolds, and on the other hand, the picture space must be inside the spheroidal surface, resting on a cup-shaped base. The solution to the problem is complicated by the fact that, unlike the linear perspective, which has a simple geometric transformation that converts the straight lines of objective space into straight image plane, transformation of objective space into spheroidal space that converts straight lines of the objective space into curved lines of the picture plane, does not have the same simple mathematical apparatus. Indeed, if the linear perspective can be made using elementary drawing tools, namely, compasses and ruler, then there simply does not exist a simple constructing of curvilinear images of straight lines of the objective space which can be accomplished with compasses and ruler.

Nevertheless, the artists of Northern Europe have found a way of representing psychophysical space on the picture plane, which does not require the development of a complex mathematical apparatus. If masters of the Italian Renaissance saw in the reflection, which can be observed in a flat mirror, a geometric model of psychophysical space, then artists of Northern Europe saw an example of the optical organization of space in the reflection, which creates a spherical convex mirror. Indeed, the Middle Ages considered art as a reflection of the divine essence of the world, and the mirror was not just a physical body, which served as a model of optical reflection of reality, but also a symbol of truth, the material embodiment of the Eye of God. This view was promoted by the fact that in the Middle Ages mirrors were given only a round shape. About 600 years ago, besides the symbol of the Eye of God, a round mirror gained another meaning: in the spherical form of a mirror, the people of the Middle Ages saw conformity with the shape of the earth surface outlined by the horizon, and in the mirror reflection – the essence of art. Indeed, if, on the one hand, mirror reflects the material forms of reality, and on the other hand, according to the ideas of medieval artists, the task of art is to reflect the divine world in pictures of earthly life, then it follows that the round mirror becomes a symbol not only of earthly life but also its reflections in art. However, not until about 350 years ago, the method of amalgamation of flat glass surfaces had been invented, all mirrors in Europe were made of metal, which was given a round convex shape. Of course, reflection in a spherical mirror contained significant optical distortions that were noticeable to people who lived in the Middle Ages, even though they did not know flat mirrors. This corresponded to the prevailing worldview, according to which the earthly world is viewed as the abode of sin and repository of vice, and therefore pictures of earthly life cannot be an exact reflection of the divine world. It follows that reflection in a spherical mirror, which creates the impression of an optical illusion, quite naturally became an example of a system of spatial constructions on the picture plane.

Thus, for artists of Northern Europe, the model of the psychophysical space was a reflection of the objective space in a spherical convex mirror. The rules of geometric constructions on the plane of paintings by artists of Northern Europe began to be defined by optical laws, according to which reflection of space in a spherical convex mirror is formed. The geometry of paintings by artists of Northern Europe was so peculiar and so different from the way of reproducing space on the picture plane developed by the masters of the Italian Renaissance, that works by artists from Germany, the Netherlands and Austria from near 650-550 years ago figured in the history of world art as the Northern Renaissance.

Once again, note that invention by masters of the Northern Renaissance of an artistic technique of reproduction of space coincided with changes in their worldview. The thinkers of the early Middle Ages considered the earthly world and the divine world as completely different, having no common points of contact between the sphere of human activity and supernatural beings. Therefore, the space of paintings by masters of the early Middle Ages is divided into the foreground, in which the divine being flows, and the background, where earthly life unfolds and does not contain an intermediate, middle distance which would link the foreground and background. Thus, the picture space undergoes a gap that is usually covered by either an architectural structure or a group of characters separating the pictorial plane of the picture from the back spatial plan.

Gradually, the divine world would lose the value of the highest standard of supernatural, incomprehensible reality, and overtime began to be viewed as a continuation of earthly life, a reflection of its light and dark sides. The fine art of the late Middle Ages responded to changes in the view of thinkers of the time about supernatural, divine essence, and in the paintings of great predecessors of the Northern Renaissance, first attempts were made to convey the middle distance providing a spatial extension from the foreground to the background.

Thus, the image of the divine world would lose its paramount importance for the predecessors of the Northern Renaissance, and over time their attention focused on the pictures of earthly life. This change in views on the subject of art was made throughout the Middle Ages and ended in the works of Hieronymus Bosch, who finally approved the idea of the supernatural world as the reflection of the material beauty of earthly life or continuation of its physical and moral imperfection. The divine world is completely supplanted by the image of a human and his daily activities which flow exclusively in the space of earthly life as a result of which the system of spatial separation of the picture into the foreground and background loses its original meaning.

Furthermore, Hieronymus Bosch was the first to perceive the reflection of objective reality in the convex surface of a spherical mirror as a system of organization of the picture space. Certainly, the artist did not know the ways to reproduce the sphericity of the celestial vault and the shape of the earth's surface in a picture with mathematical precision. However, in the convex surface of a spherical mirror, the artist saw not just reflection of the real world, but also optical laws, which create a middle distance connecting the foreground and background with a single, continuous spatial environment.

The Cure of Folly was the first picture of Hieronymus Bosch, in which the method of reflecting reality in a spherical mirror was adopted as a system for constructing space. This principle of space organization is emphasized by the fact that Hieronymus Bosch gave the picture the shape of a circle, which is regarded as an image of a spherical mirror. The application of this principle of organization of space is confirmed by the fact that the painting The Cure of Folly contains the effects of the shape of the earth space and the concavity of the horizon line. In the upper part of the picture, the horizon line and the lines of the spatial planes of the earth surface spaced from the observer for certain distances are bent from the edges of the picture to its middle so that their bulging sides are at the top; and in the lower part of the picture, the lines of the wall hiding behind the figures of the characters and the lines of the spatial plans of the earth surface bend in the opposite direction. The geometry of the picture creates the impression that the earth's surface resembles a concave bowl, and the pictorial plane of the picture expands like a spherical surface. This visual illusion is enhanced by the fact that the color gamut is lightened from the edges of the picture to its middle, and seems to end with a bright glare on a metal vessel placed in the geometric center of the picture. Also, the visual illusion is emphasized by the fact that the volume of geometric shapes increases from the edges of the picture to its middle and eventually the center of the picture is the most rounded object, namely: a metal vessel in the hands of the doctor's assistant. At the same time, spherical space is more a symbol of the earthly world than a real human environment. The curvature of the lines belonging to the foreground and background, and converging in the center of the picture, as if compressed its middle distance. If in the art of the early Middle Ages there was no middle distance, then in The Cure of Folly the middle distance appears but is not yet perceived as the real spatial environment between the foreground and background.

Other works by Hieronymus Bosch also have the form of a round mirror, for example, the painting The Seven Deadly Sins and the Four Last Things, which is painted on the tabletop. The composition is constructed in such a way that the image of the Eye of God is located in the middle, and allegorical scenes with images of human sins are unfolding around it. Each scene is a fragment of the earthly world, that is, a fragment of a spherical space in which human life flows. These ideas are reflected in the system of spatial structures: the horizon line and the lines of landscape plans are bent following the circular bends of the lower and upper edges of the ring into which the pictures of earthly life fit. At the same time, the figures of the characters are brought to the foreground, detached from the back-landscape ground, and seem to be projected onto a convex spherical surface, the shape of which takes the visual plane of the picture. Hieronymus Bosch uses optical illusion observed in reflecting reality in the convex surface of a spherical mirror, and in building interiors. For example, in the scene with the image of the allegory *Vanity*: on the left wall, both parts of the window sill are angled to each other, and the line along which the floor and wall join is gently bent, giving the impression of a wall bulge. Moreover, all straight lines, which are located in the real world perpendicular to the picture plane, are round, as if reflected in the convex surface of a spherical mirror.

A special place in the works of Hieronymus Bosch is occupied by the painting The Ship of Fools. The foreground of the picture is made in axonometry which is quite clearly observed in the image of the tabletop placed across the boat. The system of spatial constructions, based in the early works of Hieronymus Bosch, was perfected throughout his work and eventually formed into a pictorial model of world space. The middle distance of his paintings expands, the spatial environment linking the foreground and background, acquires features of the realistic landscape, and, as a result, the sense of continuity of space, which continues to infinity, increases. However, over time, the image of a round mirror ceases to be a symbol of the earth, and the picture begins to take place, not in the whole space of the world sphere, but a fragment cut out of rectangular frames of the picture. Nevertheless, the idea of the shape of the earth space continues to be present in many of the later works by Hieronymus Bosch, for example, in the central casement of the Viennese altar The Last Judgment. These ideas are reflected in the fact that the artist depicts a majestic landscape that is widely spread beyond the bridge like a concave bowl, and the central part of the surface of the earth – as a huge cavity (Figures 38-39).

The paintings of Hieronymus Bosch do not contain straight lines converging at one point taken on the horizon and leading the viewer's eyes to the depth of the picture space. On the contrary, the viewer's eye goes in a circle, from the depths to the foreground and again into the depths, as if following the concave horizon line and rounded outlines of the spatial landscape distances. For example, on the central wing of The Haywain Triptych on the right, from the depths, a cavalcade of noblemen comes to the foreground; in The Christ Carrying the Cross painting in Vienna, a crowd of guards and executioners makes a circular motion; on the middle leaf of the triptych The Garden of Earthly Delights, hosts of monsters, people and animals pass along the circular surface of the earth and are removed into the depths hiding behind the right edge of the picture, and are connected to each other in the infinite depth of the spherical world space (Figure 38).

*Figure 38. Hieronymus Bosch. The Garden of Earthly Delights. (c. 510 years ago). Triptych. Prado Museum. No. P002823. Madrid, Spain (Wikipedia. Bosch. The Garden of Earthly Delights, n/d).*

*Figure 39. Drawing the central part with the designation of elements of perspective*

The system of organizing space on the picture plane discovered by Hieronymus Bosch was developed and continued in the visual art of the Northern Renaissance: the perspective, the prototype of which was the idea of the psychophysical space as a space closed inside a spherical surface, became the basis of spatial constructions in the paintings of such outstanding masters of the Northern Renaissance, like Albrecht Altdorfer, Lucas Cranach and Pieter Bruegel.

One of the most remarkable masters of the fine arts of the Northern Renaissance is the painter of the Danube school Albrecht Altdorfer (Velchinskaya, 1977). The painting The Battle of Alexander at Issus is considered the greatest insight into the essence of the landscape ever achieved by the artist. A cosmic panorama of the world unfolds before the eyes of the viewer, rays of the setting sun gild the tops of the mountains whose chains describe the curvilinear solid of the earth. The amazing charm of the painting and its strong effect on the viewer's imagination is explained by an unusually high point of view, which gives the impression that the artist sees the Earth from Space and embodies his cosmic visions in an artistic canvas. The middle distance expands somewhat by reducing the length of the foreground, which is why the earth's surface, covered with a mosaic of human figures, seems slightly convex (Figures 40-41). Thus, Albrecht Altdorfer develops a system of spatial constructions, opened by his spiritual mentor Hieronymus Bosch, and his work is the threshold of the cosmic landscapes of Pieter Bruegel.

The works by Pieter Bruegel the Elder (1525 – 1569) are one of the peaks of the fine art of the Northern Renaissance. Already in the early drawings, Pieter Bruegel found a deep understanding of the integrity of nature as a unity and an infinite variety of its manifestations. Due to the high eye view, images of rocky peaks, valleys, sea spaces, and mountain streams create the impression of cosmic grandeur, the vast extent of space, the greatness of the surrounding nature.

The cosmographic concept of the world space, boundless and at the same time closed inside a spherical surface, Pieter Bruegel laid as the basis for the grand image of space in the painting The Fall of Icarus (Figure32-33). The picture clearly conveys the idea of the sphericity of the Earth and, thanks to the slightly convex horizon line, the rounded lines of the sea surface creates the impression that the sea spreads over a convex spherical surface and descends to a gentle coast, which seems to be a huge concave bowl. The diagonal construction of the landscape reinforces the impression of the curvature of the earth's surface and leads the viewer beyond the picture where the infinite Space extends. The majestic seascape absorbs the small figure of a peasant plowing the ground in the foreground and hides from the attention of the viewer the legs of Icarus fallen from the skies. This testifies both to the frailty of human life and the fact that any human attempts to rise above the forces of nature are doomed to defeat, which recalls the fate of conceited Icarus. From the geometrical point of view, the painting The Fall of Icarus is remarkable as it shows how the artist creates the illusion that the picture space bends above the plane of the picture and becomes similar to the surface of a convex spherical mirror. The illusion of the curvature of the terrestrial space is further enhanced by the fact that the artist places a glare in the picture reflected by the convex surface of the sea.

The painting The Fight Between Carnival and Lent (Figure 42) reveals an artistic device that gives the impression of the grandeur of the earthly sphere. This technique is that the artist has a high position on the horizon and thus causes the feeling that the earth surface is heaving and tilting towards the viewer. Since the picture does not contain images of hills or mountain peaks, it can be assumed that Pieter Bruegel imagined what the terrain might look like if it could be viewed from a bird's eye view.

The painting The Hunters in the Snow (Figures 34-35) conveys the beauty of the winter landscape with transparent frosty air, greenish-gray sky, thin lace of fragile branches of trees and shrubs, distant spaces of snow-covered fields and hills and warmth blowing from a human dwelling. Combinations of transparent color tones, warm and soft transitions between distant planes create a feeling of cosmic

*Figure 40. Albrecht Altdorfer. The Battle of Alexander at Issus (c. 480 years ago). The Alte Pinakothek, No. 688, Munich, Germany (Wikipedia. Altdorfer. The Battle of Alexander, n/d).*

*Figure 41. Drawing apparatus perspectives*

*Figure 42. Pieter Bruegel the Elder. The Fight Between Carnival and Lent (c.460 years ago). Kunsthistorisches Museum. No. GG_1016, Vienna, Austria (Wikipedia. Bruegel the Elder. The Fight Between Carnival and Lent, n/d).*

vastness, the infinite depth of space. The sense of the grandeur of the landscape is enhanced by a high eye view chosen by the artist, and a majestic chain of mountains completing the distant plan of the picture. The group of hunters in the foreground, rushing deep into the picture space, brings dynamism to the composition giving movement that resists the silence and tranquility of winter nature. The diagonal construction of the space not only emphasizes the dynamism of the picture but also provides a compositional connection between the figures of people in the foreground and mountain distances sinking in the frosty haze, which frame the concave bowl of the earth surface.

The painting The Tower of Babel (Figure 27) shows that Pieter Bruegel brought new things into the construction of space. Despite the fact that the picture conveys the state of nature at a certain point in time and is a static image, the concept of earth space as a space enclosed within a spheroidal surface turns the picture space into a dynamic space, the curvature of which increases with the distance from the foreground and closer to the background. Indeed, the lines of the distant parts of the tower are bent more than the lines of its near parts, and this gives the tower a slope that enhances the sense of the depth of the picture space and at the same time sets the tower in motion, giving it dynamism and instability.

It appears that if the linear perspective developed by masters of the Italian Renaissance seeks to convey stability and balance, then the system of spatial constructions in the pictorial art of the Northern Renaissance upsets the balance of geometric shapes and makes their image dynamic.

In many paintings by Pieter Bruegel (Figures 32-35), a spatial construction with a diagonal vanishing point is used. This technique expands the picture space and creates the illusion that it can be spread in all directions and continue in depth to an infinitely long distance. Also, the picture space seems to pass through the plane of the picture and surrounds the viewer standing in front of it, drawing it into the depicted action. It appears that the diagonal construction of the picture is another artistic device with which the impression of infinity and continuity of space can be created, as well as the viewer, can be inspired with the idea of the immensity of the Universe.

Hence, the geometry of paintings of masters of the Northern Renaissance gives graphic material, which allows us to say that the methods of artistic reproduction of space developed by them are much better coordinated with the peculiarities of psychophysical conditions of visual perception. However, the artists of the Northern Renaissance could neither provide a scientific basis for their visual sensations, nor develop a mathematical apparatus, nor formulate the rules by which geometric constructions are carried out. Moreover, the theory of perspective, which reproduces the geometric properties of the psychophysical space on the picture plane, has not yet been created and remains the subject of scientific research.

## COMPARISON OF SPACE IMAGING

Comparing the paintings of masters of the Northern Renaissance and Italian artists, it is possible to confirm the similarity of many techniques inherited from the late Middle Ages – primarily in the use of light and shade and aerial perspective. At the same time, the transfer of the "geometry" of space varies notably.

First, in the paintings of masters of the Italian Renaissance, the main character is a human, the landscape is the background, the element of the composition, but by no means a self-sufficient value. On the contrary, the works of the most prominent representatives of the Northern Renaissance are distinguished by a heightened sense of integrity of the world which has fallen and is dying after a human. The problem of personality, as an independent value, is relegated to the background of the unfolding universal drama, which has a mystical meaning and a tragic outcome. Therefore, nature acquires features of an emotionally and intensely battling spatial environment – but also the abode, where human imagination places the image of the Lord God, a remote and completely strange to the bustle of human existence. And such a psychophysical space cannot be linear!

Second, the landscape, if it appears in the paintings of masters of the Italian Renaissance, then, most often, as a view from the window of the room (Rotenberg, 1967). Italian artists limited the landscape to the window frame so that they would not depict those areas of space that the linear perspective recreates with distortion. On the contrary, artists of Northern Europe, when building a space, did not limit themselves to the transfer of its length into depth, – on their canvases the space spreads in all dimensions and seems to be a small part of the magnificent view that the viewer could observe if he were on the imaged area (Voronina, Maltseva & Starodubova, 1994).

Therefore, if the art of the Italian Renaissance images nature to the extent that it is associated with the image of a human and his inner world, then the art of the Northern Renaissance pushes the image of nature in the first place, and the role of the fighter and worker in the world is assigned to human; if the space in the paintings of masters of the Italian Renaissance is emotionally neutral, infinite, linear, then the space of the paintings of masters of the Northern Renaissance seems to be emotionally tense, limited to the "circle of being";

Now, there are all the necessary data for comparison of medieval, Italian and Northern European pictorial art, which is necessary here because of the significant difference in the ways of imaging space in the painting by masters of the Italian and Northern European Renaissance (Table 1).

*Table 1. Comparison of Medieval Painting with a Painting of Italy and Northern Europe of the Renaissance era Features.*

| Medieval pictorial art of Europe (late period) | Pictorial art of the Italian Renaissance | Pictorial art of the Northern European Renaissance |
|---|---|---|
| There are no differences between sacred and daily pictorial art in terms of techniques for conveying the depth of space | There are no differences between sacred and daily pictorial art in terms of techniques for conveying the depth of space | There are no differences between sacred and daily pictorial art in terms of techniques for conveying the depth of space |
| Space is conveyed as three-dimensional; homogeneity is not recognized; the inverse and perceptual perspectives are used, as well as constructions similar to the axonometry. Aerial perspective and light and shade are applied. A psychologically authentic "emotional space" is organized. Most of the plots are related to religious themes; by the end of the period, many secular ones appear. Structuring space is not related to the image of the World Tree. | Space is conveyed as three-dimensional, homogeneous, static; a linear perspective is used, the construction methods of which are formalized and described. The main point is located within the picture and is often connected to the center of the composition. Perspective distortions are masked. Aerial perspective and light and shade are applied. Space is neutral; it is a background for the composition of figures, as well as a neutral background for the image of the emotional state of the characters. Even plots related to Christian topics are depicted by the same means as secular ones. The image of the World Tree is interpreted only as part of the biblical storyline. | Space is conveyed as three-dimensional, heterogeneous, spherical, "cosmic", dynamic, associated with a sequence of events; curvilinear perspectives with non-formalized options of the apparatus and constructions similar to axonometry are used. The main points are moved to the edge or outside the picture. Aerial perspective and light and shade are applied. Space is drawn into the plot; it is not an emotionally neutral background. Even plots related to Christian topics are depicted by the same means as secular ones. The image of the World Tree is interpreted only as part of the biblical storyline. |
| Most often the center is present however not in a geometric sense but is determined by the plot. Cyclicity is almost invisible. | The geometric center is present and often corresponds to the center of the composition. Cyclicity is absent. | Different options of the ratio of geometric, plot and compositional centers. Cyclicity is a particular case of a holistic approach to the plot in the Universe. |
| The top and bottom, right or left side are clearly distinguishable; as a rule, the action takes place "right here and right now". | The top and bottom, right or left side are clearly distinguishable; as a rule, the action takes place "right here and right now". | The top and bottom, right or left side has "different intensities"; the action often develops in time |
| Violation of scaling and proportions is present | Violation of scaling and proportions is practically absent | Violation of scaling and proportions is determined by the "intensity" of action and distortion of perspective |
| Images of people are distinguished by psychological accuracy and pronounced individuality; images of animals are schematic; scenes of war and violence are present | Images of people are distinguished by psychological accuracy and pronounced individuality; images of animals are schematic; scenes of war and violence are present | Images of people are distinguished by psychological accuracy and pronounced individuality; images of animals are schematic; scenes of war and violence are present |
| Polymorphy is episodic | Polymorphy is absent | Polymorphy is episodic |
| Complex cult scenes occupy a significant place | Cult scenes are profaned and shown as domestic | Complex scenes with religious symbols occupy a significant place |
| Samples of writing are poorly presented in the paintings (except miniatures) | Samples of writing are almost not present in the paintings | Samples of writing are almost not present in the paintings |

## WORLD TREE IN THE RENAISSANCE PAINTING

The World Tree is found in paintings of both Italian and Northern European artists. It carries no other meaning than that recorded in the biblical story about the expulsion of Adam and Eve from paradise. Techniques typical of Italian (Figure 43) and Northern European (Figure 44) masters are used to depict the Tree. On Figure 43 the highlighted position of the Tree among the background landscape opening from the window view is exactly the indication that it is the Tree of Life that is depicted.

The image of the World Tree, characteristic of hermetic societies, was discussed in chapter 10; the Renaissance has brought nothing new to such interpretations of it.

Thus, the progressive profanation of this image, which is manifested, in particular, in purely secular (Table 1) techniques of the Tree imaging, take place.

Let us sum up these considerations in Table 2.

*Figure 43. Sandro Botticelli. The Cestello Annunciation ≈530 years ago. Uffizi, 1608 (Inventario 1890), Florence, Italy Web Gallery of Art (Wikimedia. Botticelli. The Cestello Annunciation, n/d).*

*Figure 44. Lucas Cranach the Elder. Paradise (Garden of Eden). ≈490 years ago. Kunsthistorisches Museum. No. GG 3678. Vienna, Austria Web Gallery of Art (Wikimedia. Lucas Cranach the Elder. Paradise, n/d).*

*Table 2. Comparison of the medieval and Renaissance versions of the myth of the World Tree*

| Feature | Medieval version | Renaissance version |
|---|---|---|
| 1. | The structuring of the world is symbolized by the World Tree. | The World Tree is not perceived as a symbol of the worlds' connection. |
| 2. | The use of the Tree by God to sacrifice himself and move to hell took place. Birth of God in the world by an earthly woman – the Mother of God. | The sacred meaning of the myth is replaced by the literal. |
| 3. | Connection with the cult of the dead - direct | Connection with the cult of the dead is absent |
| 4. | Very distant views of the world of the dead in the "Book of the Dead" and Bible prophecies about the resurrection. | Absent |
| 5. | The myth is restored in the entirety of representations of all previous eras; at the same time, Jesus Christ is the central figure, and the cross is only a means | Only the literal and profane meaning of the myth is recognized |

## MARKERS ANALYSIS

**Space imaging.** A series of techniques for conveying the depth of space (light and shade, aerial perspective) have received some improvements. A linear perspective was developed as a geometrically reliable and scientific way of depicting spatial forms and relationships, and a curvilinear perspective, albeit due to ideological factors, but not receiving a scientific apparatus. In both cases, although with the help of various methods, both the psychophysical space and the emotional space of the human were recreated.

Thus, in the period under review, the technique of imaging space developed within the framework of stage 4 of self-organization, and this process was all-European. Since the interest in the feelings and beauty of nature has increased significantly, the deployment of stage 5 (sense and feeling perception) has a place (Figure 44).

**The myth of the World Tree**. There is a profanation of the plot of the myth, already reduced exclusively to the biblical version: the means used to depict the events taking place in paradise (that is, the sacred place) are no different from those used, for example, for landscape sketches.

The weakening of intuition that is irreversible for most people of the era, smoothing of confrontations associated with reason and will, emphasis on spatiotemporal authenticity characteristic of the mind, profaning of symbolic images to sensual specific for sensory-emotional channels continue.

## RENAISSANCE HUMAN MIND

As it appears from the foregoing, the methods of representing the space by Italian masters unequivocally indicate that they have achieved the stage four of self-organization (rationality) and a strong movement towards the stage five (sensuality). In this sense, human consciousness is finally becoming modern.

The intuitive-mystical component is more strongly represented in the minds of masters of the Northern Renaissance. However, representations of space as three-dimensional and interest in imaging human beauty, passions and experiences suggest that they have reached the same stages of self-organization as their Italian contemporaries. This is also confirmed by further development of methods for imaging space on a plane, which has become quite scientific and localized, primarily in the countries of Northern Europe – France and Germany.

However, the Renaissance is characterized, perhaps, by an abnormal number of balanced personalities belonging, according to their potential, to its two highest subtypes – geniuses and creators (Table 1, chapter 4, *Self-organization: evolution and behavior*). They lived in Italy (for example, Leonardo da Vinci), and in the North (for example, Albrecht Durer – who spent many years in Italy). Let`s make – for the edification of the heroes of our time – their generalized psychological portrait.

## BEHAVIOR PATTERNS

Apply Tables 1-6 of chapter 4, *Self-organization: evolution and behavior*.

**Priests and Monks**: Of course, it's difficult to make such generalizations – however, it seems, the Renaissance, at least in Western Europe, is completely devoid of saints! Savonarola is too hot, Thomas à Kempis is too cold, Martin Luther and Jean Calvin are too cruel. Many priests, starting with the Popes,

*Figure 45. The state of mind of a balanced personality of the Renaissance – genius, and creator.*

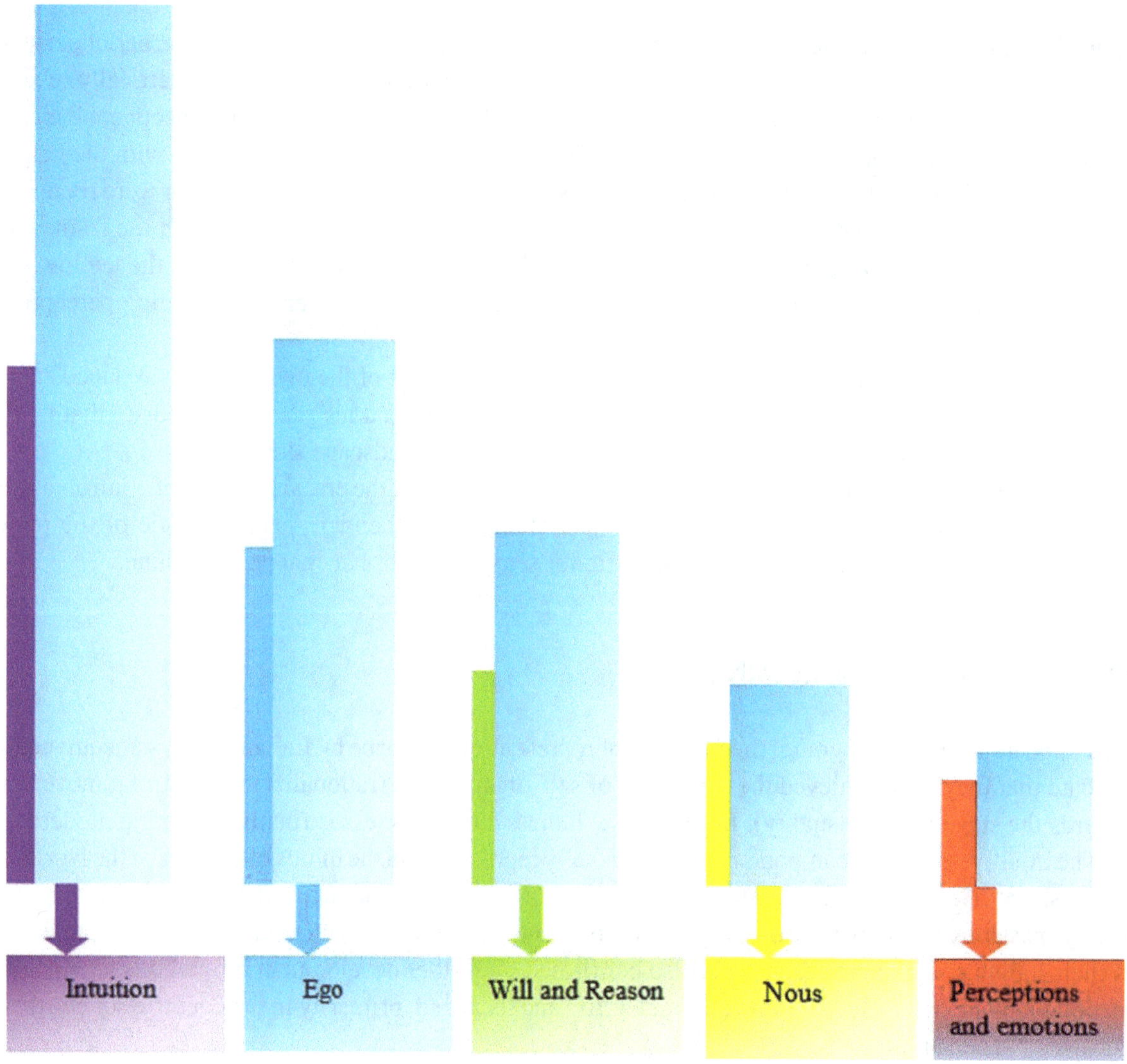


have been severely degraded, considering their vows, at best, nominally, without even trying to back up the declared salvation of the soul with ascetic feats[2].

**Rulers**: Egoists. Proud behavior, vindictiveness – especially in Italy, the inclination to violence – everywhere, benevolence – only against the background of other estates. It further depends on the potential and development of other channels.

**Knights:** disappear, being replaced by condottiere mercenaries, respectively, the warrior type is replaced by the invader type (however, they also have some honor codes in the famous battle of Anghiari described by Niccolò Machiavelli and becoming the plot for Leonardo da Vinci, in which up to 10,000 people participated, only one horseman died, who fell unsuccessfully from a horse), there are simply fighters present.

**Intelligentsia**: There is a certain aberration due to the high potential of brilliantly balanced personalities. Including entrepreneurs – such as the Medici family, becoming the basis of the "black aristocracy". "Simple" scholars and entrepreneurs are in their shadow, but numerically, of course, they predominate. A tendency to innovations and creativity in the first group, conservatism, and envy in the second.

**Commoners**: There are many tradesmen, a lot of frank marginal ones. There are a lot of exalted personalities, outraged by their failures, the blame for which they put on a world that is not too fair, becoming adepts and first victims of numerous sects, Reformations, and Counter-Reformation of this turbulent era. Their behavior is mean, depraved, superstitious, slavish and arrogant at the same time. But there is enough vital energy to become "servants of two masters".

## CONCLUSION

The connection between the technical features of pictorial art in Italy and Northern Europe is traced. The tendencies of the development of pictorial art of the Renaissance are established and their connections with the general laws of evolution of the human mind are revealed.

The quick evolutionary process for a short time - just a few centuries - led to the balance of the human mind - and this was reflected in the birth of a large number of exceptionally gifted personalities.

If abstract from an abnormally large number of people with high potential (not only in the sphere of fine arts, but also science, entrepreneurship, and navigation), it is possible to say that the Renaissance logically completes the Middle Ages. However, synergy with the culture of Byzantium, the expansion of horizons in connection with the Age Discoveries, the birth of exceptionally gifted people made it possible to make clear the coming-to-end way to new stages of evolution – final assertion of domination of intellect and release of emotions from moral oppression of church precepts.

And neither the fierce sermons of Savonarola or Jean Calvin nor mass repressions of the Inquisition could stop it.

The mind of the Renaissance man is as close as possible to the modern, closer than the mind of people of any previous era.

The most significant contribution of masters of fine art in Italy to the technique of depicting space was that the linear perspective, opened by the great masters of the Renaissance, expressed the spatial image in a strict mathematical form. This means that the linear perspective made rational and logical what seemed to be neither measured, nor computed, nor subordinated to formal geometric constructions, namely: the inner world of a human psychic sensation linking it with unified laws with the outer physical space. True, this undoubted success was achieved at the cost of violating the laws of natural human visual perception.

Masters of the Northern Renaissance were able to find another way to depict space, much closer to the laws of visual perception. However, they could not solve the problem of its rigorous mathematical justification – which, incidentally, is a problem for modern science. In later times, they also had followers, certainly, in a smaller number than supporters of the linear perspective.

The myth of the World Tree is reconstructed. The lettering and profanation of even the curtailed biblical version of the plot – "signs" instead of "symbols" are marked. Of course, this fact reflects the evolution of the human mind and has been given an appropriate interpretation.

The most interesting state of mind of the Renaissance man is reconstructed on the example of a psychological portrait of a brilliantly balanced personality. Its main characteristics are the high potential and close to the optimal ratio of channels of human interaction with the environment, which creates opportunities for creative, energetic, versatile activities so typical of the genius personalities of that era.

Patterns of the behavior of the main estates are considered. The changes that have occurred since the Middle Ages are marked and connected with the evolution of the mind. The lack of a subtype of saints is noted (with some reservations).

## FUTURE RESEARCH DIRECTIONS

The new times differ from the Renaissance era in more dryness, science, a spread of ideas of enlightenment and universal equality. This period is well documented and leaves no room for discussion about the meaning of the processes. Much more interesting is the modern period; it will be discussed in the next chapter.

State of mind and behavior patterns will also be reconstructed; however, bearing in mind the last chapter which gives an example of the practical benefits for the modern man of the methods and results of this study, the consideration will be transferred to the personal level.

## REFERENCES

Academic. Verocchio_Lorenzo de Mrdici. (n.d.). Retrieved from https://dic.academic.ru/pictures/wiki/files/86/Verrocchio_Lorenzo_de_Medici.jpg

Benesh, O. (1973). *Art of the Northern Renaissance. Its relationship with modern spiritual and intellectual movements*. Moscow, Russia: Isskustvo.

Brownson, C. D. (1981). Euclid's optics and its compatibility with linear perspective. [Springer Berlin/Heidelberg.]. *Archive for History of Exact Sciences*, *24*(3), 165–194.

Burke, P. (1990). The European Renaissance: Centre and Peripheries, 1998.

Butterfield, H. (1965). Origins of Modern Science 1300-1800. Free Press. 1965.

Cattaneo, A. (2016). European Medieval and Renaissance Cosmography: A Story of Multiple Voices. Retrieved from https://www.researchgate.net/publication/303740142_European_Medieval_and_Renaissance_Cosmography_A_Story_of_Multiple_Voices

Criminisi A., Kemp M., & Kang, S. B. (2003). Reflections of reality in Jan van Eyck and Robert Campin Measuring Art: A Scientific Revolution in Art History – Paris, 1256-1272.

Criminisi, A., Kemp, M., & Zisserman, A. (2002). *Bringing picture space to life: computer techniques for the analysis of paintings. Proceedings Computers and the History of Art* (pp. 1126–1137). London, UK: British Academy.

Debus, A. (1978). *Man and Nature in the Renaissance*. Cambridge, UK: Cambridge University Press.

Dirk De Vos. (2002). *Netherlands painting. Masterpieces of old masters*. Moscow, Russia: Bely Gorod.

Eco, U. (2004). *The evolution of medieval aesthetics*. St. Petersburg, Russia: Azbuka-klassika.

Edgerton, S. Y. Jr. (1977). The Renaissance Rediscovery of Linear Perspective. *The Art Bulletin*, *59*(2), 281–282.

Ferguson, W. K. (1951). The Interpretation of the Renaissance: Suggestions for a Synthesis. *Journal of the History of Ideas*, 483–495.

Field, J. V. (1997). *The Invention of Infinity, mathematics, and arts in the Renaissance*. London, UK: Oxford University Press.

Gershenzon-Chegodaeva, N. M. (1983). *Bruegel*. Moscow, Russia: Isskustvo.

Golos. Andrea del Verocchio. (n.d.). Retrieved from https://golos.io/vp-painting/@vp-painting/andrea-del-verrokko-velikii-master-epokhi-vozrozhdeniya

Google. Leonardo. Tajnaya-vecera. (n.d.). Retrieved from https://sites.google.com/site/leonardo123vinchi/kartinnaa-galerea/tajnaa-vecera

Grant, E. (1994). *Planets, Stars, & Orbs: The Medieval Cosmos, 1200–1687*. Cambridge, UK: Cambridge University Press.

Grashchenkov, V. N. (1988). "Judgment of the Eye" in the theory and practice of the art of the Italian Renaissance. [Moscow, Russia.]. *Soviet Art Studies*, *24*, 112.

Hause, S., & Maltby, W. (2001). A History of European Society. [Belmont, CA: Thomson Learning.]. *Essentials of Western Civilization*, *2*, 245–246.

Hieronymus, B. (1974). *Reproduction album / Text author and compiled by Fomin, G. I.* Moscow, Russia: Isskustvo.

Hockney, D. (2001). *Secret Knowledge: Rediscovering the Lost Techniques of the Old Masters*. London, UK: Viking Press.

Hockney, D., & Falco, M. (2003). Optics of the Dawn of Renaissance. *Technical Digest of the Optical Society, 87th Annual Meeting*, 852-857.

Kemp, M. (1989). *The Science of Art*. New Haven, CT: Yale University Press.

Kemp, M. (2000). *Visualizations: the nature book of art and science*. Berkeley, CA: The University of California Press.

King, Margaret L. (1991). Women of the Renaissance. 1991.

Knorr, W. R. (1992). When circles don't look like circles: An optical theorem in Euclid and Pappus. [Springer Berlin/Heidelberg.]. *Archive for History of Exact Sciences*, *44*(4), 287–329.

Kovalyev, Yu. N., & Nitsyn, A. Yu. (2009). The connection of the visual perception of space with the spiritual currents of the Renaissance in the South and North of Europe. Geometrical and computer modeling, 24. pp. 81-91. Kharkiv, 2009.Kubovy, M. (1988). The Arrow in the Eye: The Psychology of Perspective and Renaissance Art. Cambridge, UK: Cambridge University Press.

Lahr, C. D. (1997). *Mathematics and Knowledge: Models of Reality*. Hanover, NH: Dartmouth College.

Losev, A. F. (1978). *Aesthetics of the Renaissance*. Moscow, Russia: Misl.

Lvov, S. L. (1971). *Pieter Bruegel the Elder*. Moscow, Russia: Isskustvo.

Miller, N. (1993). Euclid Redivivus: A Hypothesis on a Proposition. *Sighttte des Beaux-Arts, 121*, 213–226.

Monfasani, J. (2016). *Renaissance Humanism, from the Middle Ages to Modern Times*. Taylor & Francis.

Monuments of world art: in 9 vol. / Russian Academy of Arts. Institute of Theory and History of Fine Arts. Moscow, Russia: Isskustvo, 1. Art of Italy of the XVI century / E. I. Rotenberg (1967).

Monuments of world art: in 9 vol. / Russian Academy of Arts. Institute of Theory and History of Fine Arts. Moscow, Russia: Isskustvo, 9: Art of Renaissance in The Netherlands, France, England / T. S. Voronina, N. L. Maltseva, V. V. Starodubova. (1994).

Monuments of world art: in 9 vol. (1987). Russian Academy of Arts. Institute of Theory and History of Fine Arts. Moscow, Russia: Isskustvo, 8. Italian art of late 13th - 15th centuries / I. A. Smirnova. (1987).

Nikulin, N. N. (1999). *The Golden Age of the Dutch painting. XV century*. Moscow, Russia: Publishing House AST.

Nitsyn, A. Yu. (1999). Results of the experiment on the study of the dependence of the visually perceived size of an object on the distance to the observer. Transactions of the Tauride States. Agricultural Academy. Melitopol, Ukraine: TGATA, 1999, 4. Applied geometry and engineering graphics, 7. pp. 68–70.

Nitsyn, A. Yu. (2006). Results of the experiment to determine the distance from the center of projection to the picture plane. *Geometrical computer modeling*. Kharkiv, Ukraine: KhDUKhT, 2006. *VIP, 15*, 91–94.

Panofsky, E. (2004). *Perspective as a "symbolic form". Gothic architecture and scholasticism: Trans. from German – St.* Petersburg, Russia: Azbuka-klassika.

Panofsky, E. (2006). *Renaissance and the "Renaissances" in the art of the West*. St. Petersburg, Russia: Azbuka-klassika.

Paoletti, J. T., & Radke, G. M. (2011). *Art in Renaissance Italy* (4th ed.).

Patrick, J. A. (Ed.). (2007). Renaissance and Reformation (5 vol 2007).

Prokofiev, V. N. (1985). *On art and art history. Articles from different years*. Moscow, Russia: Sov. Khudozhnik.

Rauschenbach, B. V. (2001). *The geometry of the picture and visual perception*. Saint Petersburg, Russia: Azbuka-Klassika.

Ruggiero, G. (2015). *The Renaissance in Italy: A Social and Cultural History of the Rinascimento*. Cambridge, UK: Cambridge University Press.

Shlahova, A. (2000). Problems in the Perception of Perspective in Drawing. [Oxford, UK: The Alden Press.]. *International Journal of Art & Design Education, 19*(1), 102–109.

Shlain, L. (1991). *Art & Physics: Parallel Visions in Space, Time, & Light*. New York: William Morrow and Company.

Sinisgalli, R. (2011). *On Painting. A New Translation and Critical Edition*. Cambridge, UK: Cambridge University Press.

Stephens, J. (1990). *The Italian Renaissance: The Origins of Intellectual and Artistic Change Before the Reformation*. Routledge.

Summit, J. (2011). Renaissance Humanism and the Future of the Humanities. *Literature Compass* (2012), 9-10, pp: 665–678.

Talbot, R. (2003). Speculations on the Origin of Linear Perspective. *Nexus Network Journal, 5*(1), 64–98.

Taylor, J. C. (1981). *Learning to Look: A Handbook for the Visual Arts*. Chicago, IL: University of Chicago Press.

van den Broecke, M. P. R. (1996). Ortelius Atlas Maps. An illustrated Guide. HES Publishers, 1996.

van Mander, K. (2007). *The book about artists*. St. Petersburg, Russia: Azbuka-klassika.

Velchinskaya, I. L. (1977). *Albrecht Altdorfer*. Moscow, Russia: Isskustvo.

Wikimedia. Bosh. Judgment. (n.d.). Retrieved from https://commons.wikimedia.org/wiki/File:Hieronymus_Bosch_-The_Last_Judgement.jpg

Wikimedia. Bosh. The Garden. (n.d.). Retrieved from https://commons.wikimedia.org/wiki/File:Hieronymus_Bosch_-_The_Garden_of_Earthly_Delights_-_The_exterior_(shutters).jpg

Wikimedia. Bosh. Vision of Hereafter. (n.d.). Retrieved from https://commons.wikimedia.org/wiki/File:Poliptyk_Hieronima_Bocha_Visions_of_the_Hereafter.JPG

Wikimedia. Botticelli. Simonetta. (n.d.). Retrieved from https://commons.wikimedia.org/wiki/File:Sandro_Botticelli_066.jpg

Wikimedia. Botticelli. The Cestello Annunciation. (n.d.). Retrieved from https://commons.wikimedia.org/wiki/File:Botticelli,_annunciazione_di_cestello_02.jpg

Wikimedia. Bruegel the Elder. Tower. (n.d.). Retrieved from https://commons.wikimedia.org/wiki/File:Pieter_Bruegel_the_Elder_-_The_Tower_of_Babel_(Vienna)_-_Google_Art_Project_-_edited.jpg

Wikimedia. Gans Memling. (n.d.). Retrieved from https://commons.wikimedia.org/wiki/File:Hans_Memling_Passione.jpg

Wikimedia. Leonardo. Madonna Litta. (n.d.). Retrieved from https://commons.wikimedia.org/wiki/File:Leonardo_da_Vinci_attributed_-_Madonna_Litta.jpg

Wikimedia. Leonardo. Mona Lisa. (n.d.). Retrieved from https://commons.wikimedia.org/wiki/File:Mona_Lisa,_by_Leonardo_da_Vinci,_from_C2RMF_retouched.jpg

(Wikimedia. Lucas Cranach the Elder. Paradise. (n.d.). Retrieved from https://commons.wikimedia.org/wiki/File:Lucas_Cranach_d.Ä._-_Paradies_(1530,_Kunsthistorisches_Museum).jpg

Wikimedia. Lucas Cranah. Judgment. (n.d.). Retrieved from https://commons.wikimedia.org/wiki/File:Lucas_Cranach_d._Ä._-_The_Judgment_of_Paris_-_WGA05629.jpg?uselang=ru

Wikimedia. Massccio. Trinitia. (2002). Retrieved from https://commons.wikimedia.org/wiki/File:Masaccio,_trinità.jpg

Wikimedia. Petus Cristus. (n.d.). Retrieved from https://commons.wikimedia.org/wiki/File:Petrus_Christus,_Portrait_of_a_young_girl.jpg

Wikimedia. Pierro della Francesko. (n.d.). Retrieved from https://commons.wikimedia.org/wiki/File:Piero_della_Francesca_044.jpg

Wikimedia. Raffael-058. (n.d.). Retrieved from https://commons.wikimedia.org/wiki/File:Raffael_058.jpg

Wikimedia. Savanorola. Rerrara. (n.d.). Retrieved from https://commons.wikimedia.org/wiki/File:Savonarola_(Ferrara).jpg?uselang=ru

Wikipedia. Altdorfer. The Battle of Alexander, (n.d.). Retrieved from https://ru.wikipedia.org/wiki/Битва_Александра#/media/File:Albrecht_Altdorfer_-_Schlacht_bei_Issus_(Alte_Pinakothek,_München).jpg

Wikipedia. Bosch. The Cart of Hay. (n.d.) Retrieved from https://ru.wikipedia.org/wiki/Воз_сена_(триптих_Босха)#/media/File:Haywain_central_panel_of_the_triptych_WGA.jpg

Wikipedia. Bosch. The Garden of Earthly Delights. (n.d.). https://ru.wikipedia.org/wiki/Сад_земных_наслаждений_(триптих_Босха)#/media/File:El_jardín_de_las_Delicias,_de_El_Bosco.jpg

Wikipedia. Bruegel the Elder. The Fight Between Carnival and Lent. (n.d.). Retrieved from https://ru.wikipedia.org/wiki/Битва_Масленицы_и_Поста_(картина)#/media/File:Pieter_Bruegel_d._Ä._066.jpg

Wikipedia. Bruegel the Elder-001. (n.d.). Retrieved from https://ru.wikipedia.org/wiki/Список_картин_Питера_Брейгеля_Старшего#/media/File:Pieter_Bruegel_d._Ä._001.jpg

Wikipedia. Bruegel the Elder. Soroka. (n.d.). Retrieved from https://ru.wikipedia.org/wiki/Сорока_на_виселице#/media/File:Die_Elster_auf_dem_Galgen.jpg

Wikipedia. Bruegel the Elder. The Fall of Icarus. (n.d.). Retrieved from https://ru.wikipedia.org/wiki/Падение_Икара#/media/File:Pieter_Bruegel_de_Oude_-_De_val_van_Icarus.jpg

Wikipedia. Bruegel the Elder. Hunters in the Snow. (n.d.). Retrieved from https://ru.wikipedia.org/wiki/Охотники_на_снегу#/media/File:Pieter_Bruegel_d._Ä._106b.jpg

Wikipedia. Bruegel the Elder. Triumph of Death. (n.d.). Retrieved from https://ru.wikipedia.org/wiki/Триумф_смерти_(картина)#/media/File:Thetriumphofdeath.jpg

Wikipedia. Nurenberg. Dance of Death. (n.d.). Retrieved from https://en.wikipedia.org/wiki/File:Nuremberg_chronicles_-_Dance_of_Death_(CCLXIIIIv).jpg

Wilde, C. (1994). Painting, Alberty and the Wisdom of Minerva. *British Journal of Aesthetics*, *34*, 48–59.

Wright, L. (1983). *Perspective in Perspective*. London, UK: Routledge & Kegan Paul.

Znamerovskaya, T. P. (1975). Quattrocento problems and Masaccio's work: *Abstract of a dissertation for the degree of Doctor of Art History*. Leningrad, Russia. State un-t them.

## ENDNOTES

1 Other sources also supported the interest in antiquity: Roman popes and German emperors constantly appealed to the greatness of Rome, involving Guelphs and Ghibellines into their struggle; largely based on the same has built his career Cola de Rienzo, whose banners were decorated with the abbreviation S.P.Q.R. of the times of the Roman Republic; the majestic ruins themselves created an appropriate atmosphere – the same Cola de Rienzo left his work about them: "Description of the City of Rome and its magnificence".

2 A classic example is the Borgia family (Pope Alexander VI, his sons Cesare, cardinal, and Gioffre, Duke of Gandia, and also daughter Lucretia, Countess of Pesaro, Duchess of Bisceglie, Duchess-Consort of Ferrara). They are credited with, in addition to nepotism, moneymaking and simony, incestuous ties, violence, poisoning. In fairness, let us say that modern historians have questioned many of these episodes.

# Chapter 12
# Modern Human Mind


## ABSTRACT

*The ways of depicting space in the paintings of modern artists is considered. The modern version of the World Tree myth is described. A qualitative assessment of the activity of the channels of human interaction with the environment, from the time of the Upper Paleolithic to the present, illustrated by changes in plot and expressive means of the World Tree myth was obtained. Markers of an evolutionary changes are analyzed and the most active channels are established. The results are presented in the form of generalized psychological portraits and descriptions of behavior patterns. Simple tests of personality research were selected and how to use the test data to compile an individual psychological portrait was selected; this is necessary to determine personal motivation and needs, as well as conditions of comfort.*


## BACKGROUND

A practically modern type of human mind was formed during the Renaissance. Since then, the concept of three-dimensional external space has become the object of scientific analysis of geometry, and the internal subjective space has become the subject of the study of psychology. There is much more in common between psychology and geometry than it might seem at first glance - both of them use logical analysis, and the processing and presentation of psychological data are often based on a mathematical apparatus and various geometric models.

First of all, the scientific justification was obtained for the methods of accurately displaying objects of three-dimensional space on a two-dimensional plane, such as linear perspective, axonometry, and orthogonal projections, which is due to their practical value. They have engaged in a special section of geometry - descriptive geometry, and in recent decades - computer graphics (Kovalyov, Vereschaga 2012). There is also an interest in the scientific study of curvilinear perspective, due to its property of visually more accurate transmission of the depth of space compared to a linear perspective.

Painting continued its development. Several stylistic trends have changed, for each of which extensive art research has been carried out. Relevant literature includes the publication of works, biographical data of artists, an assessment of technical techniques and stylistic features, including ways of depicting space, as well as interpretation of works and their consideration in comparison with paintings by con-

DOI: 10.4018/978-1-7998-1706-2.ch012

temporaries and in historical retrospective (Wattenmaker, 1993; Cahoone, 1996; Harvard, 1998; Childs, 2000; Dempsey, 2002; Mallén, 2003; Mullins, 2006; Artnet, 2014; Thiébaut, 2018; Artists, n/d). It can be noted that all methods of image space were used, which were by no means reduced to the classical linear perspective.

Summarizing, it is possible to state that in the preceding modern era, the period that historians call the New Time and count from the end of the Renaissance, two tendencies developed, noted in Chapter 11: the strengthening of rationalism and the increase in the value of feelings and emotions.

About 140-130 years ago, the balance between them was disrupted, and the rational motives of the artists receded into the background, freeing up space for the depiction of both immediate, "here and now" impressions and the game of imagination.

This period, which continues in our time, is defined as modern.

The tasks of determining the current state of mind, obtaining a general picture of its evolution, as well as reconstructing patterns of behavior remain relevant.

However, now is the time to discuss not only "academic," but also the practical objectives of this study.

There are many possibilities for applying a systematic approach and a scenario of self-organization to various technical and life situations. The authors wanted to focus on one consequence that is important for every person - the choice of a comfortable home.

Theorists of architecture, since antiquity, conducted extensive research on the problem of creating a comfortable home, or, in general, a comfortable living environment, which can include, for example, the urban environment. Since the 19th century, representatives of other sciences have joined in such studies - thermophysics, acoustics, specialists in safety, ecology, ergonomics and so on. As a result, the problem of creating a physiologically comfortable home has been solved - which cannot be said about a psychologically comfortable home.

Relying on their approaches and refusing in this case excessive academism (the more "hard" publication (Mkhitaryan, 2017; 2018) the authors go to the level of personality research and provide all the necessary data (include psychological tests (Urbina, et al., 1997; Robert 2003; Kaufman, 2009; American Psychological Association, n/d; British Psychological Society, n/d) for a rational choice of housing, which will be discussed in the Chapter 13.

## GENERAL CHARACTERISTICS OF THE EPOCH AND THE PREVIOUS PERIOD

The period which was immediately preceding the contemporary epoch is called the Age of Reason, thus proving the indisputable fact of the prevalence of nous among all the channels of human interaction with the environment.

However, it should be mentioned such negative events as the numerous wars and revolutions that caused the genocide of the aristocracy that degraded considerably so far.

The human race has become even more distant from its origins; the concept of obedience, asceticism, and modesty has been completely abandoned, at least it was evident within the public mind. Having no criterion of truth, the optimal scheme of social organization - meritocracy – has lost its value as a means to bring the most successful persons into power, if you don't know who is more successful? A saint person? A tyrant? A hero? A genius? A businessman who has earned more than the others? Or maybe the most exalted person who seems to be a charismatic man to people?

The Age of Reason never gave us the answer to those questions, neither it made a human person happier.

The contemporary epoch has accepted all the legacies of the previous era, but it has not found solutions for the problems – again during the revolutions they murdered many successful researchers, two World wars passed, when many fighters lost their lives, pressure from the mass media took place that prevented many prominent individuals from being successful due to the fear of oppression, along with the religious and spiritual themes that have already faded away long ago, but still used to make the quality of human mind significantly worse.

The significant reduction in potential should also be noted in this respect. Due to this reason, the mind of primitive tribes is not similar to that of the mind of the Paleolithic or Neolithic humans; the problems of most religious communities are no longer spiritual, but rather a political issue; the sovereign rulers never see themselves as the gods' sons, neither they set any special tasks; the fighters are never fighting for war, but for peace; the scientists usually focus on private tasks with business offers; business has become financially speculative, while the figures of mass culture are having success; plebeians and marginals lead, according to the standards of previous eras, the unprecedented luxurious life.

In this context, the shift from the rational world and enhanced trends of sensual perception doesn't seem to be surprising, thus, making the essence of the contemporary processes of evolution. Let's consider the typical contemporary means of space depicting.

## SPACE IMAGES IN MODERN PAINTING

First of all, it is possible to note the use of the techniques that were characteristic of primitive humans (Figure 1); there are the two-dimensional space, polymorphism of the figures, the contrast features, etc. (Chapter 6, *Neolithic: separation and war*) at the paintings of the contemporary Australian Aborigines (Figure 2) and primitive artists (Figures 3-4) (Pryimachenko, 1979).

*Figure 1. Aboriginal rock art on the Barnett River, Mount Elizabeth Station, Australia. Photographer: Graeme Churchard (Wikipedia. Aboriginal rock art, n/a).*

*Figure 2. Contemporary Aboriginal Art in Australia By Japingka Gallery, 2012 (Japinga-Original-Art, 2012).*

*Figure 3. Maria Primachenko (Maria Priymachenko). Black Beast. (approx. 85 years ago). Ukraine (Wikiart. Maria Primachenko, n/d).*

*Figure 4. Maria Primachenko. Two Titmice (approx. 35 years ago), Ukraine (Storinka-Kiev. Maria Primachenko, n/d).*

It is necessary to emphasize the difference - the absence of scenes of war and aggression in the works of contemporary artists.

Secondly, the same as ancient artists, two-dimensionality of images are used, which are sometimes influenced by light and shadow, the division of space into separate plans located at different distances from the viewer, overlapping images over distant objects (Figure 5-8).

Many works with the features of those techniques (in particular, the images are shown in Figures 5-6) possess a fabulous, mystical character, and the methods of space illustration seem to move the viewer into such locations (Isaeva, 2016; Thiébaut, 2018).

Other works possess the public and "poster-like" features (Figure 7-8), but they also impact upon the viewer as if from the "timelessness space", directly affecting the human mind and will (Bailey, Colin B., et al., 2001; Moor, 2007).

Thirdly, there is a noticeable interest in illustrating the polymorphic figures with extremely distorted proportions, often with mythological prototypes (Figures 9-11) (Descharnes, Robert & Nicolas, 1993; Mallén, 2003).

It has been already noted that the polymorphic figures are characteristic not only for the paintings of the Ancient World but also for the earlier periods, i.e. the Neolithic and even Paleolithic eras.

Fourthly, there are abstract forms that somehow remind of synoptric signs which are characteristic of the prehistoric period as well (Figures 12-13) (Néret, 2003; Artnet, 2014).

Fifthly, various forms of perspective perception have been applied, ranging from distortion of the linear perspective up to the quite systematic use of the curvilinear perspective which reminds of the technique of late antique artists and masters of the Northern Renaissance (Figures 14-15) (Machotka, 1996).

Sixthly, the images using the linear and aerial perspective as well as the effects of light and shade (Figures 16-17) have been widely used, it corresponds to the technique evident in the works of artists of the Late Middle Ages and the Italian Renaissance (Fedorov-Davydov, 1957; Ostrovsky, 1967).

*Figure 5. Ivan Bilibin. Vasilisa the Beautiful leaves Baba-Yaga with a «gift» (approx. 120 years ago). Russia (Wikimedia. Bilibin. Vasilisa the Beautiful, n/d).*

*Figure 6. Alphonse Mucha. Sokol Festival. (approx. 110 years ago). (Allpainters. Mucha .Sokol Festival, n/d).*

*Figure 7. Gustav Klimt. Portrait of Adele Bloch-Bauer II. Approx.110 years ago. Belvedere, Vienna, Austria (Wikimedia. Klimt. Portrait of Adele Bloch-Bauer II, n/d).*

*Figure 8. Dmitriy Moore. Have You Signed Up as a Volunteer? (approx.100 years ago) (Catalog9.Ty zapisalsya dobrovolcem, n/d).*

*Figure 9. Salvador Dalí. The Persistence of Memory (approx. 90 years ago) Museum of Modern Art, New York City, N°162.1934.0 USA (Wikipedia. Dalí. The Persistence of Memory, n/d).*

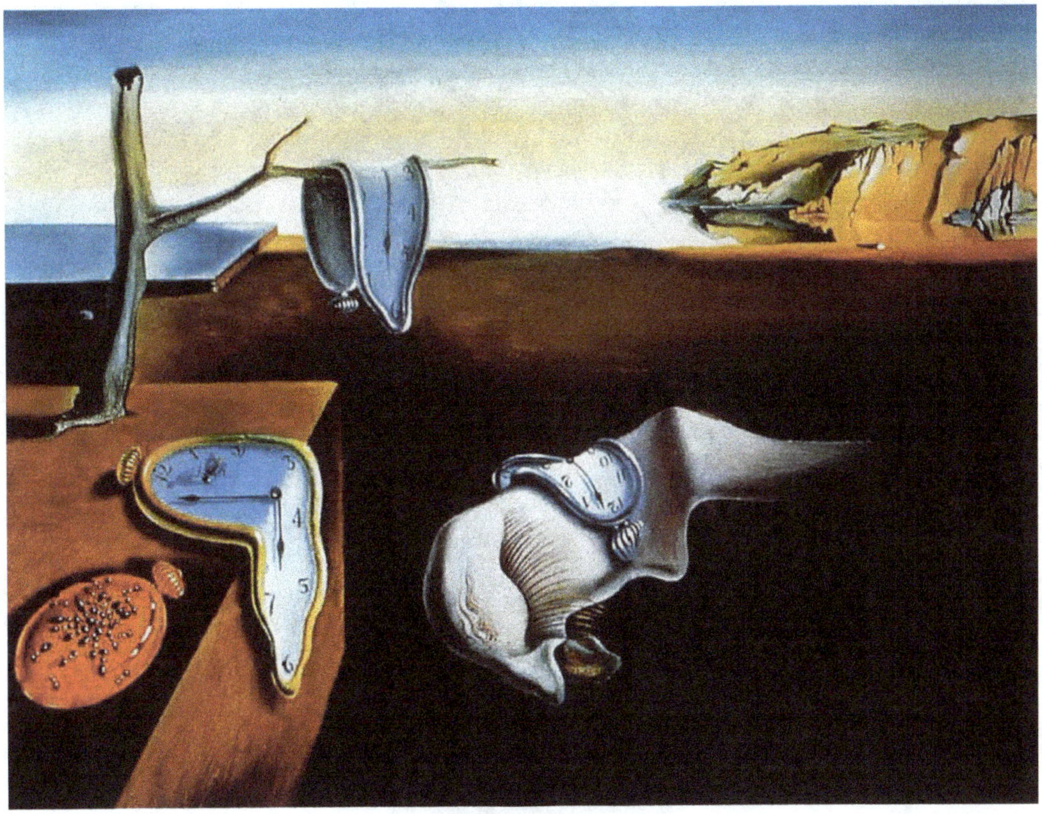

*Figure 10. Pablo Picasso. Guernica (approx. 80 years ago) The Prado Museum, Madrid N° DE00050. (Wikipedia. Picasso_Gernica, n/d).*

*Figure 11. Ivan Bilibin. Alconost (approx.100 years ago) (Wikipedia. Bilibin. Alconost, n/d).*

*Figure 12. Kasimir Malevich. Black suprematic square (approx. 105 years ago) The State Tretyakov Gallery, Moscow (Wikimedia. Malevich. Black suprematic square, n/d).*

*Figure 13. Pierre Soulages. Painting. (approx. 55 years ago) Honolulu Museum of Art. accession 4400.1. (Wikipedia. Soulages. Honolulu. Painting, n/d).*

*Figure 14. Paul Cezanne, Mont Sainte-Victoire above the Tholonet Road, approx. 125 years ago, State Hermitage Museum, Saint Petersburg, Russia. (Wikipedia. Cezanne, Mont Sainte-Victoire, n/d).*

*Figure 15. Paul Cezanne, Banks of the Marne, approx. 30 years ago, Pushkin Museum Nº 3416, Moscow, Russia (Wikipedia. Cezanne. Banks of the Marne, n/d).*

*Figure 16. Arcadiy Rylov. In blue space. (approx.100 years ago). The State Tretyakov Gallery. Moscow, Russia. (Wikimedia. Rylov. In blue space, n/d).*

*Figure 17. Joseph Bokshai. Panoramic view of Uzhgorod (approx. 70 years ago). Ukraine. (Uzgorod-novosti. Bokshai. Panoramic view of Uzhgorod, n/d).*

*Figure 18. Piet Mondrian. Composition with Color Planes and Gray Lines 1 ≈125 years ago (Wikimedia. Mondrian. Composition, n/d).*

Finally, and seventhly, the game of abstract color figures has been structured, it either arranges the color space strictly or make the free impact upon human emotions and feelings through the color perception, in both cases this principle prevails over the geometric forms (Figures 18-23) (Milner, 1992; Album "Adalbert Erdeli", 2007; Koehler, 2010).

Let's compare the mentioned methods of space illustration with the states of activity of various channels of interaction with the environment (Table 1).

However, the feature 7 hasn't been defined only, at least in its mere appearance, in the painting of the previous eras; on the other hand, such a range of space illustration has never been observed before.

The same variety has been evidenced in the styles of other art trends. For example, in architecture, the beginning of the modern era was marked by the emergence and development of Modern (Secession, Art Nouveau, Jugendstil), but this style went out of fashion soon. The same event happened to some other trends, such as constructivism, functionalism, modernism, post-modern, deconstructivism, hi-tech, conceptualism, etc. - even the attempt to present the list of the main trends runs the risk to remain incomplete due to the variety of styles.

This circumstance requires a thorough explanation.

However, before proceeding to the explanations of this kind, let's consider the situation related to the Myth of the World Tree.

*Figure 19. Samuel Lewis Francis. Untitled. ≈60 years ago, 2012 Sam Francis Foundation, California, Artists Rights Society (ARS), NY, Photographer: yigruzeltil (Wikiart. Sam Francis, 2012).*

*Figure 20. Laurent Parcelier Landscape. By: usenkomaxim. (Usenkomaxim. Laurent Parcelier, n/d).*

*Figure 21. Adalbert Erdely. View of Uzhgorod (spring) (approx. 70 years ago) Ukraine (Uzgorod-novosti. Erdely. View of Uzhgorod, n/d).*

*Figure 22. Sergey Fedotov. Self Portrait (Fedotovartist. Avtoportret, n/d).*

*Figure 23. Leonid Afremov. Coming back to the dream (Cameralabs. Afremov. Coming back to the dream, n/d).*

*Table 1. Comparison of the methods of space depicting with the activity of the channels of interaction with the environment*

| Feature | Feature Description | The activity of the channels (group distribution by the levels) |
|---|---|---|
| 1. | Two-dimensional space, flat figures, contrast with the background, polymorphic figures | 1 (weak), 2, 3 |
| 2. | Depth is transmitted by several backgrounds, figures are flat | 1 (weak), 2, 3, 4 (less) |
| 3. | Distortion of proportions, polymorphism of figures, their fantastic or mythological features | 1, 2, 3 (all weak) |
| 4. | Abstract shapes and symbols | 1, 2, 3 (all weak) |
| 5. | Perceptive and curvilinear perspective | 1, 2 и 3 (less), 4 (stronger) |
| 6. | Linear perspective, air perspective, effects of light and shape | 1- 3 (weak), 4 (even stronger), 5 |
| 7. | Color space prevailing over the shape | 1-4 (very weak), 5 (strong) |

## THE MYTH OF THE WORLD TREE TODAY AND IN RETROSPECTIVE

The World Tree is found in religious books, in the practice of modern shamans and esoterics, in scientific research, in paintings, in literature, on tattoos or wall-paper, as a symbol of mass culture.

All means of previous epochs are involved in its image or description. Our time has added to them those techniques that indicate the complete profaning of the myth in the mass consciousness.

Having characterized the current state of the myth and comparing it with the variations discussed in chapters 5–12, drawing up a graph of the changes in the activity of the channels of interaction with the environment in different historical epochs.

Lines of different colors (Figure 24) show the activity of interaction channels (grouped by level) in different periods. Of course, this is only a qualitative, not a quantitative assessment, based on the data presented in chapters 5-12.

The next line refers to all the epochs discussed in these chapters.

The next line shows the images of the World Tree characteristic of each epoch.

The last line indicates the area where the corresponding image was used.

The Figure 24 shows how from a symbol, with a characteristic ambiguous meaning, directing mind in a certain direction, but not limiting its depth, the World Tree turns into a sign with its literal meaning, "flatness", excluding any depth of understanding. And this is typical for a profane, almost devoid of intuition.

*Figure 24. Modification of the means of depicting the World Tree and changes in the channels' activity as the evolution of the human mind (the lines of activity illustrate a qualitative value only, but not a quantitative one).*

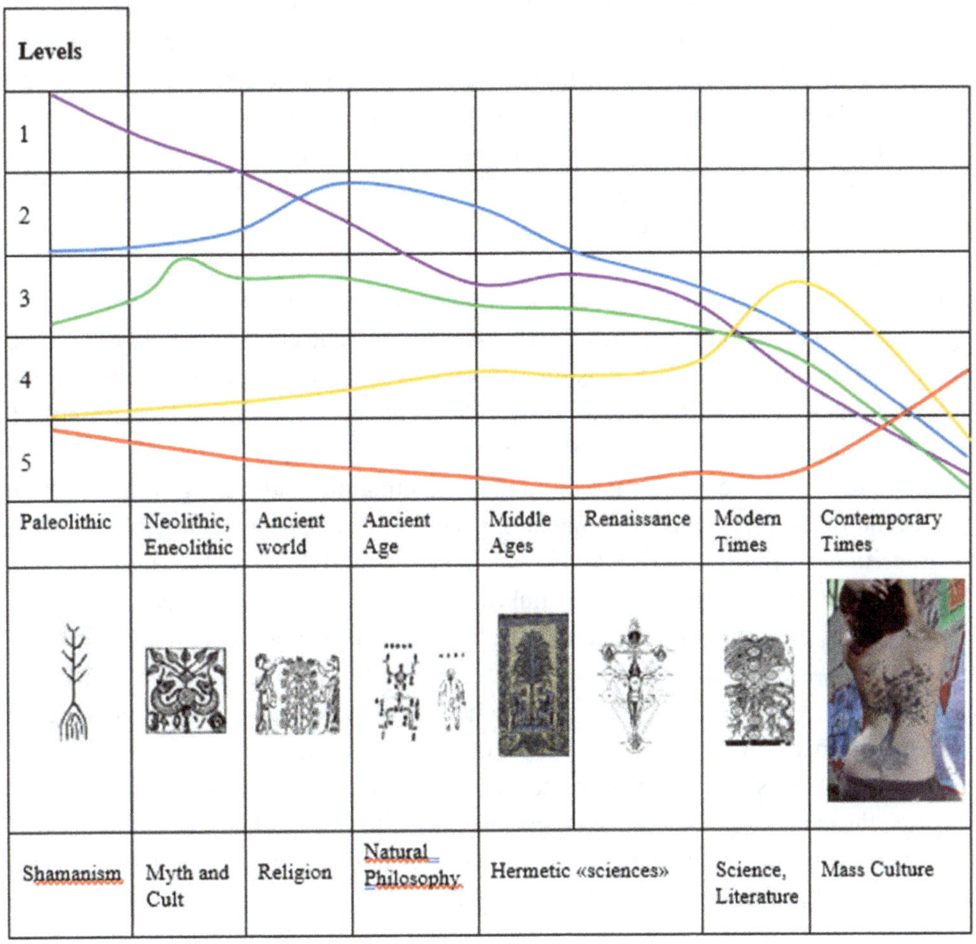

## MARKERS ANALYSIS

**Space Depicting**: All the techniques of imaging of the space depth are applied, that have been developed during the previous eras. The role of canons is decreasing, along with the growing number of styles and trends and the desire for complete freedom of expression. Feelings of violence and aggression become reduced, but there the competition between styles appears. Linear and perceptual perspectives are widely common. The sensual-and-emotional feature is significantly revealed, especially when creating the color space.

**The World Tree Myth**: Only the tribal shamans who live in remote areas and adherents of esoteric concepts consider the World Tree a concept having its operational value. Our epoch is the witness of such signs as the weakening of intuition (loss of meanings), critical weakening of the ego (suggestibility), weakening of the reason (inability to determine between the "good" and the "evil") and will (passive behavior in situations requiring decisiveness), at least, the lack of development of nous, enhancement of the subjective significance of the sensitive-and-emotional channels (weakening of the visual and hearing organs in comparison with the pre-historic times).

## STATE OF MIND

The symptoms of the end of the European-type civilization were noticed by Oswald Spengler in his book "The Decline of the West", published in 1918 (Spengler, 1993). Considering the "European centricity" of the civilizational model and the means of space illustration (chapter 10, *Renaissance: the legalization of feelings*) that has been evidenced since the period of the late Middle Ages, a proper interpretation of contemporary modern changes can either confirm or deny this point of view.

Does the reference to the ancient ways of space illustration mean a certain reversed evolution of the human mind, just as it happened in the Middle Ages?

No, it does not mean this.

For the contemporary mind (i.e. the postmodernist one), the game of images, ways, ideas without penetrating their essence - a trait that Hermann Hesse noticed ("The Glass Bead Game") (Hesse, 1943) is an important feature. Images cease to be symbols, such as, for example, icons connecting the viewer, the faces of saints and their living souls in an act of prayer, and become signs as, for example, formulas, whose connection with the designation is conventional.

But with one exception - the connection of colors with emotions and feelings remains direct, understandable, not requiring conventions or explanations. And this is direct evidence of the growing trend towards the transition of mind to the fifth level, which means - and this is sad for humanity - the loss of rational thinking, science, as the field of this thinking, and, ultimately, rationality.

Is it possible under such conditions to reverse, to intuition, to increase potential?

For humanity as a whole, no longer exists. Compared to the Upper Paleolithic "supermen", who have a large and complex brain and a strong body, our contemporaries, with a diminished and degraded brain, are weak and unable to survive alone. If the people of the Upper Paleolithic easily reached intuitive insights about the three worlds of a single Universe, our contemporaries easily reach only hysterical states, are unable to see the truth, desire primitive entertainments. They are a clear illustration of the scenario of the evolution of the mind predicted in Chapter 4: unfortunately, humanity is in a state of decrepitude, a decrease in potential, a fall into insanity and is unable to change in response to changing external

*Figure 25. State of mind of a contemporary human being*

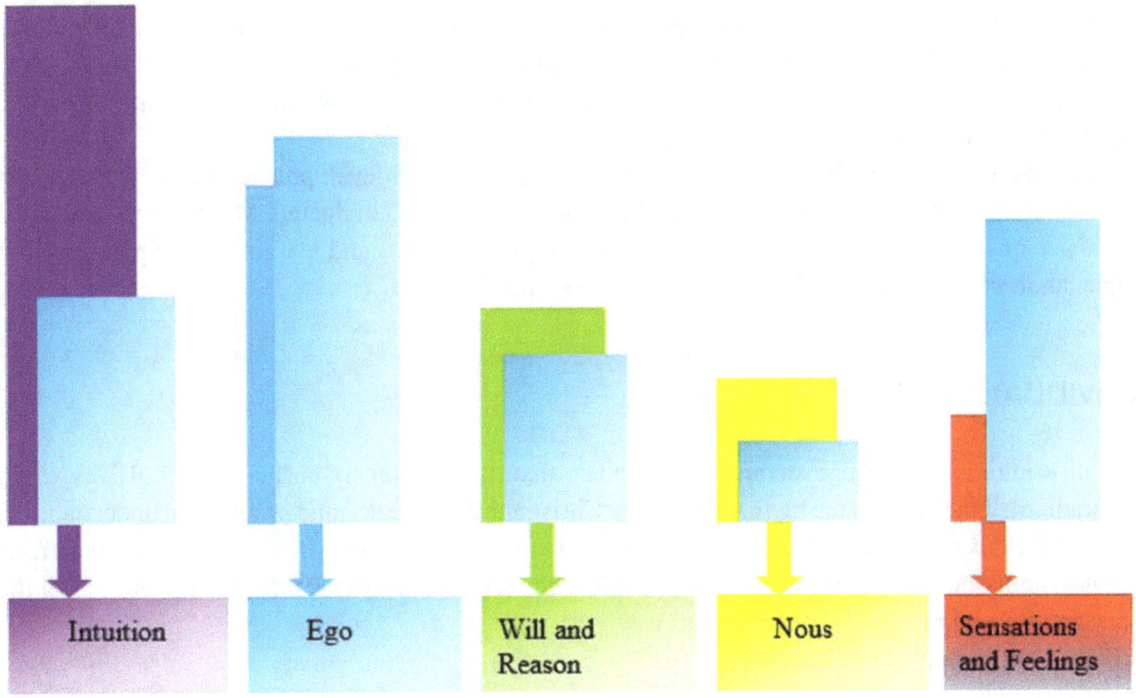


conditions. And it does not even see the need, nor the paths, nor the incentives for such changes. We conclude the above discussion with a generalized psychological portrait of a modern person (Figure 25).

## BEHAVIOR PATTERNS

Let's comment on Figure 25 using the application of Tables 1-6 of chapter 4, *Self-organization: evolution and behavior*.

**Intuition** is developed below average or, more often, frankly weak. This means the presence of a certain insight only at the household level, a vague love of nature (hence the loud theme of ecology), sociability (increased attention to public relations), and, of course, an "understanding" of feelings. Deafness to all "spiritual" impulses and senses, superstition, easy exposure to external manipulations.

**The Ego** is quite developed, however, when compared with the kings and heroes of the Ancient world, the difference is immediately noticeable. Kings and heroes felt themselves, descendants of the gods, had unusual abilities and abilities, set themselves super tasks. Our contemporaries judge everything "like gods", they want to have freedom "like gods", but they have neither the corresponding abilities, nor knowledge, nor global goals. Our confidence is only self-confidence, our pride looks ridiculous.

**The Will and the Reason** medium or low. Adventurism, perseverance, obstinacy, weakness. Who will decide, like Christ, to sacrifice himself for the sins of mankind? Who decides to go down to the Duat, undergo trial and trial, win and go out into the light? No one poses such tasks, and talk is about "the will to power" and prudence necessary for "business".

**The Nous**. It is customary to judge him by his successes in science. And yes, processors are getting faster, and smartphones are getting thinner. But computers, and flights into space, and nuclear energy, and genetics existed many decades ago. Has humanity found a new paradigm over the years in understanding the structure of the world, life, and death, consciousness, has it even learned to travel to the stars, not to mention the ways of the underworld? Not. As it could not prevent world wars, defeat disease and hunger. So, the rating is no more than average.

**The Sensations and Feelings** have a medium and high degree of development. Society in every possible way contributes to their growth; targeted propaganda is being conducted. More and more exalted artistic personalities appear, inclined to imitate meaningful activities and marginals, incapable even of simple labor. Perhaps they are the main "heroes of our time".

## INDIVIDUAL LEVEL

The path to intuitive knowledge was and remains open: albeit unconsciously, but humanity still lives within the paradigm of the World Tree myth, renewed by Christianity; published and awaiting an understanding reader, The Book of the Dead and other similar texts; there are time-tested yogic and monastic practices.

For an individual, very rare in our time, people who feel the need to take this path, there are all the necessary tips and non-zero chances of success.

But all other people need comfort.

Let's show how to do it.

### Testing

Life is the most reliable test ever. It's quite possible to consider statistical data based upon the numerous real situations when not only the abstract "reason" or "ego" are acting, but our personality as a whole, it helps us build the most objective psychological portrait. A classic example is Evariste Galois who was undoubtedly a brilliant mathematician, however, even studying with his tutors, he failed to pass the exams twice, trying to enter the Polytechnic School that was the leading highest educational institution of France in his time. The enormous stress during the examinations was too strong for him, and high intelligence could not have helped him.

However, only a few people can evaluate their strengths and weakness impartially, so this test might fail to present an objective point of view.

The next group is the projective tests that also help us evaluate the personality as a whole. Those tests include, for example, the Rorschach test as well as such tests as the "Unfinished Story" or the "Unfinished Drawing". They simulate a certain situation close to reality when a person performs a task without knowing what solution is expected of him and should act without prompts and restrictions. The specialist conducting the testing, based upon his experience and knowing the results of such tests, interprets a person's decision and makes up his psychological portrait.

The disadvantages of projective testing are also obvious - the result depends on the qualifications of the specialist, and his presence is not always desirable.

The third possibility is presented by the test questionnaires, often published even in ordinary newspapers or on the Internet. Using gaining a certain number of points after having answered the questions (the answers are chosen based upon a multiple-choice test), a person can evaluate one of his qualities or even some of them.

However, how summarize all those qualities into a whole psychological portrait? Especially in that case when the tests apply different methodological approaches and are not related to each other? These circumstances are the main disadvantages of this testing method. However, such tests are simple; anyone can take them on their own. Let's try to reduce the shortcomings by selecting tests and processing the results[1].

***Intuition Test.*** (Greatpicture. test-na-intuitsiyu. n/d).

In the written version, which is presented below, some components of the test were reduced.

Pattern.

Record the points for the answers as per each question:

- most often - 3 points;
- sometimes - 2 points;
- never - 1 point.

In the end, calculate the total number of points.

Questions.

1. You do not like to make plans and deal with organizational issues. You prefer a spontaneous approach.
2. When you are looking for a solution to a problem, you are guided by your primary sensations. In the end, your choice becomes the right one.
3. When you first meet a new person, you have a certain opinion about him, which will be confirmed later on.
4. Your intuition helped you win in some lotteries and contests. The financial resources invested in any project have been successfully paid off.
5. You know your life mission and purpose in this world.
6. When you don't listen to your inner voice, you get the wrong result and feel sorry for your actions.
7. While working, you rely on your sixth sense, and your colleagues wondering how you manage to do it.
8. You can find a way out of any situation, even if you have a few facts.
9. You are a creative person and generate interesting ideas that result in spectacular success.
10. You feel anxious or excited before some positive or negative events.
11. You have unusual ideas that become perfectly realized in real life.
12. You can feel the emotional state of a close person, even if he is not nearby.
13. Friends and the persons whom you know consider you to be a man of ideas.
14. When you solve a problem, you rely on your inner feeling and do not listen to other people's advice.

As a result, your sixth sense never lets you down.

15. You can recognize a liar and understand whether a person is sincere or not.
16. You used to predict a disaster, death or illness on a recurrent basis.

17. When you don't know how to act in a certain situation, you start to meditate and wait for the true answer inside yourself.
18. You have an unusual relation to the strangers or places where you have never been, but where you incredibly want to go.
19. You are an extremely empathic person who can feel the sensations and emotions of other people, regardless of your desire.
20. You used to escape from dangerous situations, being unaware of it.
21. If you get lost in any place, you can intuitively find your way back.

Key.

The first level of intuition (from 20 to 30 points) means an internal system that informs of an eventual danger. The intuition can send you the signals that symbolize the coming danger. Approximately 70% of the people from the world's population have this level of intuition. Some training is required.

The second level (from 30 to 40 points) means the social intuition. You use the intuition to understand the emotional state of your beloved ones. You can empathize deeply with others and support any person in a difficult situation. As a rule, this level of intuition is characteristic for those social groups where the close emotional connection exists. If you are at this level of development of the sixth sense, then you can become a successful manager (feel the motives of behavior and the people's mood). Thanks to the systematic training, you are going to learn how to hear your inner voice and will be able to trust it not only in your everyday life but also in your career, creative development, and family affairs.

The third level (from 40 to 50 points) means the creative intuition. If you are characterized as a person belonging to the third level of development of intuitive skills, then you can easily understand what is happening around you. You are surely a creative person with the skills of a successful entrepreneur and businessman. The development of the sixth sense can help you and your beloved ones to promote in life.

The fourth level (from 50 to 60 points) means the highest level of intuition. The intuition constantly helps you achieve successful results. You know that your inner voice will never let you down, and you completely trust it. Every day you concentrate on your sixth sense and react to all internal sensations that arise. Thanks to your intuition, you have achieved great success both in your career and love affairs.

Now it's necessary to enter correctly the test results in the research methods as applied for this work. Let's check the data of Table 2, chapter 4, *Self-organization: evolution and behavior*:

- according to the table, the superlative degree of skills is not evaluated by this test;
- the first level corresponds to a low degree, the second one corresponds to a medium degree, the third and the fourth (they differ as per the frequency of manifestation, but not by the level) ones mean the high degrees;
- now, it is necessary to scale the result correctly. Look at Figure 22, take the purple (intuition) column, refer a height of 100 mm to it, it will correspond to 60 points of the testing results; 0 points will correspond to zero height of the column. Scale the scored points by multiplying them by a factor of 0.6, and get a blue column - your indicator for a psychological portrait. If you gain more than 60 points, the height of the blue column should be increased.

***Egoism Test***. (Azps. Test. Egoism, n/d

Pattern.

Answer "yes" or "no" to the following questions. Count the number of points you've gained table 2).

*Table 2. Questions/points for Egoist test.*

| Answer | Number of Questions | | | | | | | | | | | |
|---|---|---|---|---|---|---|---|---|---|---|---|---|
| | 1 | 2 | 3 | 4 | 5 | 6 | 7 | 8 | 9 | 10 | 11 | 12 |
| Yes | 11 | 2 | 1 | 1 | 4 | 5 | 2 | 2 | 1 | 2 | 4 | 3 |
| No | 3 | 1 | 2 | 4 | 1 | 3 | 1 | 1 | 3 | 1 | 1 | 1 |

Questions.

1. Do you think that everything has its price?
2. Do you have a passion for any color?
3. Do you laugh at yourself?
4. Do they like you in society?
5. Do you always give in during the disputes?
6. Do you know how to conceal your failures and disease from the surrounding persons?
7. Do you write down the debts of your friends?
8. Do you like to speak loudly?
9. Do you think you know people well?
10. Do you think that a saving book is a waste thing?
11. Do you think that fate has passed you by?
12. Are get excited about the failures of your enemies?

Key.

If the total amount of the points is more than 31, you are a selfish person.

If the total amount of the points is from 27 to 31, you are a selfish person, and you have few friends because to demonstrate scarce warm feelings to people.

If the total amount of the points is from 20 to 26, you are a responsible person who is willing to help others.

If the total amount of the points is less than 20, sometimes you behave very selfishly, and sometimes your unselfishness and self-sacrifice make your friends feel rather surprised.

Lets' check the data of table 3, chapter 4:

- according to the table, the excellent and high degrees are is not evaluated by this test;
- the total amount of the points up to 26 corresponds to a low degree, while the amount from 27 to 33 points corresponds to the average degree;
- now scale the result correctly. Look at Figure 22 and take the blue column (it corresponds to the egoism) and refer a height of F*100 = 62 mm to it; 33 mm of the height of this column will correspond to 33 points of the test; 0 points will correspond to the zero height of the column. If you evaluate your egoism as "high", the height of the blue column should be equal to the height of the blue, and for the "superlative" degree the height of the blue column will exceed the height of the blue.

To improve the accuracy, you can pass some additional tests for self-improvement, ambition and the test of the kind.

***Self-determination (Reason) Test.*** (Azps. Tests. Riskman, n/d).

Pattern.

Choose one variant of the offered ones.

Questions.

1. As for you, what moves a person in his life, first of all?
   A - curiosity; B - desire; C - necessity.
2. Why do you think people move from one job to another?
   D - they are fired; E – they leave because of the big salary; F - they prefer some other work.
3. When you get into trouble, what do you do?
   A – you postpone the solution of the problem to the latest moment; B - you need to analyze how guilty you are; C - you don't even want to think about what happened.
4. You did not have time to do some work. What happens after?
   F – you declare your failure even before it becomes to be known; E - you fear to be asked about the results; D – you've been preparing for a thorough explanation.
5. When you reach some goal, how do you evaluate the news about it?
   C - with some sense of relief; B - with violent positive emotions; A - in different ways, depending on the goal, but not so violently.
6. What would you recommend to a very shy person?
   D - to avoid situations that require taking some risk; F – to get rid of it by consulting a psychologist; E - to meet people of another nature who never suffer from being shy.
7. What will you do in a conflict situation?
   B - I will talk to someone who made me involved in the conflict; A - I will write a letter;
   C - I will try to settle the conflict as an intermediary.
8. What kind of fear do you have when you are wrong?
   D – the fear that a mistake can change the order to which you've got accustomed; E – the fear of punishment; F – the fear of losing one's prestige.
9. When you talk to someone, how do you look?
   A – I look away from time to time; B – I look directly into my interlocutor's eyes; G – I look away, even when I'm addressed.
10. What is your speaking style during a conversation?
    F - the tone of the conversation usually remains calm; E - you insert some meaningless words now and then; D - you repeat, worry, your voice starts to let you down.

Key.

If the majority of answers refer to the options of C and D. You are scared not only to make decisions but even to think about them, being afraid of the coming events... Your psychological state cannot be called stable and favorable. You often expect to be criticized for your actions rather than be encouraged... Do you have any other answers? If not, then you have the problems not only with yourself but also with the surrounding people.

If almost all the answers consist of the options A and E. You are not a particularly determined person, but you cannot be called indecisive either. You don't act actively and quickly all the time, but only because you think that it's not worth it. You like brave people. But you can justify the undetermined persons often because you believe that their actions are not the results of fear, but a caring attitude. When you read books, watch films about some brave and desperate people, you often would like to be in their shoes.

If you chose the options B and F mostly. You are a determined person. Very often you neglect the things that you consider small and insignificant. In spite of this, you are appreciated as an independent and interesting person. If you also have a sense of responsibility, then you are often engaged with complex tasks, but in this case, there should be other types of people in your group who would balance your exaggerated activity ... Don't you need to think over the decisions taken by you?

Let's check the data of chapter 4, *Self-organization: evolution and behavior*, and Table 4:

- test processing options refer to the low, high and medium degrees of self-determination;
- Scale the resultы. The height of the green column in Fig.22 (corresponds to the reason) F2*100 = 38 mm. The same value will define a blue column if the third variant of the test processing takes place. Depending on a greater or a lower degree, the height should be modified accordingly.

***Willpower Test.*** (Psylist. Prakticum-00465, n/d).

Pattern.

The following 15 questions can be answered: "yes"; "I don't know" or "it happens sometimes", "it happens"; and "no". If you answered: "yes", the option refers to 2 points; "I don't know" or "it happens sometimes", the option "it happens" refers to 1 point; the option "no" refers to 0 points. You should immediately sum up the total amount of the points while answering the successive questions from 1 to 15.

Questions.

1. Are you able to complete the work you have started which doesn't seem interesting to you, even though the time and circumstances allow you to break away from it and then return to it again?
2. Did you have to overcome your internal resistance when you had to do take some unpleasant actions (for example, to go on duty on your day off)?
3. When you find yourself in a conflict situation at work (school) or at home, are you able to pull yourself together enough to give the best objective look to it?
4. If you are prescribed a diet, can you resist all cooking "temptations"?
5. Will you be strong enough in the morning to get up earlier than usual, as it was planned in the previous evening?
6. Will you remain at the scene to testify?
7. Do you give quick replies to the letters?
8. If you are afraid of the coming air flight or visiting a dentist's office, will you be able to overcome this feeling easily and not to change your mind at the last moment?
9. Will you take a very unpleasant medicine recommended by your physician?
10. Do you keep the promise that has been given on impulse, even if its fulfillment could bring you a lot of trouble, are you a person who keeps his word?
11. Are you going on a trip to the unknown city, if necessary, without any hesitation?

12. Do you strictly follow the daily routine: waking up, eating, studying, cleaning and other things?
13. Do you disapprove of bad library debtors?
14. The most interesting TV show does not force you to postpone the implementation of urgent and important work. Is it true?
15. Can you stop the quarrel and keep silent, no matter how offensive the words of the opponent party might seem to you?

Key.

The total amount of the points should be calculated. If it is within the range: from 0 to 12 points - you have a weak willpower, you are easy to convince. However, if a person knows his "weak points", he becomes stronger. You'd better work on yourself.

The range from 13 to 21 points means that your willpower is average. You act differently in different situations, sometimes showing miracles of flexibility and compliance, yet of persistence and determination. Your attractive features such as communication and original approach are significant for your activities.

The range from 22 to 30 points means that you have great willpower, you are a person of character. However, you are not alone, thus concessions, compromises, and negotiations also lead to your success in communication and affairs.

Let's check the Table 4 of chapter 4, *Self-organization: evolution and behavior*:

- test processing options refer to the low, high and medium degrees of self-determination;
- scale the results. The height of the green column in Figure 22 (corresponds to the will) is F2*100 = 38 mm. The same value will be referred to as a blue column if you score 30 points; the scaling factor is 1,27; multiply the number of points, to calculate the height of a blue column. The height of a blue column should be increased in case of a superior degree of willpower.

A comparison of the results of tests for self-determination and willpower will illustrate the personal subtypes as illustrated in Table 4. A comparison of will and mind of reason and mind (see the next test) will determine the trend to the business or scientific approach to the psycho-type of an adventurous researcher.

***Nous Developed Test.*** In terms of the notions defined in chapter 3, *Self-organization: body and mind*, and table 5 of chapter 4, *Self-organization*: *evolution and behavior*, this testing can be most adequately performed using the Raven Progressive Matrices Scale. The test is simple, it doesn't require much time for passing it, but it is large enough, and therefore not given here.

The materials and the order of processing the results (Psicabi. Test Ravena, n/d).

Let's demonstrate the scaling order of the test results only. So, using focusing on Figure 25, determine the height of a yellow column (it corresponds to reason): F3*100 = 24 mm, it will correspond to the level of high intelligence and 140 points. The scale factor is 24/140 = 0,17. It should be multiplied by the points by this number, to find the height of a blue column.

***Emotionality (Sensitivity) Test.*** (Psicabi. Test na Chuvstvitelnost, n/d).

Pattern.

You are offered a series of statements. If you agree with the statement, write "yes" next to its number, if you do not agree, then write "no". Every affirmative answer refers to 1 point.

Questions.

1. Can you blush strongly because of embarrassment or shame, so that you feel that your cheeks are blushing and the tears are coming up in your eyes?
2. Have you ever turned pale for fear or sorrow?
3. Do you often feel embarrassed when you are shy?
4. Can you cry easily for resentment, unhappiness, empathy or even because of joy?
5. Can you cry because of aesthetic pleasure when listening to music or reading poetry?
6. Have you ever got sweaty in unpleasant or difficult situations?
7. Do you have a dry mouth because of strong excitement?
8. Is your voice getting strained?
9. In moments of great excitement or embarrassment, do you feel the stiffness of the limbs when the legs become "rubbery" and grow weak?
10. Did you notice the trembling of the fingers in case of strong excitement or embarrassment, do you have an internal tremor and a state of "shivering down"?
11. Do you worry before each performance when you think as if you had forgotten everything?
12. Can you lose your train of thought, get confused and stop talking while answer at the exam or delivering the public speech?
13. Do you often get irritated and outraged?
14. Can you get angry with a child and punish him?
15. Do you quarrel with your beloved ones if you see the injustice of their actions?
    17. Does it often end with your tears, despair, and remorse?
    18. Can't you divert your attention from the troubles and sorrows, trying not to think of them or does bad mood completely get you involved for a long time?
    19. Do you become fussy in the moments of excitement or embarrassment?
    20. Do you feel any pain in the area of the solar plexus due to anxiety?

Key.
The greater the number of points, the higher the emotionality is:

0 - 3 points - low;
4 - 6 points - reduced;
7 - 9 points - average;
10 - 12 points - increased;
13 - 15 points - high.

The scaling is performed in the same way as in the previous cases: the height of a red column (it corresponds to the sensitivity) 100*F4 = 15 corresponds to 15 points; under a factor of "1", each point corresponds to 1 mm of a blue column. If emotionality reaches its exaltation, the blue column will become higher than the red one.

In particular, it's necessary to apply another type of testing, to control the emotional state at the time of testing, such as a stress-resistance test, anxiety test, and the test of the kind. However, taken into consideration that the tests for determining abilities are performed in a quiet and home environment, you can do without it.

## Compilation and Interpretation of a Psychological Portrait

It is necessary to summarize all the columns together, as illustrated in Figure 22. It is convenient to place the blue columns (test results) on the multi-colored background (i.e. to display the priorities of the channels for a balanced personality).

To determine the conditions of further comfort/inconvenience and, accordingly, the requirements for a dwelling (chapter 13, *Practice as a criterion of truth: the choice of a comfortable dwelling*), it is necessary to determine the needs and motivations. For each of the levels (i.e. the psycho-types), they are listed in table 7 of chapter 4, *Self-organization: evolution and behavior.* If you have two or more levels, respectively, there will be more priority needs and motivations - up to five requirements for a balanced personality.

Let us give an example of such a definition. For example, for an "egoist" with a highly developed analytical type of intelligence, it might look something like the mentioned scale (Figure 26). The ego and the mind prevail, the intuition and will, the ability for sensory perceptions in the redistribution of the norms are weakened. Use the data in Table 7 of chapter 4, *Self-organization: evolution and behavior.* The complex of needs and motivations will be as indicated in the boxes under the arrows.

Such a person will not be easy to become successful, because meeting the needs and realization of the motivations require the active actions, intuition is unlikely to tell you what to do, while the will and self-determination are insufficient.

*Figure 26. Presentation of the results of the personal study.*

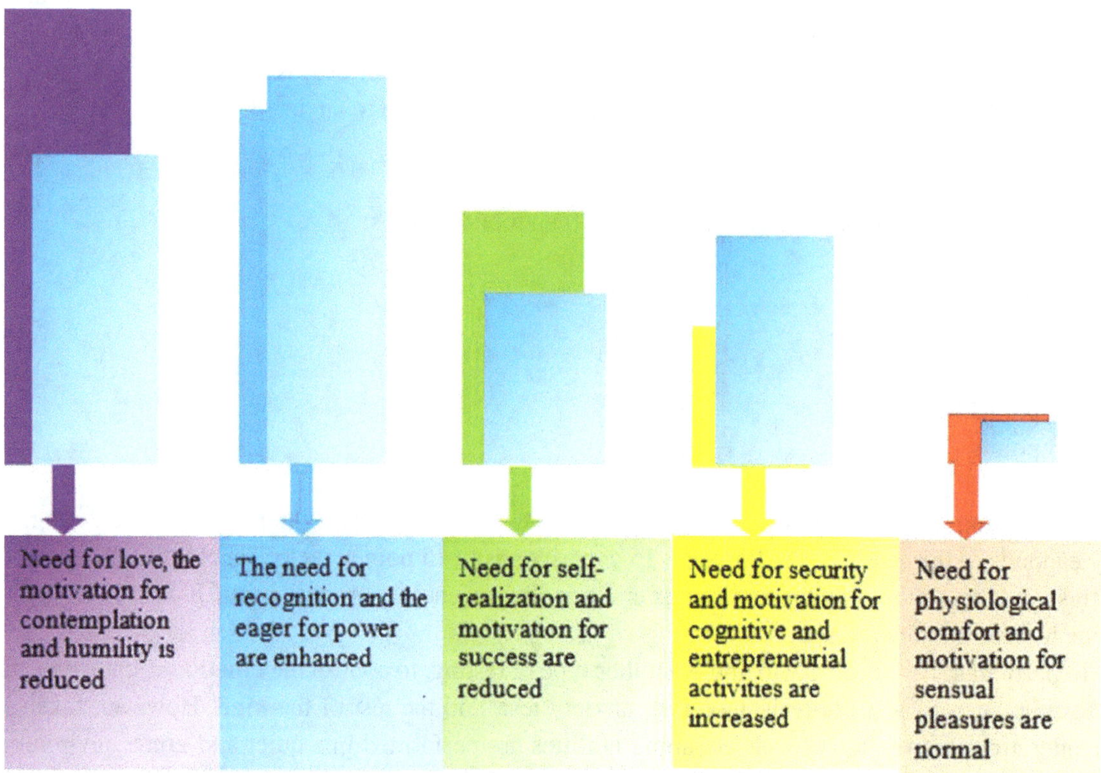

Further on, it's necessary to make adjustments as per the age (table 8, chapter 4, *Self-organization: evolution and behavior*) and temperament (the data have been provided in the same chapter). Let's continue the consideration of the example illustrated in Figure 23. Let's suppose that it's an ancient person with phlegmatic temperament. Then it will be even more difficult to meet the needs (the question will arise all the time - do I need it?); however, the possible failures will be felt, though being not so acute as compared to the years of his youth.

## CONCLUSION

The main ways of the space depicting in the modern epoch are identified; all of them, except one, were used in previous centuries. The construction of the color space, when colors prevail over forms or exist without them at all, is how our epoch differs from the previous ones.

Another difference is the separation of form from meaning, a sign from a symbol. This process, which began in ancient times and is associated with a weakening of intuition, "sanctified" by Aristotle's logic and rooted in nominalists, has reached its culmination in the postmodern consciousness.

Very important Figure 24 shows how the potential, activity of channels, features of the space imaging and the form of cognitive activity of a person changed from Upper Paleolithic for the present time. It is seen how the knowledge obtained initially based on intuitive experience is "translated" from epoch to epoch into its inherent "language", the expressive means of which depend on the channel's activity.

Currently, the meaning and content of the phenomena of science and scientific knowledge are determined by the logical form of thinking. Probably the next step will be the rejection of the logical form of thinking as a criterion of truth in favor of emotion as a criterion of truth.

Ironically speaking, geometry will become possible were only straight lines of the same color and, perhaps, sounds of the same tone will be considered a parallel. In practice, this will lead to the death of civilization of the type we are familiar with - Homo sapiens will simply turn into Homo and cease to be a reasonable person.

Testing this sad forecast with the state of the World Tree myth does not bring peace to our souls (this is a parody of the future emotional style). The same profanity is observed - "signs" instead of "symbols".

As in the previous chapters, a psychological portrait of a certain average person of our era is built. Its main characteristics are low, relative to the Renaissance, potential, as well as an imbalance in the channels of human interaction with the environment in favor of sensory perceptions and emotional reactions.

Also considered are the behavioral patterns of modern man depending on his place in the social structure.

The data are necessary for constructing a personal psychological portrait, determining needs and motivations based on it, as well as making refinements related to the stage of the life cycle and temperament are given.

## FUTURE RESEARCH DIRECTIONS

In the last chapter, all the results related to the determination of the state of mind and the characteristics of a contemporary person's behavior will be used practically - in order to determine a convenient and comfortable dwelling.

## REFERENCES


Album "Adalbert Erdeli". (2007). Uzhhorod: "Vidavnitstvo Oleksandri Garkushi", 2007. Azps. Test. Egoism. (n.d.). Retrieved from http://azps.ru/tests/stest/egoism.html

Allpainters. Mucha. Sokol Festival. (n.d.). Retrieved from https://allpainters.ru/wp-content/uploads/paintings/6th-sokol-festival-1912.jpg

American Psychological Association. (n.d.). Retrieved from https://www.apa.org/science/programs/testing/index

Artists. (n.d.). Retrieved from https://www.the-artists.org

Artnet. (2014). Pierre Soulages. Happy to Stay in the Dark, Ben Davis, June 19, 2014.

Azps. Tests. Riskman. (n.d.). Retrieved from http://azps.ru/tests/stest/riskman.html

Bailey, C. B., & Collins, J. B. (2001). Gustav Klimt: modernism in the making. New York: Harry N. Abrams in assoc. with National Gallery of Canada.

British Psychological Society. (n.d.). Retrieved from https://ptc.bps.org.uk

Cahoone, L. E. (1996). *From Modernism to Postmodernism: An Anthology*. Cambridge, MA: Blackwell.

Cameralabs. Afremov. Coming back to the dream. (n.d.). Retrieved from https://cameralabs.org/6587-krasivye-i-optimistichnye-kartiny-leonida-afremova

Catalog9.Ty zapisalsya dobrovolcem. (n.d.). Retrieved from http://catalog9.ru.net/ty-zapisalsya-dobro-volcem-kartinka.html

Childs, P. (2000). *Modernism*. London, UK: Routledge. doi:10.4324/9780203131169

Dempsey, A. (2002). *Art in the Modern Era: A Guide to Schools and Movements*. New York: Harry A. Abrams.

Descharnes, R. N. (1993). Salvador Dalí. New York: Konecky & Konecky, 1993.

Fedorov-Davydov, A. A. (1957). Picture A. Rylov "In the blue space". *Art*. 1957. No. 8. Fedotovartist. Avtoportret. (n.d.). Retrieved from http://fedotovartist.beon.ru/photos/1944-629-sergei-fedotov-avto-portret-jekspressionizm.html

Greatpicture. test-na-intuitsiyu. (n.d.). Retrieved from http://www.greatpicture.ru/test-na-intuitsiyu-naskolko-u-vas-razvita-intuitsiya/

Gregory, R. J. (2003). *"The History of Psychological Testing" (PDF). Psychological Testing: History, Principles, and Applications*. Allyn & Bacon.

Harvard, H. (1998). *History of Modern Art: Painting, Sculpture, Architecture, Photography*. New York: Harry N. Abrams.

Herman, H. (1943). *Das Glasperlenspiel*. Henry Holt and Company.

Isaeva, K. (2016, Aug. 16). *Russian fairy tales in the works of Ivan Bilibin*. Retrieved from www.rbth.com

Japinga-Original-Art. (2012). Retrieved from https://japingkaaboriginalart.com/articles/contemporary-aboriginal-art-in-australia/

Kaufman, A. S. (2009). *IQ testing 101*. Springer Pub.

Koehler, R. (2010, June 10). The Painter Sam Francis. Variety. Retrieved June 22, 2010.

Kovalyov, Y. N., & Vereschaga, V. M. (2012). Applied Geometry: Descriptive Geometry, Engineering, and Computer Graphics, Modern Trends. Kiev, Ukraine: Omega-L.

Machotka, P. (1996). *Cézanne: Landscape into Art*. United States: Yale University Press.

Mallén, E. (2003). *The Visual Grammar of Pablo Picasso*. New York: Peter Lang.

Maria, P. (1979). *Sixteen Paintings in the National Museum of Ukrainian Folk Art*. Leningrad: Shown on A Set of Sixteen Color Postcards Published by Aurora Art Publishers.

Milner, J. (1992). *Mondrian*. London, UK: Phaidon.

Mkhitaryan, N. M. (2017). *Man and Dwelling*. Gangemi Editore International Publishing.

Mkhitaryan, N. M. (2018). *Habitat: images of comfort*. Kiev, Ukraine: Naukova Dumka.

Moor, P. (2007). Archived 21 December 2007 at the Wayback Machine.

Mullins, C. (2006). *Painting People: Figure Painting Today*. New York: D. A. P.

Néret, G. (2003). Kazimir Malevich and Suprematism 1878-1935, Taschen, 2003.

Ostrovsky, G. (1967). Joseph Bokshay. Moscow, Soviet artist, 1967.

Psicabi. Test na Chuvstvitelnost. (n.d.). Retrieved from https://psycabi.net/testy/507-test-na-chuvstvi-telnost-diagnostika-emotsionalnosti-metodika-v-v-suvorova

Psicabi. Test Ravena. (n.d.). Retrieved from https://psycabi.net/testy/717-test-ravena-progressivnye-matritsy-raven-progressiv-matrices-metodiki-dlya-diagnostiki-intellekta-vzroslykh

Psylist. Prakticum-00465. (n.d.). Retrieved from https://psylist.net/praktikum/00465.htm

Spengler, O. (1993). The Decline of the West. Essays on the morphology of world history. M.

Storinka-Kiev. Maria Primachenko. (n.d.). Retrieved from https://storinka-m.kiev.ua/product.php?p_id=2646

Thiébaut, P. (2018). *Mucha et l'Art Nouveau*. Paris, France: Éditions du Chêne. (in French)

Urbina, S., & Anastasi, A. (1997). *Psychological testing*. (7th ed.). Upper Saddle River, NJ: Prentice Hall.

Usenkomaxim. Laurent Parcelier. (n.d.). Retrieved from http://usenkomaxim.ru/2013/01/03/laurent-parcelier-sovremennyiy-frantsuzskiy-impressionizm/

Uzgorod-novosti. Bokshai. Panoramic view of Uzhgorod. (n.d.). Retrieved from http://uzhgorod.in/novosti/2015/dekabr/uzhgorod_v_iskusstve_naslazhdaemsya

Uzgorod-novosti. Erdely. View of Uzhgorod. (n.d.). Retrieved from http://uzhgorod.in/novosti/2015/dekabr/uzhgorod_v_iskusstve_naslazhdaemsya

Vitruvius. (1936). *Ten books about architecture*. Moscow, Russia: Publishing House of the Academy of Architecture.

*Wattenmaker, R. J. (1993).* Great French Paintings from the Barnes Foundation: Impressionist, Post-impressionist, and Early Modern. New York: Alfred A. Knopf.

Wikiart. Maria Primachenko. (n.d.). Retrieved from https://www.wikiart.org/ru/mariya-primachenko/black-beast-1936

Wikiart. Sam Francis. (2012). Retrieved from https://www.wikiart.org/en/sam-francis/untitled-yellow-1961

Wikimedia. Bilibin. Vasilisa the Beautiful. (n.d). Retrieved from https://commons.wikimedia.org/wiki/File:Vasilisa.jpg

Wikimedia. Klimt. Portrait of Adele Bloch-Bauer II. (n.d.). Retrieved from https://commons.wikimedia.org/wiki/File:Gustav_Klimt_047.jpg

Wikimedia. Malevich. Black suprematic square. (n.d.). Retrieved from https://commons.wikimedia.org/wiki/File:Чёрный_супрематический_квадрат._1915._ГТГ.png

Wikimedia. Mondrian. Composition. (n.d.). Retrieved from https://commons.wikimedia.org/wiki/File:Mondrian,_Composition_with_color_planes_and_gray_lines,_1918.jpg

Wikimedia. Rylov. In blue space. (n.d.). Retrieved from https://commons.wikimedia.org/wiki/File:Rylov_Blue_Expanse.jpg?uselang=ru

Wikipedia. Aboriginal rock art. (n.a.). Retrieved from https://en.wikipedia.org/wiki/File:Aboriginal_rock_art_on_the_Barnett_River,_Mount_ElizabethStation.jpg

Wikipedia. Bilibin. Alconost. (n.d.). Retrieved from https://ru.wikipedia.org/wiki/Файл:Bilibin_Alkonost.JPG

Wikipedia. Cezanne. Banks of the Marne. (n.d.). Retrieved from https://en.wikipedia.org/wiki/File:Paul_Cézanne_104.jpg

Wikipedia. Cezanne, Mont Sainte-Victoire. (n.d.). Retrieved from https://en.wikipedia.org/wiki/File:Paul_Cézanne_210.jpg

Wikipedia. Dalí. The Persistence of Memory. (n.d.). Retrieved from https://ru.wikipedia.org/wiki/Файл:The_Persistence_of_Memory.jpg

Wikipedia. Picasso_Gernica. (n.d.). Retrieved from https://ru.wikipedia.org/wiki/Файл:PicassoGuernica.jpg

Wikipedia. Soulages. Honolulu. Painting. (n.d.). Retrieved from https://en.wikipedia.org/wiki/File:%2717_December_1966%27_by_Pierre_Soulages,_Honolulu_Museum_of_Art.JPG

## ENDNOTE

1 A few links for a better understanding of the psychological research of personality problems: (Urbina, et al., 1997; Robert 2003; Kaufman, 2009; American Psychological Association, n.d.; British Psychological Society, n.d.).

# Chapter 13
# Practice as a Criterion of Truth:
## The Choice of a Comfortable Dwelling


## ABSTRACT

*The conditions of comfort for representatives of all psychotypes and subtypes are determined. They can be extended to the main estates of modern society. The same conditions are fundamental in determining the goals of designing a dwelling, including its relationship with the environment, style, functionality, transformability, and controllability, space-planning solutions, color solutions, selection of materials and equipment, and subject content. Peaceful relations in the family, in society, with the environment, are an essential condition for comfort. Possible conflicts and the opportunities to resolve them by using the dwelling are given and classified. Rational spending of money to get the most comfortable dwelling is a very essential component of our comfort/discomfort! And the solution to these problems, based on the theory of self-organization, is also given in this chapter.*


## BACKGROUND

In the previous sections, the creation of a wave model of self-organization scenarios proceeded from the abstract to the concrete. First, axiomatics was considered, then general scenarios, then a scenario of the evolution of the mind and human behavior, which was checked using anthropological and historical data, as well as special markers. Very long periods were considered.

Can this theory be used "here and now"? Its usefulness cannot be limited by a better understanding of human evolution or worldview, but it should bring practical benefits to our contemporaries.

And here a very interesting task is the choice (design, construction) of the dwelling. Data on the channels of interaction with the environment, psychotypes, social groups, behavioral patterns, needs and motivations - these are ready-made and very important information for the correct determination of housing requirements for different focus groups. And, as dealers say, the key to sales success and the guarantee that spending a lot of money, a person will receive exactly what will make his life as comfortable and happy as possible (Kolcaba, 2003; Miller, 2009).

DOI: 10.4018/978-1-7998-1706-2.ch013

Complementation the general approach with individual testing data (Chapter 12), it can identity of a particular customer, and make the choice of dwelling even more accurate. The technique used there was based on popular tests (there are the numerous alternatives provided in the scientific literature and popular online services). It is only important that their results be included in the process of a psychological portrait compiling, which was discussed in chapter 12.

Associative thinking is the most important professional quality of people of art. And the architect is looking for material embodiment of timeless ideas and imperishable beauty that he feels (Ikonnikov, 1985; Iovlev 1996; Potaev, 2008; Tetior, 2008; Vityuk, 2015; Kavtaradze, 2017). At the same time, style, functionality (Panero & Zelnik, 2006), transformability, space-planning solutions, color solutions (Buymistru, 2008), materials, furniture and so on are some associative connections between the needs and motivations of a person and environmental conditions. The potential of such a system is the financial and other resources involved in its organization.

Below, numerous examples of such associations will be considered, and this will enable the potential customer to explain to the architect what exactly he needs.

Special attention to the fact that the needs and motivations of a person do not remain unchanged, changing both during life (Morris, et all, 2013) and shorter periods. These changes and their periods are studied by chronobiology and chronomedicine (Zidermane, 1988). Accordingly, it is necessary to pay attention to the potential for the transformation of the home, as well as the possibility of regulating environmental parameters (Gagge, Stolwijk & Hardy1967; Leroux, 1970; Shimko, 1990; Reimers, 1994; Godish, 2001; Panero & Zelnik, 2006; Alexander, Ishikawa & Silverstein, 2014; Humphreys, 2016; Visloguzov, 2016; Colin Ellard, 2017). Modern technologies of internet of things, smart house, and smart city allow to do this. (D'Ulizia, Ferri, Grifoni, & Guzzo, 2010; Acharjya, Geetha, eds.; 2017; Maddox, 2018; Pal, Funilkul, Charoenkitkarn, & Kanthamanon, 2018).

The correct construction of relations returns to one of the branches of psychology - conflict studies, and even mathematics - the theory of catastrophes; however, it is possible to consider this problem in the paradigm of the theory of self-organization (Mkhitaryan, 2017).

And finally, the optimization problem. Mathematicians (Shtoer, 1992), economists (Podinovskiy & Nogin, 1982), psychologists (Grant, Halvorson & Higgins, 2014; Sadiq-Khan, 2017), designers (Norman, 2006; Alexander, Ishikawa & Silverstein 2014) pay great attention to it. The theory of self-organization is no exception (Mkhitaryan, Badeyan & Kovalyov, 2004); the approaches developed within its framework were used to solve this problem.

## COMFORT / DISCOMFORT CONDITIONS AND DESIGN AIMS

### The Concepts of Comfort and Discomfort

In determining comfort state, it is necessary to proceed from the fact that a human, being a complex system and obeying the laws of self-organization, at the same time remains a part of the "human-dwelling-environment" system. A comfortable state of a human will be called a condition characterized by the following features:

- harmony of human, dwelling, and environment;
- bodily and mental health;

- sufficient stock of resources for adaptation;
- "youth" of all components of the human-dwelling-environment system.

Let's consider every of these features separately.

*Harmony of a human, dwelling, and environment.* Violation constancy of the internal environment of the body will inevitably lead to physiological and mental disorders that will be perceived as the presence of discomfort. Moreover, if constancy of the internal environment can cannot be restored by regeneration and processes of exchange for a long time the state of person become pathological. This is the most important feature relating to the first level of organization of the system.

Lack of symmetries at all or some levels of the organization, inharmonious distribution of potentials between levels and elements, and impossibility of adaptation are specific features of violation of the integrity.

Without external factors and their harmonization with internal factors, no comfort can be achieved.

*Physical and mental health* of a human is interpreted as a simultaneous state of constancy of physiological indicators and mental equilibrium. This feature emphasizes importance of internal factors. It should be emphasized that disturbances of mental equilibrium are sometimes useful for activation of certain types of activity ("throes of creation"), but they are still perceived as uncomfortable.

Restoration of health occurs due to such influences of the external environment or dwelling, which help to restore the integrity of a human.

*Sufficient stock of resources* for adaptation is necessary to adapt to changes in external conditions, such as seasons, state of atmospheric pressure, time of day, etc.

The "youth" of the system components, which implies the presence of adaptability, as well as the existence of a reserve of existence time, is also a characteristic feature of a comfortable state.

On the contrary, an uncomfortable state of a human refers to a state characterized by one or a combination of the following features:

- lack of harmony of human, dwelling, and environment;
- bodily and mental illness;
- insufficient to adapt the stock of vitality;
- "old age" of the system components.

The psychological component of comfort and discomfort is correlated with the types of perceptions and states of consciousness.

For level 1, a feeling of chance, luck, favorable course of events, harmony with the environment will be an indicator of comfort.

An uncomfortable state is characterized by a feeling of a "black luck", unfortunateness, and disharmony.

For level 2, a feeling of self-sufficiency, health, peace, joy, optimism, faith in the future is an indicator of comfort. This is a result of inner harmony, as well as a sense of beauty and eternity of the outside world.

An uncomfortable state is characterized by a sense of illness, internal conflict, dissatisfaction, pessimism, imminent death, ugliness of the surrounding landscape, people, frailty of the world, etc.

For level 3, comfort is characterized by a sense of freedom, as well as sense of superiority, confidence in attainment of a goal, in other words, "psychology of the winner". The will of a human plays a dominant role in such perceptions.

Feeling of discomfort is characterized by a feeling of weakness, lack of self-confidence, fear of not achieving a goal.

At level 4, there are two groups of indicators corresponding to sensory perceptions and nous. Therefore, a feeling of comfort will include both a sense of space-time order, "correctness" of the outer and inner world, as well as their intellectual awareness of both natural and knowable.

The feeling of chaos, unknowability, chance, danger, depression, disturbance, anxiety, insecurity, and weakness will be uncomfortable.

For level 5, the number of indicators of comfort is increased. A comfortable state is characterized by a sense of balance, strength, sensory comfort, "harmony of elements" at the level of intellectual constructions and direct perception of the environment.

The feeling of imbalance, exhaustion of forces, sensory fatigue, "enmity of the elements", etc., is uncomfortable.

Since every level corresponds to the calculated potential, one can, knowing the priority and coefficient of perceptual conditions, quantify the contribution of any feature to the perception of comfort/discomfort.

## Estimate of the Relative Weight of Different Indicators of Comfort/Discomfort

A. The locality or globality of an indicator's effect is determined by the fact that it characterizes several levels or only one of them. Accordingly, the weight of the indicator is estimated as the sum of the potentials of the involved levels of human and environment interactions, expressed relative to the potential of the first level. Specifically, the weight of the indicator is expressed by coefficient $k_1$, for either of the layers involved. Here there is an alternative - use $k_1 = 0.618$ for a "balanced personality", or choose it based on the test results;

B. The coefficient of perception conditions $k_2$ expresses the degree of difficulty in perceiving an indicator; for either layer it is in the range 0...1;

C. When summing up the indicators, assign "+" sign, if they are indicators of comfort, otherwise - "-" sign.

Because of the above, the "**formula of comfort**", which characterizes the sum of comfort indicators of all levels p expressed in conventional units, is:

$$p = \sum_{i=1}^{n} \cdot \sum_{j=1}^{m_i} \pi_{ij} \cdot k_{1ij} \cdot k_{2ij}, \quad (1)$$

where $n$ – is the number of levels; $m_i$ – the number of indicators of *i-th* level. The highest value $p$ here corresponds to the best comfort.

The extent of importance of the channel can be associated with its potential and expressed in words reflecting its development (chapter 4, *Self-organization: evolution and behavior*), establish correlations with psychological states, assessed as comfortable or uncomfortable. Then, aims of designing are defined (Ttables 1-6).

*Table 1. The degrees of intuition, correlation with the states of comfort/discomfort and aims of designing for contemplators*

| Degrees of the channel (the highest potential is at the top, the lowest – at the bottom) | State of comfort/discomfort | Aim of designing |
|---|---|---|
| Communion with God | Comfort | Harmony with spiritual world |
| Love to God | | |
| Contemplation | | Harmony with nature |
| Insight | | |
| Associativity | | |
| Perception | | Harmony with people |
| Love of neighbor | | |
| Sociability | | |
| Dullness | Discomfort | Peace and social adaptation |

*Table 2. The degrees t of ego, correlation with the states of comfort/discomfort and aims of designing for egoists*

| Degrees of the channel (the highest potential is at the top, the lowest – at the bottom) | State of comfort/discomfort | Aim of designing |
|---|---|---|
| Autism | Discomfort | Peace and social adaptation |
| Pride | Comfort | Self-sufficiency, safety, prestige |
| Complacence | | |
| Self confidence | | |
| Self esteem | | |
| Lack of faith in one's strength | Discomfort | Peace and social adaptation |
| Self-deprecation | | |

*Table 3. The degrees of the will/reason, correlations with the states of comfort/discomfort and aims of designing for fighters*

| Degrees of the channels (the highest potential is at the top, the lowest – at the bottom) | | State of comfort/discomfort | Aim of designing |
|---|---|---|---|
| Will | Reason | | |
| Aggression | Wisdom | Comfort | Transformability |
| Expansion | Determination | | |
| Adventurism | Evaluation | Comfort/ Discomfort | |
| Perseverance | | | |
| Stubbornness | Indecision | Discomfort | Peace and social adaptation |
| Weakness | Nondiscrimination | | |

*Table 4. The degrees of reason, correlations with the state of comfort/discomfort and aims of designing for enterprising researchers*

<table>
<tr><th>Degrees of the channel (the highest potential is at the top, the lowest – at the bottom)</th><th>State of comfort/discomfort</th><th>Aim of designing</th></tr>
<tr><td>Faith in the cognoscibility of the world</td><td rowspan="6">Comfort</td><td rowspan="6">Orderliness, proportionality, tectonics</td></tr>
<tr><td>Feeling the harmony of the world</td></tr>
<tr><td>Ability to find regularities</td></tr>
<tr><td>Ability for abstracting</td></tr>
<tr><td>Ability to analyze and synthesize</td></tr>
<tr><td>Ability to define relations</td></tr>
<tr><td>Confusion of ideas</td><td rowspan="2">Discomfort</td><td rowspan="2">Peace and social adaptation</td></tr>
<tr><td>Thoughtlessness</td></tr>
</table>

*Table 5. The degrees of perceptions and emotions, correlation with the states of comfort/discomfort and aims of designing for bohemian personalities*

<table>
<tr><th colspan="2">Degrees of the channels (the highest potential is at the top, the lowest – at the bottom)</th><th rowspan="2">State of comfort/discomfort</th><th rowspan="2">Aim of designing</th></tr>
<tr><th>Perceptions</th><th>Emotions</th></tr>
<tr><td>Extrasensory</td><td>Frenzy</td><td rowspan="2">Discomfort</td><td rowspan="2">Peace and social adaptation</td></tr>
<tr><td>Ability to distinguish more than 7-9 shades and tones</td><td>Exaltation</td></tr>
<tr><td>Ability to distinguish between 7-9 tones</td><td>Sensitivity</td><td>Comfort</td><td>Sensory comfort</td></tr>
<tr><td>Loss of sight, hearing, etc.</td><td>Cold</td><td>Discomfort</td><td>Peace and social adaptation</td></tr>
</table>

*Table 6. The degrees of balanced channels with the state of comfort/discomfort and aims of designing for balanced personalities*

<table>
<tr><th>Degrees of the channels (the highest potential is at the top, the lowest – at the bottom)</th><th>State of comfort/discomfort</th><th>Aim of designing</th></tr>
<tr><td>Ability to make fundamental discoveries</td><td rowspan="4">Comfort</td><td>Opportunities for creativity: scientific;</td></tr>
<tr><td>Ability to create original works</td><td>Opportunities for creativity: artistic;</td></tr>
<tr><td>Ability to do any difficult work well</td><td>Opportunities for creativity: inventive and engineering;</td></tr>
<tr><td>Ability to do any simple work</td><td>Opportunities for creativity: craft</td></tr>
</table>

## DESIGN AIMS AND DWELLING SPECIFICATIONS

Let's go on to the next stage. It is necessary to determine the correlations between the aims of designing following tables 1-6 and characteristics of dwellings that relate to various levels of organization and interaction of the system:

1. Relationship with the environment;
2. Style;
3. Functionality, transformability and adjustability;
4. Space-planning solutions;
5. Color solutions, acoustic, tactile and other sensory comfort, and interior elements.

For all these levels, one has to achieve aims of designing, promoting a state of comfort, as well as eliminate any conditions that create discomfort.

Let's summarize the results in Tables 7-12.

*Table 7. Correlation of the aims of designing and parameters of the dwelling for contemplators*

| Aims of designing | Levels | Specifications |
|---|---|---|
| Harmony with the spiritual world | 1 | Openness to inspirations and adepts/protection against profane and deadly effects of the environment |
| | 2 | Caves, huts, sketes, cells, and monasteries |
| | 3 | Minimum life support, no transformations |
| | 4 | The simplest |
| | 5 | Spiritual books, objects of worship, the rest does not matter |
| Harmony with the nature | 1 | Openness to the beauties of nature/minimal protection |
| | 2 | Country houses, cottages with elements of eco-and bio-styles, in harmony with the nature |
| | 3 | Safety, normal life support |
| | 4 | Open planning, terraces, and large openings |
| | 5 | Natural colors, natural materials, sensory comfort |
| Harmony with humans | 1 | Openness to the beauties of the urban environment |
| | 2 | Townhouses, penthouses, condominiums with elements of eco-and biostyle, in harmony with the urban environment |
| | 3 | Safety, normal life support, and communication |
| | 4 | Open planning, terraces, large openings |
| | 5 | Natural colors, natural materials, and sensory comfort |
| Peace and social adaptation | 1 | Openness to communicate with their peers/restricted access territory; beautiful, peaceful area |
| | 2 | Rehabilitation centers, country houses with elements of eco-and bio-styles that do not stand out from the terrain |
| | 3 | Safety, normal life support, communication in closed networks with social trainers and others |
| | 4 | Open planning, terraces, large openings |
| | 5 | Natural colors, natural materials, sensory comfort |

*Table 8. Correlation of the aims of designing and parameters of the dwelling for egoists*

| Aims of designing | Levels | Specifications |
|---|---|---|
| Peace and social adaptation | 1 | Relatively isolated terrain with high reputation |
| | 2 | The dwelling stands out against the terrain |
| | 3 | Maximum life support and security, lack of transformability, communication in closed networks with social trainers and others |
| | 4 | Closed layout, minimum communication with the environment |
| | 5 | Unique items, equipment for creativity, libraries |
| Self-sufficiency, safety, prestige | 1 | Administrative, financial, business centers/prestigious suburban areas/"islands for millionaires", rich resorts |
| | 2 | City and country houses, estates, prestigious hotels, standing out against the surrounding background, classic or modern "rich" styles |
| | 3 | Territory security, maximum life support |
| | 4 | Overall solutions, many rooms, closed or open planning, terraces, large openings, depending on paragraphs.1-3 |
| | 5 | Gold, purple, expensive materials and furniture, attributes of power and wealth, adjustable environmental parameters |
| Peace and social adaptation | 1 | Relatively isolated terrain |
| | 2 | The dwelling is different from other environment objects |
| | 3 | Maximum life support and security, lack of transformability, communication in closed networks with social trainers and their ilk |
| | 4 | Closed layout, minimum communication with the environment |
| | 5 | Equipment for work, developing literature |

*Table 9. Correlation of the aims of designing and parameters of the dwelling for fighters*

| Aims of designing | Levels | Specifications |
|---|---|---|
| Transformability | 1 | City/country Zones |
| | 2 | Country houses for warriors that do not stand out on the terrain (samurai's dwelling), city houses, military bases contrasting with the environment, deconstructionism for invaders and fighters |
| | 3 | Normal life support |
| | 4 | The possibility of easy transformation |
| | 5 | Elements of sentimentality for warriors, entourage of "camp life" for invaders and fighters |
| Peace and social adaptation | 1 | Isolated terrain |
| | 2 | The dwelling is hidden by the elements of the environment |
| | 3 | Normal life support and security, transformability, communication in closed networks with social trainers and others |
| | 4 | Closed layout, minimum communications |
| | 5 | Equipment for work, developing literature |

*Table 10. Correlation of the aims of designing and parameters of the dwelling for enterprising researchers*

| Aims of designing | Levels | Specifications |
|---|---|---|
| Orderliness, proportionality, tectonics | 1 | Cities/townhouses/country zones |
| | 2 | Pragmatic styles: minimalism, constructivism, functionalism, etc. |
| | 3 | Normal life support, communication |
| | 4 | The ability to ensure repair and upgrade |
| | 5 | Color minimalism, laconic forms, technically advanced equipment |
| Peace and social adaptation | 1 | Sparsely populated area |
| | 2 | Social housing |
| | 3 | Normal life support |
| | 4 | Minimalist layouts |
| | 5 | Equipment for work, developing literature |

*Table 11. Correlation of the aims of designing and parameters of the dwelling for Bohemian personalities*

| Aims of designing | Levels | Specifications |
|---|---|---|
| Peace and social adaptation | 1 | Relatively deserted, calm, beautiful countryside |
| | 2 | The dwelling fits into the locality |
| | 3 | Modern, Art Deco, Classicism |
| | 4 | Open planning |
| | 5 | Equipment for creativity, libraries, workshops |
| Sensory comfort | 1 | City "Hollywood"/resort areas |
| | 2 | City and country houses, estates, prestigious hotels that stand out against the surrounding background, classic or modern outrageous styles |
| | 3 | Territory security, maximum life support |
| | 4 | Excessive dimensions, many rooms, closed or open plan, terraces, large openings depending on paragraphs1-3 |
| | 5 | "Theatricality", expensive materials and furniture, attributes of self-importance, adjustable environmental parameters |
| Peace and social adaptation | 1 | Relatively isolated, tranquil area |
| | 2 | Special requirements for persons with poor eyesight, hearing, restricted mobility, etc. |
| | 3 | Normal life support concerning disabilities, security, communication with staff, social trainers, etc. |
| | 4 | Depend on the type of disability; simple and transparent anyway |
| | 5 | Equipment for work and rehabilitation, furniture, taking into account the limited possibilities |

*Table 12. Correlation of the aims of designing and parameters of the dwelling for balanced personalities*

| Aims of designing | Levels | Specifications |
|---|---|---|
| Opportunities for scientific creativity | 1 | Not far from the places of stay of the scientific and cultural elite (university cities, laboratories, etc.) |
| | 2 | City apartments, townhouses, country houses in both classical and modern styles without any "excesses" |
| | 3 | Normal life support |
| | 4 | Tectonic, simple, self-similar |
| | 5 | Libraries, communication equipment with colleagues and laboratories |
| Opportunities for artistic creation | 1 | Openness to the beauties of nature/minimal protection |
| | 2 | Urban and country houses, townhouses, summer cottages in classical or modern, especially eco-and bio-styles, in harmony with the terrain |
| | 3 | Safety, normal life support |
| | 4 | Open plan, terraces, large openings |
| | 5 | Natural colors, natural materials, sensory comfort, bohemian surroundings, workshops |
| Opportunities for inventive and engineering creativity | 1 | Openness to the beauties of the urban environment |
| | 2 | Townhouses, penthouses, country houses, in harmony with the environment |
| | 3 | Safety, normal life support, communication |
| | 4 | Strict and orderly |
| | 5 | Physiological comfort, special equipment in the workshops |
| Opportunities for craft creativity | 1 | Relatively uncrowded, possibly rural |
| | 2 | Mostly country houses far from neighbors |
| | 3 | Safety, normal life support |
| | 4 | Simple layouts |
| | 5 | Physiological comfort, equipped workshops |

One should also take into account age characteristics, temperament, climatic conditions and influence of traditions.

Let`s give examples of realization of designing aims and specifications for dwellings (for permanent and temporary residence, apartments, country houses, etc.) of people belonging to different psychological types.

## DWELLINGS FOR CONTEMPLATORS

Despite the common ability of all representatives of this psychological type for direct perception of the surrounding world, depending on the scope of the potential, the object of perception and the conditions of comfort will vary (Table 7).

For advanced mystics (and this group, albeit in small numbers, is represented in all epochs), the habitat should provide conditions for meditation and contemplation - be quite uninhabited, difficult to reach and quiet (Figure 1). Requirements directly to the dwelling are minimal - protection against climatic and other disturbing factors and very few amenities. Cells, huts, dugouts will be quite enough (Figure 2).

*Figure 1. Inaccessibility, desolation, silence of Mount Athos (Religions-union. Afon, n/d)*

*Figure 2. Minimum life support requirements - monks' cells on Mount Athos. Pravoslavie. Mount Athos, n/d)*

For beginning mystics, mutual assistance is important, both from the more advanced fellows and strictly organized routine of life and religious ritual. Accordingly, a monastery will be an optimal type of dwelling (Figure 3). The requirements for the location of the dwelling are the same as in the previous case. An integral part of the interior will be a religious entourage that continuously affects the minds of the monks (Figure 4).

Secular people **united with nature** prefer nature to human society. Accordingly, the best would be the isolated location of dwellings in a beautiful, open to contemplation of the area (Figure 5). However, the requirements for comfort are very high, and their implementation requires the use of advanced technical solutions.

Sometimes the desire for openness takes on very exotic forms (Figure 6).

*Figure 3. Patala (the official residence of Dalai Lama in Tibet) (Estateline. Dosug. Pictures, n/d)*

*Figure 4. Religious symbols - an attribute of monasteries and cells. By Yana Bayturina (Knews. Mnenie. Religia, n/d)*

Another group of contemplators - talking about nature - are afraid of losing at least a small part of their comfort. For them, urban dwellings in ecostyle, with a characteristic desire to move a piece of nature inside will be the most preferred (Figure 7). Fortunately, there are many variations of eco-style that can satisfy every taste. The interior also has matching surroundings (Figure 8).

*Figure 5. Frank Lloyd Wright: Fallingwater (Edgar J. Kaufmann Sr. Residence), Bear Run, Pennsylvania, USA. (Faqindecor. Frank Lloyd Wrigh, n/d)*

*Figure 6. Transparent dwelling project The Glass House by Santambrogiomilano Company. Milano, Italy. (Archidea. Steklyannyy dom v Milane, n/d)*

*Figure 7. Vincent Callebo. Asian pyramids (project). Changzheng, China (Telegraf. Aziatskie pyramidy, n/d)*

*Figure 8. Eco-style elements in the interior. (Boxtoner. Celling-Decoration, n/d)*

## DWELLING FOR EGOISTS

Prestige, self-sufficiency, pathos - these are qualities that all egoists would like to see in their dwellings. However, the severity of these qualities may be different - several subtypes are also distinguished here. There are two main ones: aristocrats and those who are called self-made man, people who became rich thanks to their efforts.

*Figure 9. Elements of Baroque*

Classical styles, such as Baroque, are better suited for aristocrats, with the consistency of aesthetics of all the details inherent in this style - from the dwellings themselves to household items, clothing and even book fonts (Figure 9).

For self-made man, an empire or Victorian style is an acceptable alternative.

Modern styles for egoists, of course, differ from historical ones. However, the general pathos, luxurious forms, and expensive materials remain unchanged attributes.

Consider these dwellings in more detail.

Location. Traditional locations of the elite near the centers of state and financial power, in prestigious resorts, etc., are complemented by exotic, but equally prestigious locations (Figure10).

Expressiveness on the background of the surrounding area and buildings. The United Arab Emirates has many modern and unique facilities. But even against their background, the hotel Kempinski&Residences Palm Jumeirah (Figure 11) stands out, located on the Palma Island-1 and bringing "The Thousand and One Nights" to mind.

Self-sufficiency and safety. The hotel has a full range of safety and comfort facilities, such as a spa, gym, several brand shops and free shuttle service to the Mall of the Emirates, a Mediterranean restaurant, a bar with a wide selection of cocktails and cigars, pools, etc. (Figure 12)

*Figure 10. Photograph of the new islands "the universe" as well as the new design of "Palm deira" in Dubai Author: Tobias Karlhuber. (Wikimedia. The_Universe, n/d)*

*Figure 11. Kempinski & Residences Palm Jumeirah, Palms Islands, UAE.*

*Figure 12. Swimming pool of Kempinski&Residences Hotel, Palm Jumeirah*

*Figure 13. Mansions on the territory of the hotel*

*Figure 14. Options of luxury rooms interior*

Freedom of self-realization. All these components allow every visitor to freely organize his/her stay at the hotel. However, the organization of personal space is the most important element of self-realization, and here the hotel also provides the most extensive opportunities. Fully self-contained are the mansions located on the hotel, which can be rented for a long time or sold in private ownership (Figure 13).

Not only apartments, but also suites have a balcony, dining/lounge area, and kitchen, and are equipped with DVD player, 40-inch TV, and Hi-Fi acoustics. The visitor can choose the stylistic and color scheme of the room (Figure 14). Only the atmosphere of luxury remains unchanged, and its required high-tech components not only do not interfere but also contribute to the feeling of homely comfort.

Space-planning solutions. Figures 15 gives an idea of how thoughtful are the space-planning solutions of the hotel, where are spacious and proportional rooms of various forms and free layouts, spacious transit zones, a developed system of terraces and balconies, sunscreens, landscaping, etc.

*Figure 15. Elements of space-planning solutions*

Sensory comfort. The openness of the environment is combined with the maintenance of optimal thermal and moisture parameters, light levels, color, and acoustic comfort. The splashing of the waves and sound of the sea wind are not muffled by anthropogenic noises, and the natural colors of the sky, water, and sand are combined with light tones of finishing materials. But, as the hotel itself stands out against the background of the island and buildings, so the individual details focused on different target groups emphasize both the status of the hotel and the viability of its guests (Figure16).

*Figure 16. Elements of interior*

## DWELLING FOR FIGHTERS

Let's start a summary of the dwellings for this psychological type with **warriors**. But who are they?

Warriors are people for whom the struggle has become a means of self-development, in particular, spiritual one. It was they who developed a certain view on the place of a human in the universe and the purpose of his life, however strange it may seem to other people. Reading Bushido: "The path of a warrior means death. When there are two ways to choose, choose the one that leads to death. Do not argue! Direct the thought to the path you preferred, and go!" or "Think every morning about how to die. Every evening refresh your mind with thoughts of death. And may it always be so. Train your mind. When your thought constantly revolves around death, your life path will be straight and simple. You will do your duty; your shield will turn into a steel shield". There is even a specific only for Japan type of literature "Death Poems", which was considered a special achievement for the warriors. A properly organized environment should give them opportunities for development, and be safe for the outsiders.

Warriors love to change this world by their ideas - and dwelling should provide them with this opportunity, laying the quality of transformability into the basis of design solutions. However, it is also necessary to promote the harmonization of consciousness, because participation in constant conflicts acts destructively. It is very instructive to combine these conflicting aims in traditional Japanese dwelling.

First, efforts are being made to ensure harmony with nature. The house fits into a sanitized landscape; water, stones, trees - Feng Shui in action. Doors, windows, terraces are arranged so that you can admire the most beautiful parts of the garden.

Secondly, the style of the dwelling reflects the ideas of the Great Void[1] to some extent (lack of interior walls, windows, familiar furniture, in modern terms, studio layout), change (partitions can be easily moved and cleaned) and frailty of life (the house is easy to destroy, but just as easy to recover). Also, the transformability of the building is the most important correlation with the conflicting nature of the fighter. Also, there are practical considerations of earthquakes, hurricanes, wars that do not cause the owner of critical losses.

Thirdly, space-planning and design solutions are simple: a wooden rectangular frame in a plan, most often just one capital wall, other external walls are removable lightweight structures without windows between the columns, a gable roof thatched with a straw, wooden platform that goes beyond the walls. And inside - tatami, removable and sliding partitions, and doors, usually made of rice paper on a light frame. There were simply no water supply, sewage, or heat supply systems (except for a clay oven). Depending on the season, one could easily remove external partitions, and depending on the situation, it was easy to change the elements and area of the interior. Local materials were widely used in construction.

Fourthly, the elements of the interior are also unusual for a human of Western civilization. There are no habitual pieces of furniture: mattresses are used instead of beds and sofas and removed after sleeping in special niches or, in modern homes, built-in wardrobes, pillows are used instead of chairs, and dining tables have low portable tables. There is no similarity and symmetry. Drawings are not repeated.

In all this severity, there is also some sentimentality: ancient, time-honored items are very valuable: darkened teapots and plates covered with a net of cracks and plates, inherited dolls, paper lanterns, engravings, etc. They are stylistic accents against the background of empty space - in spirit, shape, color, purpose. Traditional Japanese aesthetics includes the categories Wabi-sabi (imperfect, impermanent, incomplete beauty of things), sibus (simple, subtle and unobtrusive beauty), yugen (subtle, elusive, up to the mystical feeling, the depth of things) and a number of others that characterize the many facets of the nature of a true warrior.

*Figure 17. Arbol Design. House in Nishimikuni, Japan (Magazindomov. Gorodskoj dom v yaponii, n/d).*

One has to admit that the traditional Japanese dwelling - Minka - has anticipated many stylistic, space-planning and constructive decisions of our days.

Now is an example of a modern dwelling located in cramped urban conditions (Figures 19-22). Even in such conditions, dwellings typical of country houses and comfortable for warriors (security, transformable empty spaces, "points of sentimentality", means of providing physiological comfort) can be realized.

Low-rise buildings, surrounded by high-rise buildings, stand out against this background by themselves. Therefore, calm solutions are possible, similar to the "eye of the storm" in a quiet area in the center of a city hurricane (Figure 17). The sense of peace is achieved thanks to simple shapes, sizes comparable to humans, wooden wall decorations, and vegetation.

The transformability of space also takes place (Figure 18): glass light-mounted partitions and doors are used, and the play of light and shade of vertical boards, plants, and interior objects creates a shaky border to internal volumes.

Space-planning solutions (Figure 19) are also simple and well thought out; there is an almost complete absence of windows - the above-mentioned detail of a traditional Japanese dwelling.

With a total area of 92 $m^2$, the set of premises is quite decent, including even two small gardens.

The interior design (Figure 20) is not cold and quite comfortable: the geometric rigor of the premises is compensated by the possibilities of transformation; regular forms - crowns of trees and open sky, visible through transparent partitions; alternation of white surfaces and wooden floors in harmony with traditional furniture and tatami; unnecessary items that are put in boxes do not interfere with the sense of emptiness, "sentiment points" are presented, for example, with old dishes.

The activity of the **invader** and **fighter**, of course, is different - the first get something valuable for his works, the second can say about himself with the words of Porthos: "I fight because I fight". As a result, the first is less constrained in funds than the second. However, the consequences of the activities of both bring destruction and chaos into the world, and there is also a certain aesthetics expressed

*Figure 18. Arbol Design. House in Nishimikuni, Japan. Transformability of space. (Magazindomov. Gorodskoj dom v yaponii, n/d).*

*Figure 19. Arbol Design. House in Nishimikuni, Japan. House plan. (Magazindomov. Gorodskoj dom v yaponii, n/d).*

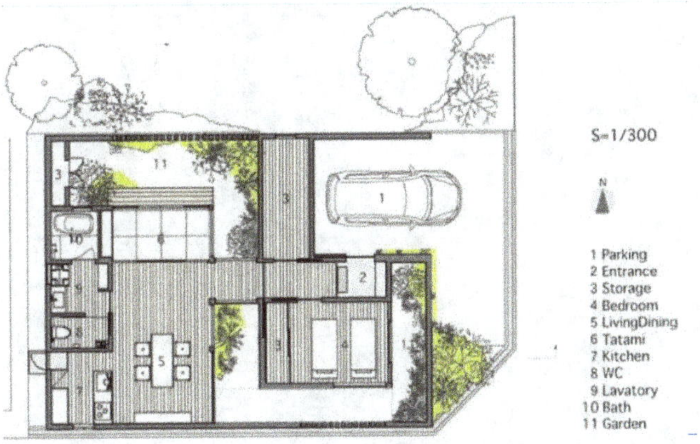


by deconstructionism. This style best suits the needs and motivations of these subtypes. Let's give an example (Figures 21-23).

The house is located in a beautiful area near the ski resort and fits into the surrounding landscape, reminding a mountain cliff. Mountains have always served as a source of additional energy, but if a relatively orderly row of mountain peaks is suitable for quiet meditation, then a jumble of rocks, cliffs and steep slopes is a good place just for invaders and fighters.

The exterior of the house, characterized by granular, disharmonious forms and openings, compared with other buildings, creates the effect of dissimilarity and even audacity (Figure 21).

Transformability, even destruction of space, is an essential feature of deconstruction, demonstrating the victory of the fighter's will over the ordering mind. In such conditions, even the use of screens, partitions, and other similar means are not necessary.

Space-planning solutions are expressed with the same idea (Figure 24).

The finishing touches are brought in by the interior: granular shapes, acute angles, level drops, semi-enclosed spaces. As a contrast to all of the above - standard form of the furniture and rather calm combinations of colors (Figure 23).

*Figure 20. Arbol Design. House in Nishimikuni, Japan. House interiors. (Magazindomov. Gorodskoj dom v yaponii, n/d)*

In general, the arsenal of deconstructionist methods of organizing the interior is quite wide, including the use of distorted forms, destruction of color harmony, a dominance of unnatural, aggressive colors, disturbance of natural lighting, the design of furniture, lamps, plumbing, various accessories, clothing and so on.

The dwelling of the last subtype (troublemaker, with a bohemian raid, or hooligan, with the domination of marginality) is as unpleasant as such people and does not deserve a separate example.

## DWELLING FOR ENTERPRISING RESEARCHERS

Here transparency and rationality should be ensured. Functionalism, rationalism, constructivism, minimalism, hi-tech and their combinations (fusion) will be the most appropriate stylistic decisions.

Let's try to retreat from the dwelling demonstration scheme used in the three previous cases, and focus on age-related features, considering dwellings for the student and elderly family. This will provide an opportunity to show two more types of homes - campuses and townhouses.

*Figure 21. Patkau Architects. Private house, Whistler. Canada. (Archdaily. Patkau Architects. Private house, Whistler, n/d)*

*Figure 22. Patkau Architects. Private house, Whistler. Canada. Plan and section. (Archdaily. Patkau Architects. Private house, Whistler, n/d).*

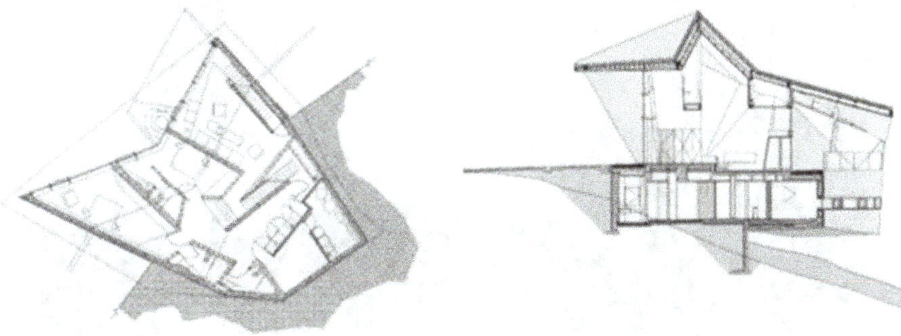

Let's consider what should be the living conditions of students on the example of Tietgen campus, Copenhagen, Denmark (Figures 24-27).

Pay attention to a safe, quiet and clean place of construction, canals and lawns, bike paths; surrounding infrastructure has all the necessary infrastructure (Figure 24).

Tietgen itself is a round residential building with a large courtyard (Figure 25). It has 30 kitchens. There is a reading room inside where students can study, with scanners, and a photocopier in it, a room equipped with computers with printers, and parking for bicycles. Interiors are shown in Figure 26.

There are 360 rooms on seven floors with a total area of 26,800 sq.m. Each room takes 26-33 square meters. 30 rooms for couples with two bedrooms and a total area of 45 sq.m. The planning of all rooms is the same: a large window, bathroom, and toilet (Figure 29).

*Figure 23. Patkau Architects. Private house, Whistler. Canada. Interiors. (Archdaily. Patkau Architects. Private house, Whistler, n/d)*

Now let's consider a townhouse, representing a compromise between a country house and a city apartment for a pragmatic enterprising researcher (Figures 28-30).

That is how the struggle for environmental safety looks like for some representatives of the intelligentsia, and they see their contribution to saving our planet in buying very expensive apartments with a full ecological package (solar panels, heat recovery system, green roofs, rainwater harvesting for household needs).

*Figure 24. Lundgaard & Tranberg Architects. Campus Tietgen. Copenhagen, Denmark. Location. (Arhidaily. Lundgaard & Tranberg Architects. Campus Tietgen, n/d).*

*Figure 25. Lundgaard & Tranberg Architects. Campus Tietgen. Copenhagen, Denmark. Building plan. (Arhidaily. Lundgaard & Tranberg Architects. Campus Tietgen, n/d).*

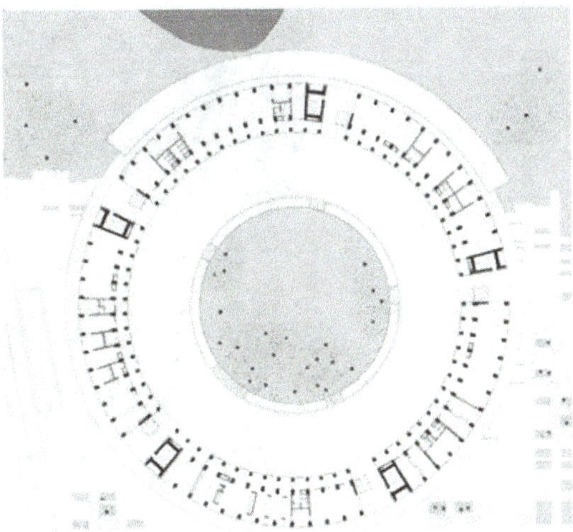

*Figure 26. Lundgaard & Tranberg Architects. Campus Tietgen. Copenhagen, Denmark. Interiors. (Arhidaily. Lundgaard & Tranberg Architects. Campus Tietgen, n/d).*

*Figure 27. Lundgaard & Tranberg Architects. Campus Tietgen. Copenhagen, Denmark. Room. (Arhidaily. Lundgaard & Tranberg Architects. Campus Tietgen, n/d).*

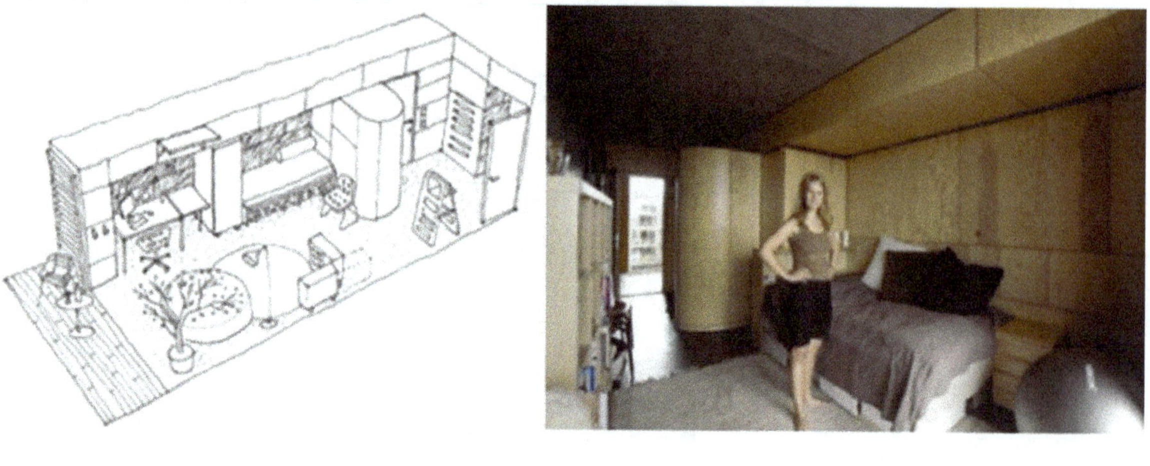

*Figure 28. C.F. Møller Architects. Project Sustainable Town Houses Stockholm, Sweden. Location and Exterior. (Archdaily. Møller Architects. Project Sustainable Town Houses Stockholm, n/d)*

*Figure 29. C.F. Møller Architects. Project Sustainable Town Houses Stockholm, Sweden. Plan and side facade. (Archdaily. Møller Architects. Project Sustainable Town Houses Stockholm, n/d)*

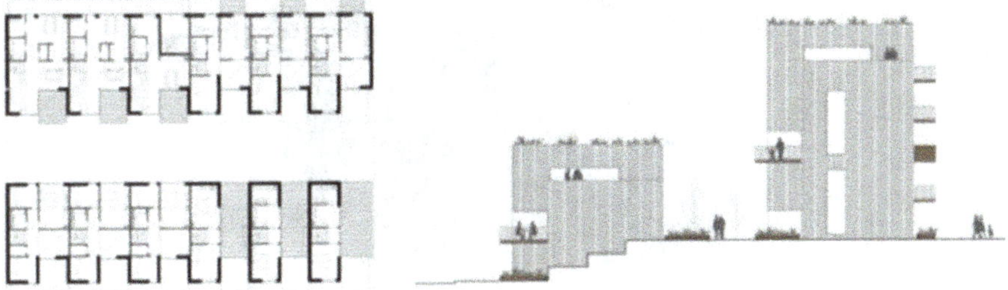

*Figure 30. C.F. Møller Architects. Project Sustainable Town Houses Stockholm, Sweden. Interior project. Nordic environmental minimalism. (Archdaily. Møller Architects. Project Sustainable Town Houses Stockholm, n/d)*

## DWELLING FOR ARTISTIC PERSONS

Here, the principle is opposed to the previous one - no laws, abundance of forms, materials of colors, which correspond to the modern or art deco styles. The diversity of subtypes introduces its corrections, and the examples given below will show that the theory is not dogma at all (especially for artists).

let's give two examples and focus on social differences - on the one hand, the dwelling of Russian show business stars Alla Pugacheva and Maksim Galkin (Figure 31), on the other - nameless dwelling of outcasts in a typical provincial town in Russia.

The castle (almost untypical for Russia), as it should be, stands on a hill surrounded by a high wall. The pond is also there. The exterior combines medieval and modern forms and materials (Figure 32).

*Figure 31. The appearance of the starts to the exultant people. (Glamurchic.Pugacheva, Galkin, n/d).*

*Figure 32. House of A. Pugacheva and M. Galkin. Gryaz, Russia. Location, exterior, details. (Trollno. Zhizn v zamre, n/d).*

*Figure 33. House of A. Pugacheva and M. Galkin. Interior. (Trollno. Zhizn v zamre, n/d).*

Transformability is not laid constructively, on the contrary, superstitious stars seek protection against the most incredible dangers inside solid walls. But the extensive functionality contributes to the diverse use of the house: for privacy, communication with relatives, receiving a large number of guests, PR-campaigns, etc.

Space-planning solutions were not published for security reasons, but it is known that the house has a fireplace room, dining room and kitchen with a basement and storerooms, winter garden, sports complex with a pool, bedrooms and host rooms, guest rooms, observatory, library, home theater, toilets, bathrooms, and garage.

The design of individual rooms, on the contrary, is often exposed during various types of presentations and (Figure 33). The author of the project called it Victorian, but the reader will easily find elements of eclecticism and kitsch. As the envious public would like to have…

The next subtype, **outcasts**, refers to bohemian personalities only because in both cases sensations and emotions prevail among the ways of interacting with the outside world. But it is distinguished by low

*Figure 34. Construction of a Soviet city with social dwellings*

potential, rude senses, great suggestibility, and even greater stubbornness. This seems paradoxical, but it is explained quite naturally: the suggestion occurs gradually, imperceptibly, an outcast perceives the desire invested in his heart like his own, which he cannot but follow; whereas the adoption of a clearly expressed opinion of others leads a human who does not see any alternative, downright to the collapse of the worldview, which he opposes with all his might.

Such people do not have enough strength for self-development (and there is no such need); if there is strong ideological pressure from the government, inspired by ideas of enlightenment and progress, they may acquire some signs of civilization. But as soon as the pressure stops - everything disappears: so "the heaviest readers in the world" the just a couple of decades after the death of the Soviet Union, this population seems to not need books at all. Alas, it is not only a post-Soviet phenomenon.

But let's go back to the dwelling. Of course, here it is possible to talk only about social dwelling. Figure 36 shows typical buildings of provincial Soviet cities, which were often formed around a large industrial facility.

Space planning solutions, due to limited budgets, were simple but reasonably worked out (Figure 35).

Social dwelling consumers would save on everything; so, he is his designer. Because of this, the interiors showed the deepest properties of his nature, usually defined by the word "philistine" (Figure 36).

Modern designers have offered a lot of interesting ideas for reconstruction of such apartments here, but are they in demand? After some time, the apartment will still acquire its original view, in the sense of Figure 36.

*Figure 35. The layout of 5-story houses of 1-335 series. Small kitchens and rooms, but the furniture can be arranged conveniently. Low ceilings. No elevators. (Yaroslav. Hrudchevka, n/d).*

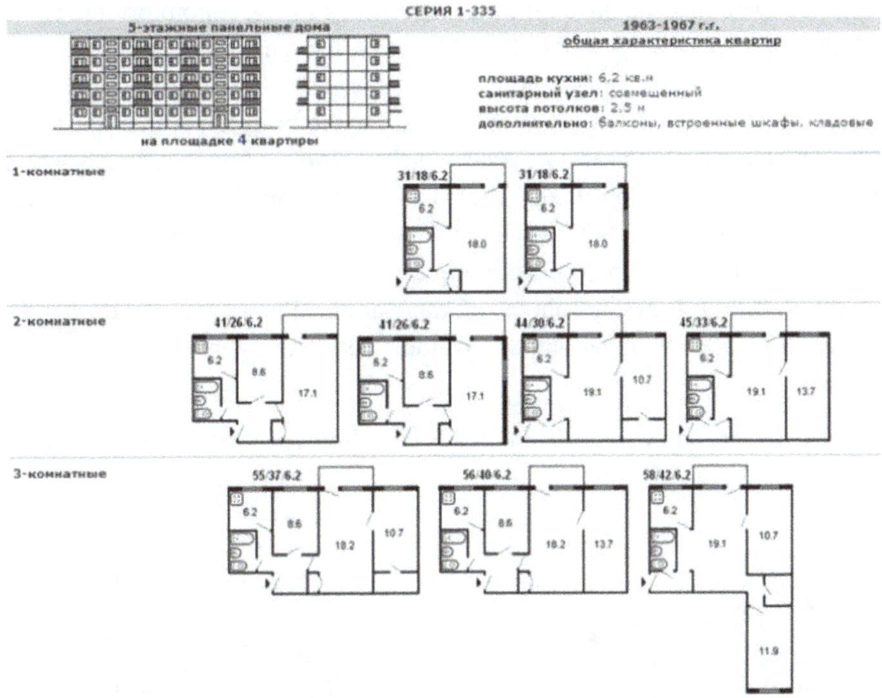


*Figure 36. Mixing of incompatible things in a confined space. Used paintings by Aleksandr Laktionov. (Narod. USSR. Laktionov, n/d) and Yuriy Patsyrev. (Artlib. Uid=2952, n/d).*

## DWELLING FOR BALANCED PERSONS

The creation of such dwellings is a difficult task – one needs to create comfortable conditions for the operation of all the channels while keeping their harmony, and encourage the creative activity of the individual, which can be realized in different forms.

Let`s give a personal example - just as it was done for the brightest woman of the Middle Ages, Aliénor d'Aquitaine (chapter 10, *The Middle Ages: a failed return*). And the same bright personality of the necessary type in our time Nikola Tesla (1856-1943). The great inventor, adventurer, well-known billiard player, the true son of the Art Nuovo era, when creators, like insane gods, played with images and symbols, forgetting about meanings and consequences…

There is no data on the dwellings of Imhotep or, for example, Pythagoras, but the hotel where Nikola Tesla spent the best years is well known - this is Waldorf Astoria (Figures 37-40) in New York.

*Figure 37. Towers of Astoria and Waldorf Hotel (till 1929), New York, USA. (Wikimedia. Waldorf Hotel, n/d).*

*Figure 38. Astoria Hotel and Waldorf Hotel. Floor plans. (until 1929), New York, USA. Author: William T. Comstock. (Wikipedia. Astoria Hotel, n/d).*

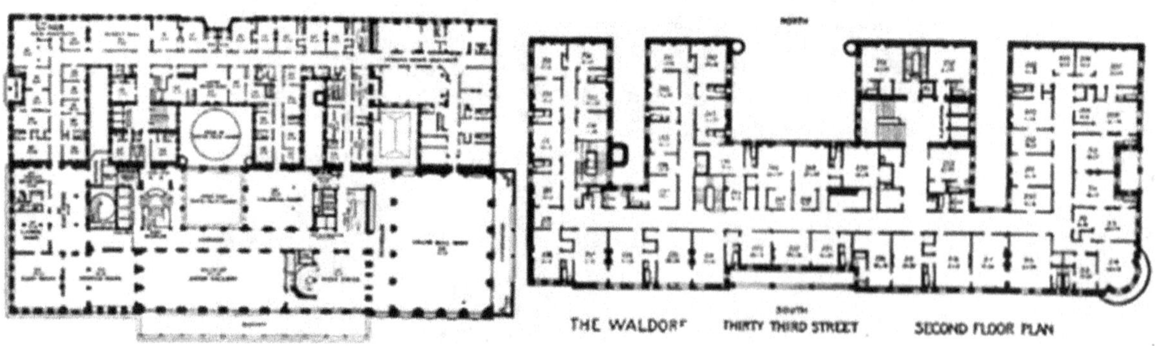

*Figure 39. The interiors of Waldorf Astoria Hotel. Reception and Restaurant. (Wikipedia. Waldorf-Astoria, n/d).*

*Figure 40. Interiors of Waldorf Astoria Hotel. Theatre and reading room. (Wikipedia. Waldorf-Astoria, n/d).*

Tesla, prone to artistry, loved the atmosphere of luxury, cuisine, various technical tricks (pneumatic mail, calls, light bulbs, staff call), generously implemented in this most sophisticated and aristocratic hotel of the time – as long as he had enough money. And he died in the hotel too.

Let's point out Tesla's inability to accumulate money - this is another sign that distinguishes a balanced personality from an enterprising researcher - and move on to the typology and functionality of the dwelling, based on Tesla's experience, but at a new technological level.

Thus, such options are possible:

- a high-class hotel as a place of permanent residence;
- apartments (a part of a hotel, allocated for apartments without access of ordinary visitors to them);
- elite apartment or house in the suburbs with a well-trained and unobtrusive servant.

And all this equipped with the most advanced systems, "smart house".

Of all the options, the apartments seem to be optimal - the opportunity to organize exquisite food at the base of the hotel's restaurant, as well as entrust its staff with any household staff, plus relative privacy, but with the ability to hold noisy parties in the restaurant combine the virtues of the two other types of dwellings and are free from their drawbacks.

## DWELLING AS A PROTECTOR, ADVISER, AND DOCTOR

The next group of means to increase comfort refers to the facilities and devices designed to meet the needs of the owner, taking into account the stage of his life cycle, seasonal, monthly and daily rhythms, health conditions - physical and psychological, the need for socialization and rehabilitation.

The requirements related to the organization of the living space and providing light and color comfort are solved using architecture and design (Panero, Zelnik, 2006; Buymistru, 2008). Heating, ventilation, air conditioning, water supply, and drainage systems are used for physiological comfort (Bessoudo, 2008; Visloguzov, 2016). The dynamics of the habitat are ensured by the smart house management system (Morris, et al., 2013). So far, such systems can manage kitchen equipment, regulate the temperature in the rooms, identify house residents and foreigners with security systems, etc. But this is not bad - the cost of energy supply can be reduced by up to 30%.

But, of course, soon the functionality of such systems will be expanded significantly, and the interface will become more flexible and unobtrusive. Let's consider several possibilities that in the future would become part of a smart house in the form of separate units.

1. Mood correction. Imagine that you are coming back home on a late winter evening, after a hard and nervous working day. The system captures the expression of your face and a few slow reactions. The "Relaxing" program is activated: the temperature of the living room and bedroom rises, no drafts, light food is offered, the lighting is muted and the color is changed, melodious music is turned on, soothing aromas are added, relaxing bath is prepared, etc.
   And if your desire to work is recorded in the morning, the "Courage" program is activated - the temperature is reduced, vigorous music is on, some oxygen is added to the air, etc.
   Such means of adjusting or maintaining one's mood are known to both ergonomists and designers, and the main thing is to combine them into experimentally tested programs and create a qualitative unit for analyzing human behavior, allowing for easy customization.
2. Expert system. Modern means of communication allow accessing any large library or archive, joining a group of interest, simply contacting administration, police or an expert in any field. Again, it's about using artificial intelligence to combine all these possibilities, identify the owner's interests and organize the required connections and inquiries, saving him from long searches and expectations.
   A separate topic - concerns related to the operation, payment, and repair of the house. It would be right to let an expert system solve these troublesome tasks.
3. Developing system. There are many educational programs in the world and the network - from sophisticated training to remote education in the best universities. The natural development of the expert system would refer to the identification of the user's interests and organization of the required education courses for him. Here both age-related peculiarities and discipline of a person could be taken into account, if necessary, the system can also become a strict trainer, or, if you will, a collector!

4. Socialization system. When analyzing the conditions of comfort for various psychological types, the topic of socialization sounded quite often. Many people, due to the peculiarities of mental development or physical disabilities, have difficulties in communicating with society. And here a smart house can make life easier. For example, having organized access to the closed social networks where autists, for example, communicate. Or by providing people with impaired vision or hearing access to "extended reality". Some disadvantages can be eliminated by hard training - and here a smart house will play the role of a social trainer. To organize contacts with an employer for a designer or programmer who prefers to work at home, or a craftsman who has a 3D printer, to ensure the purchase of materials and much more - these are the tasks for such systems, which also require the development of artificial intelligence.
5. Home doctor. Scientific activities, communication, movement, just life support of genius of our time - Stephen Hawking – were supported with technical devices for many years. It is also preferable to entrust such functions as the organization of diagnostics and consultations, purchase and timely taking of medications, regulation of environmental parameters, use of unusual means of treatment, such as aromatherapy, to the smart house. Simple monitoring of vital parameters and prompt reaction to their change, with notification of the attending physician, are tasks that are quite feasible for electronic assistants – in many cases, it can save lives.
6. Personal servant. It's nice when someone wakes you up, brings coffee to bed, and sneakers to your feet, invites you to dinner, reminds you of a meeting, surrounding you with care and attention in every possible way, staying patient to your grievance and bad mood. But people are less and less willing to accept such work. And the robot, under the supervision of the smart house, is quite capable of being at your service 24 hours a day. There is no doubt that such a system will be in demand.

All that has been said is just a brief sketch of the prospects for the development of a dwelling, following the trend to make it more dynamic and controllable.

## DWELLING AS A PEACEMAKER

Usually, the choice of dwelling gives rise to conflicts: the husband likes one, and the wife likes the other and the children like something else. There is a situation of incompatibility. To prevent this from happening, one should analyze the nature of incompatibility and understand how to use a dwelling that plays the role of a regulator in the "human-dwelling-environment" system to prevent it. Consider the problem of compatibility, within the paradigm of this study, at the interpersonal level – within a family, between friends and in general, "in one's circle".

The contradictions are based on unequal potentials and different development of channels of interaction with the environment, with imposed difference of temperaments and ages, cultural and social differences, different needs, motivations and life experience (partly due to the difference of channels, and partly – due to the living conditions). Do not forget about sex differences.

The construction of a dwelling should start with planning, which, in its turn, should be based on facts - psychological portraits of family members obtained as a result of testing (chapter 12, *Modern human mind*, an example of a psychological portrait – Figure 23) or life path analysis based on natural observation.

Then, one should compare the portraits and decide for what purpose and how will compatibility be implemented? Such options are possible:

1. **Development of Priority Channels**: if all family members have similar portraits, we select a dwelling that is comfortable for priority channels, as described above. Note that one should choose a place for a house or apartment in a comfortable environment, or create such an environment, using the techniques described in (Mkhitaryan, 2017), and then select the style, presence or absence of transformability, space-planning solutions, interior design, equipment, and furniture - then the habitat will be complete and there will be no cognitive dissonance. As a result, "strong" channels will become even stronger, gradually shaping the corresponding habits, character, and the whole trajectory of the development of family members.
2. **Redistribution of Potentials** from priority channels to those development thereof is desirable is an option, again possibly with the similarity of psychological portraits. For example, enterprising researchers have wished to become warriors. Then they should carefully study Japanese style, and following the recommendations outlined above, carefully reproduce the appropriate habitat, enroll in the martial arts group or enlist in the army. Or the egoists decided to become monks, but they don't want to go to the monastery - then recommendations on the choice of homes for contemplators, plus enhanced communication with various "wanderers" and "people of God" will do. The tragic fate of Tsar Nicholas II[2], his family and the entire Russian Empire shows that such experiments are far from always appropriate, and before you decide on them, you need to consider the consequences thoroughly. Although a human psyche is very adaptive, the proverb "do not sit in someone's else sleigh" is appropriate here as nowhere else.
3. **Complimenting Personalities**: Let's say, the wife has good intuition, and a husband is a person with high intelligence, and their small child is a very emotional being. Each of these personalities is not balanced and has many "gaps" in interaction with the environment and society. But they complement each other, mutual love smoothed dissimilarity and the family is very balanced as a whole. So is it worth changing anything here? In this case, the style can be neutral (for example, minimalism, eco- or bio-style), not annoying to anyone and not stimulating particularly. This is especially important for transit zones, living rooms, dining rooms and other gathering places for the whole family. On the contrary, in their premises, stylistic variations, aimed at stimulating intuition, intellect, and emotions for the wife, husband, and child, are appropriate, respectively.
4. **Synchronization**: Let's not forget that the priority of this or that channel is connected with a certain historical epoch. In the previous example, the wife is in the Paleolithic, the husband is stuck in the Age of Enlightenment, and only the child is our contemporary. Of course, it is desirable to be adapted to modern realities, and why not to learn something from your child? Under the slogan "all the best for children", it would be appropriate to shift the stylistic accents somewhat towards modernity, postmodernism, or even light kitsch.
5. **Neutralization of Incompatibilities**: No wonder they say that opposites converge. A husband and a wife can be quite dissimilar to each other. The difference in age and worldview gives rise to the problem of "fathers and children", where you can safely add more "grandchildren". And if there are "owls" and "larks" among them, fans of football and classical music, then one can be concerned about the destiny of this family, provided there are strong characters in there.

In this case, only territorial delimitation will help - the floors of "fathers" and "children", several bedrooms, bathrooms, study rooms, televisions, etc. for husband and wife, separate rooms for "grand-children" etc. - each room in its own style interfaced with the individual characteristics of the owner. Plus, a neutral style for transit zones and canteens. Yes, there will be eclecticism, perhaps, a horrendous one for guests, but comfortable for family members, at least, a decay that does not provoke its inevitable in other solutions.

In reality, of course, other options are possible - especially if the spouses belong to mixed psychological types. But a careful analysis will make it possible to find at least not an annoying decision, even it does not satisfy everyone.

## OPTIMIZATION TASK

At the end of the chapter, let's talk about money - after all, any option of the dwelling change requires considerable expenses. To prevent the conversation from becoming abstract, let`s focus on the needs of the person whose psychological portrait is shown in Figure 25 chapter 12. For convenience, let`s copy this portrait (Figure 41).

Thus, our character decided to build the most comfortable dwelling for himself. How should he act?

Any budget is limited (let it be, for example, \$1,000,000), and one has to spend money with the maximum return. For our character, financial resources shall be directed, first of all, to ensure prestige, self-sufficiency, and security of the dwelling, secondarily, ensure the rationality of architectural and design solutions, and only then - meet any other conditions of comfort.

*Figure 41. A psychological portrait of personality - the basis of optimization*

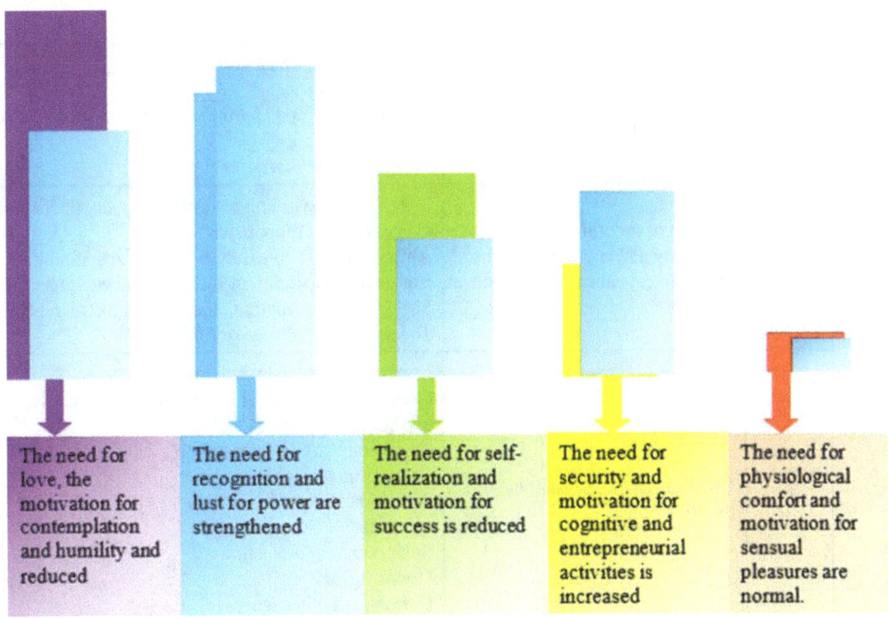

In what proportion shall he distribute his funds? In a perfect scenario, according to the height of the blue columns in Figure 44. In reality, this will not work, due to the disproportionate cost of different works, but one should still strive for such a distribution.

Taking into account this circumstance and roughly distributing the expenses among priority targets (see the column "targets for levels" in Table 13), let`s move on.

The realization of targets for levels involves clarifying conditions of comfort and preventing any discomfort. This clarification is decomposed into physiological, psychological and biorhythmic indicators. Some of them, mainly physiological, are regulated by national building codes and regulations.

So, it is a list of quality criteria. One can do it oneself, using the advice of friends, following regulatory documents, etc. Let`s make their selection and distribution by levels as shown in the column "individual indicators" of Table 13.

*Table 13. Summary of consumer's indicators of the dwelling in different levels*

| Target | Levels | Targets for levels | Individual indicators |
|---|---|---|---|
| Maximum comfort ($1,000,000) | 1 | A sense of harmony with the outside world, good luck, indestructible health, a favorable course of events, optimism about the future ($220,000) | Compliance with traditions (religion, culture, lifestyle)<br>Compliance with climate<br>Compliance with the environment (topography, availability of water sources, nature of vegetation, cleanliness)<br>The harmony of the system city - district - ensemble - building - dwelling<br>The possibility of improving the environment<br>Convenience of location<br>*The list can be shorter or longer* |
| | 2 | The sense of self-sufficiency, self-esteem, prestige, security, joy, and optimism ($270,000) | 4. Expression and purity of style ($36,000)<br>5. Expressiveness on the surrounding background ($45,000)<br>5. The ability to attract and catch attention ($45,000)<br>5. The prestige of the location ($45,000)<br>5. Compliance with the status of the owner ($45,000)<br>3. Advanced security features ($27,000)<br>2. Autonomy ($18,000)<br>1. The consistency of exterior and interior ($9,000)<br>*The list can be shorter or longer.* |
| | 3 | The sense of freedom, superiority, confidence, and possibility of activity ($160,000) | Maximum functionality<br>Transformability;<br>"Smart home" and appropriate program used to regulate its parameters<br>*The list can be shorter or longer.* |
| | 4 | A sense of orderliness of the regularity and cognoscibility of space-time, beauty, and harmony ($200,000) | Functional, constructive, aesthetic, symbolic feasibility of forms<br>Consistency of proportions<br>Anthropometric consistency of the sizes of premises<br>Ergonomic feasibility of furniture and equipment<br>Scale to human, rhythm, nuance, and metric repetitions<br>*The list can be shorter or longer.* |
| | 5 | Comfort sensory sensations ($150,000) | Light and black and white plastic<br>Color - the harmony of combinations, any unusual combinations, and stimulating effect<br>Tones and textures of materials<br>Temperature-moisture comfort<br>Acoustic comfort<br>No drafts<br>Solar radiation protection<br>Use of flavors<br>Harmony of other senses<br>*The list can be shorter or longer.* |

Now it is necessary to distribute the budget among private indicators. This is done in two steps.

A. Let's determine the importance of particular indicators for ourselves at every level.
In order not to complicate the calculations, it can be recommending the following scale:
5 - critical;
4 - very important;
3 - important, but you can "get along" with a disadvantage;
2 - not important;
1 - desirable, but you can accept the lack;
0 - can be ignored safely.
Let either of 8 particular indicators score the number of scores shown in the right column to the left of the indicator name for the highest priority target (level 2).

B. Do the following calculations.
   1. Summarize the scores. The total is 30.
   2. Divide the budget by the number of scores: 270 000/30 = $9 000 for 1 score.
   3. Determine the budget for every indicator, multiplying 9,000 by the number of scores obtained.

The result is given to the right of the indicator name. The same actions are performed for all particular indicators. As a result, this is a painted budget of "our wishes".

Now the most difficult, creative part is to find correlations between "our wishes" and real architectural and design solutions, as well as the actual cost of work. Figures 1-43 will help to make a statement of work for architects and designers and evaluate their counter offers. As for the cost of work, then you should trust professionals. And life experience suggests that the budget will be *always* exceeded. Therefore, our character would better have a reserve of $500,000, and even better $1,000,000 for unforeseen expenses!

## CONCLUSION

The conditions of comfort for all channels of human interaction with the environment, all psychological types, and subtypes. Based on the obtained results, the aims of dwelling designing been determined, including its relations with the environment, style, functionality, transformability and adjustability, space-planning solutions, color solutions, selection of materials and equipment, and interior items. Examples of dwellings with such recommendations implemented have been provided.

Ways of accounting any changing needs during the life cycle, annual, daily and other cycles, as well as in states of stress and illness have been considered. Their implementation is possible with the improvement of smart house technologies. The directions for such improvement have been indicated. Types of conflicts and possibilities of a dwelling to solve, redirect or smooth them have been given.

An example of rational use of money to obtain the most comfortable dwelling, that is, solving a limited multicriteria optimization problem with heterogeneous optimization criteria based on the theory of self-organization has been given. Thus, the practical use of the theory developed in this publication has got a logical finishing.

## FUTURE RESEARCH DIRECTIONS

The study has the following perspectives:

- definition and accurate analysis of the needs and motivations of target groups with their social context, categories of dwelling, budget parameters, etc.;
- selection of the most appropriate design solutions concerning the needs and motivations of the target groups, in particular, main indicators of the dwelling comfort;
- development of new technologies, in particular - smart house.
- optimization of expenses of the customer and construction company to achieve maximum efficiency of capital investments.

## REFERENCES

Acharjya, D. P., & Geetha, M. K. (Eds.). (2017). *Internet of Things: Novel Advances and Envisioned Applications*. Springer.

Alexander, K., Ishikawa, S., & Silverstein, M. (2014). *Template language*. Cities. Building. Construction.

Archdaily. Møller Architects. Project Sustainable Town Houses Stockholm. (n.d.). Retrieved from https://1dom.wordpress.com/2010/09/20/ekologicheskie-taunxausy-v-shvecii/

Archdaily. Patkau Architects. Private house, Whistler. (n.d.). Retrieved from http://www.archdaily.com/610622/hadaway-house-patkau-architects

Archidea. Steklyannyy dom v Milane. (n.d.). Retrieved from https://archidea.com.ua/ideas/innovations/359382-steklyannyy-dom-v-milane

Arhidaily. Lundgaard & Tranberg Architects. Campus Tietgen. (n.d.). Retrieved from https://www.archdaily.com/474237/tietgen-dormitory-lundgaard-and-tranberg-architects/52f3041de8e44eb12300006a-tietgen-dormitory-lundgaard-and-tranberg-architects-photo

Artlib. (n.d.). Retrieved from http://www.artlib.ru/?id=11&fp=2&uid=2952

Bessoudo, M. (2008). Building Facades and Thermal Comfort: The impacts of climate, solar shading, and glazing on the indoor thermal environment. VDM Verlag, 2008.

Birkmire, W. H. (n.d.). *Waldorf floor plan.* New York: J. Wiley. Retrieved from https://en.wikipedia.org/wiki/File:Waldorf_floor_plan.jpg

Boxtoner. Celling-Decoration. (n.d.). Retrieved from https://boxtoner.ru/the-ceiling/decorating-with-own-hands-from-gypsum.html

Buymistru, T. (2008). *Coloring. Color is the key to beauty and harmony*. Moscow, Russia: Niola-Press.

D'Ulizia, A., Ferri, F., Grifoni, P., & Guzzo, T. (2010). Smart homes to support elderly people: innovative technologies and social impacts. In *Pervasive and Smart Technologies for Healthcare: Ubiquitous Methodologies and Tools* (pp. 25–38). Hershey, PA: IGI Global.

Ellard, C. (2017). *Habitat: how architecture affects our behavior and well-being*. Moscow, Russia: Alpina Publisher.

Estateline. Dosug. Pictures. (n.d.). Retrieved from http://estateline.ru/dosug/pictures/12102/

Faqindecor. (n.d.). *Frank Lloyd Wright*. Retrieved from http://faqindecor.com/frjenk-llojd-rajt-istorija-arhitektury-napisannaja-zolotymi-bukvami/

Gagge, A. P., Stolwijk, J. A. J., & Hardy, J. D. (1967). Comfort and thermal sensations and associated physiological responses at various ambient temperatures. *Environmental Research*, *1*(1), 1–20.

Glamurchic. (n/d). *Pugacheva, Galkin*. Retrieved from https://glamurchik.tochka.net/197883-pugacheva-povezla-galkina-kupatsya-v-livadiyu/

Glazychev, V. (2017). *Urbanistics*. Europe Publishing House.

Godish, T. (2001). *Indoor Environmental Quality*. Boca Raton, FL: CRC Press.

Grant, H., Halvorson, T., & Higgins, T. (2014). *Psychology of motivation. How deep installations affect our desires and actions*. AST.

Humphreys, M. (2016). *Adaptive thermal comfort: foundations and analysis*. Abingdon, U.K.: Routledge.

Ikonnikov, A. V. (1985). Art, environment, time: Aesthetic organization of the urban environment. Moscow, Russia: Soviet artist.

Iovlev, V. I. (1996). *Ecopsychology for architects: process and form*. Ekaterinburg, Russia: Architecton.

Kavtaradze, S. (2017). *Anatomy of architecture. Seven books on logic, form, and meaning*. Higher School of Economics.

Knews. Mnenie. Religia. (n.d.). Retrieved from https://knews.kg/2018/03/13/mnenie-religiya-okazyvaet-bolshoe-vliyanie-na-molodezh-kyrgyzstana/

Kolcaba, K. (2003). Comfort Theory and Practice: A Vision for Holistic Health Care and Research.

Leroux R. (1970). Human Ecology - the science of housing. Moscow, Russia: Izd. construction literature.

Lin, Mike V. (2010). Modern Design. Walkthrough. Moscow, Russia: Astrel, AST.

Maddox, T. (2018). Smart cities: 6 essential technologies. Tech Republic; Retrieved April 1, 2018.

Magazindomov. Gorodskoj dom v yaponii. (n.d.). Retrieved from http://www.magazindomov.ru/2014/07/30/gorodskoj-dom-v-yaponii-67/

Miller, D. (2009). The Comfort of Things.

Mkhitaryan, N. M. (2017). *Man and Dwelling*. Gangemi Editore International Publishing.

Mkhitaryan, N. M., Badeyan, G. V., & Kovalyov, Y. N. (2004). *Ergonomic aspects of complex systems*. Kiev, Ukraine: Naukova Dumka.

Morris, M. E., Adair, B., Miller, K., Ozanne, E., Hansen, R., Pearce, A. J., ... & Said, C. M. (2013). Smart-home technologies to assist older people to live well at home. *Journal of aging science, 1*(1), 1-9.

Narod. USSR. Laktionov. (n.d.). Retrieved from http://webstarco.narod.ru/ussr/laktionov.html

Pal, D., Funilkul, S., Charoenkitkarn, N., & Kanthamanon, P. (2018). Internet-of-Things and Smart Homes for Elderly Healthcare: An End User Perspective.

Panero, J., & Zelnik, M. (2006). *Fundamentals of ergonomics. Man, space, interior. Handbook of project standards*. Astrel.

Potaev, I. (2008). *Urban planning and territorial planning*. Phoenix.

Pravoslavie. (n.d.). Mount Athos. Retrieved from https://pravoslavie.ru/100213.html

Prokhorov, M. I. (1946). *City Square*. Moscow, Russia: State Architectural Publishing House.

Reimers, N. F. (1994). Ecology. Laws, principles, hypotheses. Moscow, Russia.

Religions-union. (n.d.). Afon. Retrieved from https://religions.unian.net/afon/1488789-7-tayn-i-samyih-bolshih-zagadok-svyatoy-goryi-afon.html

Sadiq-Khan, D. (2017). *Battle of the city: how to change our streets. Revolutionary ideas in urban planning*. Moscow, Russia: Olympus Business publishing house.

Shimko, V. T. (1990). *Architectural Formation of the Urban Environment: A Manual for Architectural Specialties of Universities*. Moscow, Russia: Higher. Sc.

Shtoer, R. (1992). Multi-criteria optimization: theory, calculations, and applications. Moscow, Russia: Radio and communication.

Telegraf. Aziatskie pyramidy. (n.d.). Retrieved from https://telegraf.com.ua/ekologiya/450729-aziatskie-piramidyi-fotogalereya.html

Tetior, A. N. (2008). *Urban ecology*. Moscow, Russia.

Trollno. (n.d.). *Zhizn v zamre*. Retrieved from https://trollno.com/zhizn-v-zamke-kak-vyglyadit-dom-ally-pugachevoj-i-maksima-galkina/

Visloguzov, A. (2016). *Features of Modern Design of Heating, Ventilation, and Air Conditioning Systems in Public, Multi-Storey, and High-Rise Buildings: A Tutorial*. SKFU.

Vityuk, E. (2015). *Search of an Ideal City*. Architecton.

Wikimedia. (n/d). The_Universe. Retrieved from https://commons.wikimedia.org/wiki/File:The_universe.jpg

Wikimedia. Waldorf Hotel. (n.d.). Retrieved from https://commons.wikimedia.org/wiki/File:Demolished_Astoria_hotel,_NYC.jpg?uselang=ru

Wikipedia. (n.d.). Astoria Hotel. Retrieved from https://en.wikipedia.org/wiki/File:Astoria_Hotel_-_First_Floor_Plan.png

Wikipedia. (n.d.). *Waldorf-Astoria.* Retrieved from https://en.wikipedia.org/wiki/Waldorf–Astoria_(1893–1929)

Williams, N. D. (2006). *Design of familiar things*. Moscow, Russia: Publishing House.

Yaroslav. (n.d.). Hrudchevka. Retrieved from https://yaroslavl.n1.ru/articles/yaroslavskaya_hruschevka_carstvo_ili_myshelovka-50334181/

Zidermane, A. A. (1988). Some issues in chronobiology and chronomedicine: Zinatne.

## ENDNOTES

1 Shunya is one of the key ideas of Buddhism, along with the dharmachakra - the Wheel of Law, cycle, teaches to perceive life as a short moment between diving into the void. Such a feeling permeates the famous medieval novel "Taira Monogatari", about the death of the Taira clan in the fight against the warriors of Minamoto, as well as many legends and poems.

2 The last Russian Tsar was unduly influenced by his devout wife and the pseudo-holy Grigory Rasputin. Under their influence, he abdicated from reign during the February Revolution, not willing to shed blood. He was immediately arrested and later shot by the Bolsheviks with the whole family during the civil war, which cost millions of lives.

# Conclusion

The book tells about the evolution of the human mind and patterns of behavior from ancient times to the present.

The author's hypothesis: since a human is a complex system, the evolution of his mind and behavior is considered as the implementation of the general laws of self-organization of complex, open, emerging systems. Testing this hypothesis was the purpose of the publication.

Let`s consider how successful the was testing of the hypothesis and what results were obtained due to its application.

The most important changes in the process of human evolution are bipedalism, the release, and improvement of hands, binocular vision, the predominance of vision over other senses, changes in the structure of the larynx, hair loss.

Even more important is the complication of the structure of the brain, the increase in its volume and energy consumption (in an adult, the share of brain metabolism of total energy needs reaches 20-25% during mental work, which is much more than in primates and mammals, with a brain mass of 1.5-2% body weight).

The paradox of the changes is that the positive effect manifests itself through many generations and is initially useless or even harmful to human survival.

Thus, vertical walking and the liberation of hands after at least a million years led to the beginning of the production of tools of the Olduvai type. Only 300 thousand years after its appearance, the Paleoanthropus switched to a more advanced Mousterian instrument manufacturing technique.

The same applies to symbolic human activity: the first Homo sapiens, at least 130-150 thousand years, in no way showed the superiority of their more perfect brain over Neanderthals, and only then did their painting and stone processing go far ahead. And precisely at this time - and this is another paradox - the human brain began to degrade, simplifying and decreasing from the Upper Paleolithic to the present day by 400-450 $cm^3$.

These features create the impression of the existence of unknown evolutionary factors. These factors should be sought in the more general laws of evolution of complex open systems because man is also a complex system.

However, existing methods of modeling complex open systems have in their axiomatic basis contradictions with the properties of real systems. Therefore, it is necessary to develop new methods and models where this contradiction is eliminated.

The axiomatic wave model of S-space proposed by the authors corresponds to the well-known properties of open complex systems. The model includes waves and solitons of various modalities, nonlinear operations, and has the properties of openness and emergence. The theory of self-organization describes scenarios for the evolution of S-space, different depending on the initial conditions and external factors. When modeling real systems, based on preliminary data, external influences, the type of scenario, the distribution of parameters by system levels, and units of measurement of parameters are determined. Missing parameters can be calculated based on theoretical principles.

This apparatus was proposed to be used to research the evolution of the human mind and behavior. The above facts, related to the problem of human origin, were used to verify it: if their adequate description and explanation are possible, then the apparatus is suitable for solving the tasks.

Within the theory of self-organization, man, his systems and organs are waves and solitons that have a certain potential and modalities, exist in a dissipative medium for a limited time and evolve under the influence of non-systemic factors.

So, the potential for upright posture and the release of the hands appeared as a result of a genetic mutation (actualization of the wave and its modality of excitation) quickly manifested itself and consolidated (the wave passed into a soliton). There was a potential possibility to use hands for making tools. And after a long time, as a result of another external impulse, it became actual. Similar external impulses preceded other anatomical changes.

The same applies to the improvement and then degradation of the human brain - an external impulse led to a genetic mutation and continued to operate, as a result of which the brain grew and became more complicated from generation to generation, and the body adapted to “this parasite”, and only after a long time the brain could manifest its usefulness. And then the impulse dried up, and it was time for degradation.

This process can be compared with the life of an individual man: in childhood, the body grows “on its own”, for no apparent reason and without conscious effort. Studying is just as easy, lacking strength in sports and so on. There comes a time of maturity, and a person should already spend his energy on maintaining his intellectual and physical form. Old age is approaching - again, for no apparent reason, everything is beginning to deteriorate - neither training nor medication helps. The body is degrading, creative activity is dying, and nothing can prevent the inevitable death. The transition of a wave with expansion modality (childhood) into a stable soliton state (maturity) and a wave with weakening (old age) and discontinuation (death) modalities describes this process.

So, in the framework of the theory of self-organization, the paradoxes of human origin are adequately described and explained theoretically. Thus, the developed apparatus is suitable for studying the evolution of the human mind and patterns of behavior. The time frame of the study is determined by the criteria for verifying the results and includes the period from the Upper Paleolithic to the present day.

In terms of the theory of self-organization, the formulation of the author’s hypothesis is as follows: the course and features of the evolution of a person, his mind and behavior are the implementations of one of the scenarios of self-organization - (1S, 1O).

The first results of using the theory are the construction of a model of human-environment interactions and the determination of the structure of the subjective space of a human.

It turned out to be possible to integrate, in the framework of the model, different-quality characteristics obtained by different methods (psychotype, motivation, and needs, temperament, age, social status).

The self-organization scenario is impersonal, being a “great gestalt” that determines the innate ability to interact with the environment. Based on this understanding, the following answer to the main epistemological question can be proposed:

1) The world is cognizable only to the extent that for man and the Universe the same laws of self-organization take place.
2) The human perceives the surrounding world due to the symmetry of the channels of perception structure and the structure of the Universe; spontaneous symmetry breaking leads to the inherently metabolic process of perception.
3) The process of cognition develops as a process of self-organization; first, intuition, which determines unity with the subject of knowledge is activated, then the ego establishes the "subject-object" relationship, then the object is affected and its reactions evaluated (will and reason), reactions and evaluations are related to the categories of space and time (nous), and finally complemented by sensory impressions. Thus, the process of cognition corresponds to the scenario of the S-space bundle. The previous levels establish the context within which the data of the channels located at the next levels are interpreted.
4) There is a reverse process corresponding to the convolution scenario - when sensory impressions are ordered by the mind and evaluated, based on previous experience, by the mind operating in binary categories.
5) Different people with significantly different experiences recognizes the world relatively equally, since the processes of bundle and convolution are invariant, although they differ quantitatively for different people.

Within the framework of the human-environment model, such an approach to determining behavior models is possible.

The relations of society with the environment are related by the same laws of self-organization as for a human. Therefore, they are structured similarly: levels support unity relations, relations of separateness, relations of action and evaluation of environmental reactions, relations of ordering in space and time, and, finally, sensory relations.

Each of these levels corresponds to a certain estate, representatives of which have one dominant channel; it can be: intuition, ego, will, etc. The formation of estates is closely connected with the evolution of the human mind (interdependence is determined by the axiomatic statement 2.1 of the wave model of S-space). Therefore, the dominant channel of a particular man determines his social position. Consequently, his behavior corresponds to the pattern of one of the estates - priests, rulers, bourgeois, etc., well described in history and literature and can be understood from these descriptions.

It is substantiated that a marker of the human mind state is a method of space imaging in painting. Artists always depict the forms and relations of the subjective space, and the ways of representing the space correspond to the forms and relations of that part of the subjective space that is relevant for the self-organizing mind.

Also, the data for the mind state estimation can be obtained by analyzing the myth of the World Tree, considered as the additional marker.

This approach allowed us to track the stages of self-organization of the human mind, linking them with the technique of space depicting, as well as the development of the World Tree myth.

Further research was conducted according to the following plan (chapters 5-12):

- For each of the periods of the development of mankind, starting from the Upper Paleolithic, samples of painting and data were selected to reconstruct the state of the World Tree myth.

- Analyzed the features of the space imaging and the features of the myth version. Comparisons of monuments of different cultures were made, and the results of the analysis with the previous period were compared.
- Conclusions about the state of mind, patterns of behavior and the direction of their evolution were drawn.
- In section 12, after analyzing the current state of the human mind and behavioral patterns, the course of their evolution was generalized and presented in Fig. 24.

So:

- Paleolithic techniques correspond to channels of 1 and 2 levels (intuition, ego).
- Neolithic – to channels of 2 and 3 levels (ego, will, reason).
- This is followed by a long transitional period to channels of 4 and 5 levels (nous, emotions), which began during the Eneolithic and ended in late Antiquity.
- The Middle Ages turns the mind, for a short time, back to mysticism (intuition, ego; levels 1 and 2), but does not reach them to the extent that it was in the Paleolithic).
- But already in the Renaissance, the mind finally goes to levels 4 and 5 - to the rationalism and sensual pleasures.
- The modern type of human mind was mainly formed during the Renaissance. In the past about a hundred years, there has been a tendency to transition of mind to level 5.

Now more detail.

Having reached the peak of its development by the beginning of the Paleolithic, the human brain was built and "optimized" for the intuitive perception of extra-systemic factors. Hence the redundancy of its complexity compared with the needs of survival. Subjective space at this time was mostly intuitive. Subsequently, due to the wave nature of the man-environment system, the human brain degraded, and consciousness passed to descending levels. This is not surprising - the development of structures at any level and an increase in their potential is accompanied by a decrease in the potential of the previous level and, possibly, the subsequent level, and this is only partially compensated by exchanges with the environment. In later times, this process continued.

The sacred painting was a mean of preventing separation and stimulating the reverse process (convolution). The primitive painting was such in a large part.

If in the Paleolithic, the weakening of intuition and the "loss of gods" associated with it caused discomfort in the spiritual realm and palliative means for "treatment" in the form of sacral painting and ritual practices were sufficient, in Neolithic it was already manifested much deeper and in many aspects of interaction with the environment. And to overcome it, all the achievements of the Neolithic revolution were required.

If in the Neolithic period, the weakening of intuition and the sensation of comfort caused numerous wars and called for great inventions that changed the lifestyle of hunters to farmers or cattle breeder's lifestyle, in the Eneolithic "longing for a lost paradise" resulted in the creation of cities and complex rituals.

When comparing the depicting of space by the masters of ancient Egypt and the masters of previous eras, there is immediately a difference is founded. This is the interest of the Egyptians to the imaging of the relations of exactly three-dimensional, geometric space and finding ways to do it. Of course, among the drawings, two-dimensional "Neolithic" images make up the majority, but already exist, moreover, have canonized techniques, two methods of representing the depth of space.

The share of secular painting is gradually increasing, it begins to infiltrate into the sacral region, profaning the relationship of the pharaoh-god and his wife-goddess to the attitude of ordinary people, in an unsuccessful attempt to describe the symbolism of sacred marriage with the help of geometric and color relations.

And this already indicates a new state of mind. This state grows due to the activation of the channel of the nous (the fourth level of self-organization), which perceives space-time relations, and gradually becomes dominant. It was this process that subsequently gave rise to modern civilization and science.

Further, the connection between the technical features of ancient painting is traced with the corresponding characteristics of the preceding and synchronous images. Established trends in the development of ancient art and identified their connection with the general laws of the human mind evolved in the period of the existence of ancient civilization.

Reconstruction of the myth of the World Tree and related myths was carried out for all the above eras. The conclusions made on its basis about the state of mind of a person confirm the features of its evolution. Separate consideration is the variation of the myth in Christianity.

Also, the behavior patterns of the main estates were reconstructed and their origin was discussed.

Further, the connection between the space depicting medieval paintings in Europe, the Islamic world, China with the corresponding characteristics of the previous images is traced. The tendencies of the development of medieval paintings are established and their connections with the general laws of the evolution of human consciousness are revealed. It is confirmed that evolution has a universal character.

A reconstruction of the myth of the World Tree was done. It is noted that, except for the pagan cultures of Scandinavia in Europe, China in the East, it was preserved and rethought in various hermetic communities.

The typical state of the human mind of the Middle Ages epoch was reconstructed; its main feature is a rigidly binary worldview, intransigence, slow progress of the nous, the emergence of interest in the senses. In marginal circles, however, sensual behavior is fundamental. The behavior patterns of the main estates are also considered. The Middle Ages are the standard of estate behavior; it was then that their typical traits developed to an extreme degree.

The connection between the technical features of the Renaissance painting for Italy and Northern Europe is traced. The tendencies of the development of the painting of the Renaissance are established and their connections with the general laws of the evolution of the human mind are revealed.

If abstract from an abnormally large number of people with high potential (not only in the sphere of fine arts but also science, trade, banking, business, navigation), one could say that the Renaissance logically completes the Middle Ages. However, synergy with the culture of Byzantium, the expansion of horizons in connection with the Great Geographical Discoveries, the birth of exceptionally gifted people made it possible to make clear the coming-out way to new stages of evolution - the final assertion of the domination of the rationalism and the release of feelings from the moral oppression of church precepts.

A consciousness of a person of the Renaissance as close as possible to the modern, closer than the consciousness of people of any previous era.

The most significant contribution of the masters of Italy to the technique of depicting space was that the linear perspective expressed the spatial image in a mathematical form. This means that the linear perspective made rational and logical what seemed to be neither measured nor computed nor subordinated to formal geometric constructions, namely: the inner world of a person's psychic sensations, linking it with unified laws with the outer, physical space. True, this undoubted success was achieved at the cost of violating the laws of natural human visual perception.

The masters of the Northern Renaissance were able to find another way to depict space, much closer to the laws of visual perception. However, they could not solve the problem of its rigorous mathematical justification. In later times, they also had followers, of course, in a smaller number than supporters of the linear perspective.

Development of the World Tree myth was traced. The profanation of even the curtailed, biblical version of the plot - "signs" instead of "symbols" is marked. Of course, this fact reflects the evolution of the human mind and has been given an appropriate interpretation.

The state of mind of the creative person of the Renaissance was reconstructed. Its main characteristics are the high potential and close to an optimal ratio of channels of human interaction with the environment, which makes opportunities for creative, energetic, versatile activities so typical of the genius personalities of that era.

The behavior patterns of the main estates were considerate. The changes that have occurred since the Middle Ages are marked and connected with the evolution of the mind.

The main ways of depicting space in the modern era are defined; almost all of them were used in previous centuries. Building a color space, when colors prevail over forms or even do without them - this is what distinguishes our epoch from previous ones.

A psychological portrait of a certain average personality of modernity has been built. Its main characteristics are low potential and imbalance of channels of human interaction with the environment due to sensory perceptions and emotional reactions. The behavior patterns of such a person are also considered.

The data were necessary for building a personal psychological portrait, determining its basic needs and motivations, as well as making clarifications related to the stage of the life cycle and temperament are given.

Recall the conditions for the correct understanding of this portrait:

1) The subjective space of any person includes all levels, but mind resides predominantly on one of them (for this reason different psychotypes of people differ), making the corresponding channel of interaction with the environment relevant.
2) The mind of a particular person can evolve relatively free from the "spirit of the epoch", both rising to the heights of the first level (as few saints and mystics have shown) and falling to the sensual joys of the fifth level, which is observed incomparably more often.
3) Scientific methods can be used to study both the mind of a particular individual and the generalized mind of people of a particular period, acquiring, under the influence of the "spirit of the time", typical forms of social worldview and behavior.

Summarizing the above, it can be stated that the hypothesis adopted in the study, as well as the predictions made on its basis, are confirmed by the data presented.

Knowledge of the laws of evolution of the human mind and its state in the modern epoch allows us to create a comfortable habitat and is the key to creating a variety of techniques for harmonizing and improving personality.

In the last section, which has practical significance, the conditions of comfort for all channels of human interaction with the environment, all subtypes of each psychotype are defined.

Based on the obtained results, the aims of dwelling design were determined, including its relationship with the environment, style, functionality, transformability, and adjustability, space-planning solutions, color solutions, selection of materials and equipment, subject filling. Examples of dwellings implementing these recommendations are given.

Ways of accounting for changing needs during the life cycle, annual, daily and other cycles, as well as in states of stress and illness are considered. Their implementation is possible with the improvement of smart house technologies. The directions for such improvement are indicated.

Possible conflicts and opportunities for a dwelling for their resolution, redirection or smoothing are given and classified.

An example is given of rational spending of funds to obtain the most comfortable dwelling, that is, solving a limited multicriteria optimization problem with heterogeneous optimization criteria based on the theory of self-organization.

Thus, the topic of practical use of the theory, developed in this publication, received a logical completion.

# Compilation of References


(1955). World history: Vol. 1. *Stone age*. Moscow, Russia: Gospolitizdat.

(Wikimedia. Lucas Cranach the Elder. Paradise. (n.d.). Retrieved from https://commons.wikimedia.org/wiki/File:Lucas_Cranach_d.Ä._-_Paradies_(1530,_Kunsthistorisches_Museum).jpg

A monument of the Okunev Art as sketched by the Aspelin expedition (n.a.). Retrieved from Alt-Altaische Kunstdenkmäler: Briefe und Bildermaterial von J.R Aspelins Reisen in Sibirien und der Mongolei 1887-1889 herausgegeben von Hrsg. Von Hjalmar Appelgren-Kivalo. Hfors 1931. 47 s. + 76 kuvas. Authors: Aspelin, Johann Reinhold and Appelgren. https://ru.wikipedia.org/wiki/Файл:Из_Alt-Altaische.png

Abbott, L., & Sejnowski, T. J. (Eds.). (1999). *Neural Codes and Distributed Representations. Foundations of Neural Computation, Bradford Book*. Cambridge, MA: MIT Press.

Abramova, Z. A. (1971). *Primitive Art*. Novosibirsk, Russia: NSU.

Academic. Verocchio_Lorenzo de Mrdici. (n.d.). Retrieved from https://dic.academic.ru/pictures/wiki/files/86/Verrocchio_Lorenzo_de_Medici.jpg

Acharjya, D. P., & Geetha, M. K. (Eds.). (2017). *Internet of Things: Novel Advances and Envisioned Applications*. Springer.

Actistudy.info-vashhestvo. (n.d.). Retrieved from http://www.activestudy.info/pishhevye-veshhestva-i-ix-znachenie-dlya-organizma/

Adams, L. S. (2001). *A History of Western Art* (3rd ed.). Boston, MA: McGraw Hill.

Adkins, L., & Adkins, R. (2001). *The Little Book of Egyptian Hieroglyphics*. London, UK: Hodder and Stoughton.

Adolf, E. (2008). Life in Ancient Egypt. Moscow, Russia: Centrpoligraf.

Adolf, E. (2008). State, Army, and Society of Ancient Egypt. Moscow, Russia: Centrpoligraf.

Aif-archive161698. (n.d.). Retrieved from http://www.aif.ru/archive/1671698

Albedil, M. F. (1991). *A Forgotten Civilization in the Indus Valley*. Saint Petersburg, Russia: Nauka.

Album "Adalbert Erdeli". (2007). Uzhhorod: "Vidavnitstvo Oleksandri Garkushi", 2007. Azps. Test. Egoism. (n.d.). Retrieved from http://azps.ru/tests/stest/egoism.html

Alex-aka-jj. (n.d.). Retrieved from https://alex-aka-jj.livejournal.com/234163.html

Alexander, K., Ishikawa, S., & Silverstein, M. (2014). *Template language*. Cities. Building. Construction.

Allpainters. Mucha. Sokol Festival. (n.d.). Retrieved from https://allpainters.ru/wp-content/uploads/paintings/6th-sokol-festival-1912.jpg

Altai-go. Turochak. (n.d.). Retrieved from http://altai-go.ru/index/turochakskaja_pisanica/0-109

Amaringo. (n.d.). Retrieved from http://cannabismoon.net/ru/gallery/psycho/amaringo.html

American Psychological Association. (n.d.). Retrieved from https://www.apa.org/science/programs/testing/index

Ancient-origins. Artifacts. (n.d.). Retrieved from https://www.ancient-origins.net/artifacts-ancient-writings/reading-between-lines-decrypting-scripts-minoans-and-mycenaeans-008180

Andreae, B. (1977). *The Art of Rome*. New York: H. N. Abrams.

Andrews, P. W. (2001). "The psychology of social chess and the evolution of attribution mechanisms: Explaining the fundamental attribution error" (PDF). *Evolution and Human Behavior*, *22*(1), 11–29. doi:10.1016/S1090-5138(00)00059-3 PMID:11182572

Anokhin, P. K. (1978). *Selected Works. Philosophic Aspects of Functional System Theory. Moscow, Russia:* Nauka.

Anonymous, & Budge, E. A. (2014). *Wallis (2014).* The Egyptian Book of the Dead. Published by Createspace Independent Pub.

Anropogenez. (n.d.). Retrieved from http://antropogenez.ru/article/771/

Antclub. Muravi. Stroenie-muravya. (n.d.). Retrieved from http://antclub.ru/muravi/stroenie-muravya

Antikvar. Tripolskie ornament. (n.d.). Retrieved from https://antikvar.ua/tripolskie-ornamenty/

Antropogenez. (n.d.). Retrieved from http://antropogenez.ru

Appenzeller, T. (2012). Human migrations: Eastern odyssey. Humans had spread across Asia by 50,000 years ago. Everything else about our original exodus from Africa is up for debate. *Nature*, *485*(7396). PMID:22552074

Archaic-total-arch. (n.d.). Retrieved from http://archaic.totalarch.com/dwelling_and_settlement

Archdaily. Møller Architects. Project Sustainable Town Houses Stockholm. (n.d.). Retrieved from https://1dom.wordpress.com/2010/09/20/ekologicheskie-taunxausy-v-shvecii/

Archdaily. Patkau Architects. Private house, Whistler. (n.d.). Retrieved from http://www.archdaily.com/610622/hadaway-house-patkau-architects

Archidea. Steklyannyy dom v Milane. (n.d.). Retrieved from https://archidea.com.ua/ideas/innovations/359382-steklyannyy-dom-v-milane

Arhidaily. Lundgaard & Tranberg Architects. Campus Tietgen. (n.d.). Retrieved from https://www.archdaily.com/474237/tietgen-dormitory-lundgaard-and-tranberg-architects/52f3041de8e44eb12300006a-tietgen-dormitory-lundgaard-and-tranberg-architects-photo

Arimoya. Iskusstvo Islamskogo Mira. (n.d.). Retrieved from http://forum.arimoya.info/threads/Искусство-исламского-мира.4021

Arnheim, R. (1974). *Art and Visual Perception. Moscow, Russia*: Progress.

Arnold, O. R. (2018). People and animals: myths and reality. Alpina non-fiction.

*Art of nations and nations of the world.* (Vol. 1). (1967). Moscow, Russia: Short Art Encyclopedia.

Art. Biblioclub. Grobnitsa-Sen_n. (n.d.). Retrieved from http://art.biblioclub.ru/picture_32228_grobnitsa_sen_n/

Artists. (n.d.). Retrieved from https://www.the-artists.org

Artlib. (n.d.). Retrieved from http://www.artlib.ru/?id=11&fp=2&uid=2952

Artnet. (2014). Pierre Soulages. Happy to Stay in the Dark, Ben Davis, June 19, 2014.

Aryans of the Eurasian Steppes. (2014). *Bronze and Early Iron Epoch in the Eurasian Steppes and in Contiguous Lands*. Barnaul: Altai State University.

Asch, S. E. (1955). Opinions and social pressure. *Scientific American*, *193*(5), 31–35. doi:10.1038cientificamerican1155-31

Askizon. Okunevskaya kultura. (n.d.). Retrieved from (https://askizon.ru/item/915-okunevskaya-kultura-konets-3-nachalo-2-tys-do-n-e-epokha-rannej-bronzy-khakasiya-khakassko-minusinskaya-kotlovina

Astra-lit. Ukr-ment. (n.d.). Retrieved from https://www.astra-lit.com/ukr-ment-ru/ukr_ment_3.htm

Augustinavichute, A. (2016). Socionics. Moscow, Russia: Black squirrel, 455- 464.

Azps. Tests. Riskman. (n.d.). Retrieved from http://azps.ru/tests/stest/riskman.html

Baby-lessons_zima. (n.d.). Retrieved from http://www.babylessons.ru/zima-detskij-risunok-alisa-35-goda/#more-1171

Bailey, C. B., & Collins, J. B. (2001). Gustav Klimt: modernism in the making. New York: Harry N. Abrams in assoc. with National Gallery of Canada.

Balagan. Battle of Kadesh. (n.a.). Retrieved from https://balagan.info/the-battle-of-kadesh-in-david-rohls-new-chronology

Baron, R. S., & Kerr, N. L. (2003). *Group Process, Group Decision, Group Action*. New York: McGraw-Hill.

Beauregard, M. (Ed.). (2003). *Consciousness, emotional self-regulation, and the brain*. Amsterdam, The Netherlands: Benjamin.

Becker, K., Becker, M., & Schwarz, J. (2007). String Theory and M-theory: A Modern Introduction. Cambridge University Press. Biology-stack-exchange. (n.d.). Retrieved from https://biology.stackexchange.com/questions/53062/why-is-the-human-ear-most-sensitive-to-4000-hz-tones

Beckwith, J. (1964). *Early Medieval Art: Carolingian*. Ottonian, Romanesque, Thames, & Hudson.

Benedante. Tomb internal chariot. (2018). Retrieved from http://benedante.blogspot.com/2018/10/the-tomb-of-infernal-chariot.html

Benesh, O. (1973). *Art of the Northern Renaissance. Its relationship with modern spiritual and intellectual movements*. Moscow, Russia: Isskustvo.

BenitoJ.-M. (2006). Retrieved from https://commons.wikimedia.org/wiki/File:Techo_de_Altamira_(replica)-Museo_Arqueológico_Nacional.jpg?uselang=ru

Benton, J. R. (2002). *Art of the Middle Ages. World of Art*. London, UK: Thames & Hudson.

Bergson, A. (2001). Consciousness and Life. *Moscow, Russia*: ROSSPEN. Entanda-audiology-news. (n.d.). Retrieved from https://www.entandaudiologynews.com/features/ent-features/post/optimising-hearing-aid-processing-for-music-appreciation

Bernhard, J. (Ed.). (2001). *Ordering Medieval Society: Perspectives on Intellectual and Practical Modes of Shaping Social Relations*P. Selwyn, Trans.). Philadelphia, PA: University of Pennsylvania Press.

Bessoudo, M. (2008). Building Facades and Thermal Comfort: The impacts of climate, solar shading, and glazing on the indoor thermal environment. VDM Verlag, 2008.

Bestiary. Bog-Ra. (n.d.). Retrieved from https://www.bestiary.us/images/bog-ra-v-oblike-kota-zajca-ubivaet-zmeja-apopa

Bestiary. Zmienoobraznaya-Boginja. (n.d.). Retrieved from https://www.bestiary.us/images/zmeinoobraznaja-boginja-meritseger-v-korone-drevneegipetska-stella-novoe-carstvo

Betancourt, P. P. (2007). *Introduction to Aegean Art*. Philadelphia: INSTAP Academic Press.

Beyin, A. (2011). Upper Pleistocene Human Dispersals out of Africa: A Review of the Current State of the Debate. *International Journal of Evolutionary Biology*, *2011*(615094), 1–17. doi:10.4061/2011/615094 PMID:21716744

Biblioclub. Zolotoy Kodeks. (n.d.). Retrieved from http://art.biblioclub.ru/picture_12453_zolotoy_kodeks_iz_kenter-beri_evangelist_matfey/

Binski, P., & Messing, A. (2009). The Westminster Abbey Retable. History, technique, conservation.

Biofile. (n.d.). Retrieved from http://biofile.ru/his/24488.html

Birkmire, W. H. (n.d.). *Waldorf floor plan*. New York: J. Wiley. Retrieved from https://en.wikipedia.org/wiki/File:Waldorf_floor_plan.jpg

Blednova, N. S. (1998). Primitive Art. Problem of Origin. Kemerovo, Russia: KPI. (Dostoyanie planeti.

Blednova, N. S., & … . (1998). *Primitive Art. Problem of Origin*. Kemerovo, Russia: KPI.

Bodnar, O. Ya. (2005). Golden section and non-euclidian geometry in science and art. Ukrainian technologies. Lviv.

Boehm, C. (1999). *Hierarchy in the Forest: The Evolution of Egalitarian Behavior*. Harvard University Press.

Bohm, D. (1990). A New Theory of the relationship of Mind to Matter. *Philosophical Psychology*, *3*(2-3), 271–286. doi:10.1080/09515089008573004

Bongard-Levin, G. M., & Ilin, G. F. (1971). *Indiya v drevnosti*. Moskow, Russia.

Borinskaya, S., & Yankovsky, N. (2006). People and genes: the threads of fate, Vek 2.

Boxtoner. Celling-Decoration. (n.d.). Retrieved from https://boxtoner.ru/the-ceiling/decorating-with-own-hands-from-gypsum.html

Boyd, R. (2017). *A Different Kind of Animal: How Culture Transformed Our Species*. Princeton University Press. doi:10.2307/j.ctvc7799z

Breasted, D. A. (1915). *History of Egypt. From the Earliest Times to the Persian Conquest: in 2 Vol. Moscow, Russia:* Publishing House of M. and S. Sabashnikov.

British Psychological Society. (n.d.). Retrieved from https://ptc.bps.org.uk

Brown, P., Sutikna, T., Morwood, M. J., Soejono, R. P., & Saptomo, E. WDue, R. (2004). A new small-bodied hominin from the Late Pleistocene of Flores, Indonesia. *Nature*, *431*(7012), 1055–1061. doi:10.1038/nature02999 PMID:15514638

Brownson, C. D. (1981). Euclid's optics and its compatibility with linear perspective. [Springer Berlin/Heidelberg.]. *Archive for History of Exact Sciences*, *24*(3), 165–194.

Brunet, M., Guy, F., Pilbeam, D., Mackaye, H. T., Likius, A., Ahounta, D., ... Zollikofer, C. (2002). A new hominid from the Upper Miocene of Chad, Central Africa. *Nature*, *418*(6894), 145–151. doi:10.1038/nature00879 PMID:12110880

Burke, P. (1990). The European Renaissance: Centre and Peripheries, 1998.

Burlak, S. V. (2011). *The origin of language. Facts, research, hypothesis*. CORPUS.

Butterfield, H. (1965). Origins of Modern Science 1300-1800. Free Press. 1965.

Buymistru, T. (2008). *Coloring. Color is the key to beauty and harmony*. Moscow, Russia: Niola-Press.

Cahoone, L. E. (1996). *From Modernism to Postmodernism: An Anthology*. Cambridge, MA: Blackwell.

Cameralabs. Afremov. Coming back to the dream. (n.d.). Retrieved from https://cameralabs.org/6587-krasivye-i-optimistichnye-kartiny-leonida-afremova

Campbell, L. (1998). *The Fifteenth Century Netherlandish Paintings, National Gallery Catalogues (new series)*. New Haven, CT: Yale University Press.

Capra, F. (1999). *The Tao of Physics*. Boston, MA: Shambhala.

Carter, R. (1998). *El nuevo mapa del cerebro*. Barcelona, Spain: RBA Ediciones.

Carter, R. (2014). *The Human Brain*. New York: DK Publishing.

Catalog9. Ty zapisalsya dobrovolcem. (n.d.). Retrieved from http://catalog9.ru.net/ty-zapisalsya-dobrovolcem-kartinka.html

Cattaneo, A. (2016). European Medieval and Renaissance Cosmography: A Story of Multiple Voices. Retrieved from https://www.researchgate.net/publication/303740142_European_Medieval_and_Renaissance_Cosmography_A_Story_of_Multiple_Voices

Cavalli-Sforza, L. L., & Feldman, M. W. (2003). The application of molecular genetic approaches to the study of human evolution. *Nature Genetics*, *33*(S3), 266–275. doi:10.1038/ng1113 PMID:12610536

Chalupa, L. M., & Werner, J. S. (Eds.). (2004). The Visual Neurosciences, 2 vol. Cambridge, MA: Bradford Book.

Chambers, F. M. (1941). Some Legends Concerning Eleanor of Aquitaine. *Speculum: A Journal of Medieval Studies*, *16*(4), 459–468. University of Chicago Press. .

Chatal-Kyyuyuk. (n.d.). Retrieved from http://dostoyanieplaneti.ru/2732-chatal-khyuyuk

Cheger. Livejournal. (n.d.). Retrieved from https://cheger.livejournal.com/485661.html

Chernish, E. K. (1982). Eneolithic of the Right-Bank Ukraine and Moldova. Eneolithic USSR. Moscow, Russia, 1982. 166—347.

Childs, P. (2000). *Modernism*. London, UK: Routledge. doi:10.4324/9780203131169

Chip, W. (2013). Last Ape Standing: The Seven-Million-Year Story of How and Why We Survived. Walker Books, 2013.

Chubova, A. P. (1966). Roman pictorial art. Album reproductions. Leningrad, Russia.

Chubova, A. P. (1970). Art of Europe of 1-3 centuries, Moscow.

ChungS. J. (2014). Parallels between Confucian Philosophy and Quantum Physics.

Chung, S. J. (2018). A review of the relationship among self, mind, and brain in functional magnetic resonance imaging study: Tree-pattern image of the semantic map in human brain viewed from the Ultron-Logotron Theory. *Open Journal of Philosophy*, *8*(04), 408–427. doi:10.4236/ojpp.2018.84028

City plan kannelura. (n.d.). Retrieved from http://kannelura.info/?p=8997

Civilka. Humanity. Ashel. (n.d.). Retrieved from http://civilka.ru/humanity/ashel.html

Civilka. Humanity. Mustie. (n.d.). Retrieved from http://civilka.ru/humanity/mustie.html

Civilka. Humanity. Verhniypaleolit. (n.d.). Retrieved from http://civilka.ru/humanity/verhniypaleolit.html

Civilka. Samurskiy medved. (n.d.). Retrieved from http://civilka.ru/wp-content/uploads/2015/08/samuskiy_medved.jpg

Civilka. Sosud-neolit. (n.d.). Retrieved from http://civilka.ru/wp-content/uploads/2015/08/sosud_neolit.jpg

Clarke, R. J., & Tobias, P. V. (1995). Sterkfontein Member 2 Foot Bones of the Oldest South African Hominid. *Science*, *269*(5223), 521–524. doi:10.1126cience.7624772 PMID:7624772

Clayton, E. (2003). *Juniper Fuse: Upper Paleolithic Imagination & the Construction of the Underworld Paperback*. Wesleyan University Press.

Constable, G. (1995). The Orders of Society, chap. 3 of Three Studies in Medieval Religious and Social Thought. Cambridge, UK: Cambridge University Press, 1995, pp. 249–360.

Cook, R. M. (1986). Greek Art. Penguin.

Cormack, R. (1985). *Writing in Gold, Byzantine Society and its Icons*. London, UK: George Philip.

Coulson, D., & Campbell, A. (2001). *African Rock Art*. Harry N. Abrams, Inc.

Criminisi A., Kemp M., & Kang, S. B. (2003). Reflections of reality in Jan van Eyck and Robert Campin Measuring Art: A Scientific Revolution in Art History – Paris, 1256-1272.

Criminisi, A., Kemp, M., & Zisserman, A. (2002). *Bringing picture space to life: computer techniques for the analysis of paintings. Proceedings Computers and the History of Art* (pp. 1126–1137). London, UK: British Academy.

Crusado, U. Livrjournal. (n.d.). Retrieved from https://ultima-cruzado.livejournal.com/photo/album/417/?page=6

Cueva-museo-Altamira. (n.d.). Retrieved from http://www.cuevamuseoaltamira.com/

D'Ulizia, A., Ferri, F., Grifoni, P., & Guzzo, T. (2010). Smart homes to support elderly people: innovative technologies and social impacts. In *Pervasive and Smart Technologies for Healthcare: Ubiquitous Methodologies and Tools* (pp. 25–38). Hershey, PA: IGI Global.

Dale Gutrie, R. (2005). *The Nature of Paleolithic art*. Chicago, IL: University of Chicago Press.

Damasio, A. (1999). *The Feeling of What Happens: Body and Emotion in the Making of Consciousness*. Florida: Harcourt, Inc.

Danylenko, V. M. (1986). *Kamyana Mohyla*. Kyiv, Ukraine: Naukova Dumka.

David, R. (2002). *Religion and Magic in Ancient Egypt*. Penguin Books.

Debus, A. (1978). *Man and Nature in the Renaissance*. Cambridge, UK: Cambridge University Press.

Dempsey, A. (2002). *Art in the Modern Era: A Guide to Schools and Movements*. New York: Harry A. Abrams.

*Descartes , R. (1989-1994).* Works in 2 (Vol. M). Thought.

Descharnes, R. N. (1993). Salvador Dalí. New York: Konecky & Konecky, 1993.

Didenko, B. (1999). *The civilization of cannibals*. Humanity as is.

Dirk De Vos. (2002). *Netherlands painting. Masterpieces of old masters*. Moscow, Russia: Bely Gorod.

Dodson, A., & Hilton, D. (2004). *The Complete Royal Families of Ancient Egypt*. Thames & Hudson.

Dodwell, C. R. (1993). The Pictorial Arts of the West: 800–1200. Pelican History of Art. New Haven, CT: Yale University Press.

Dons-maps-lamoute. (n.d.). Retrieved from http://donsmaps.com/lamouthe.html

Dosoavtor-2018-11. (2011). Retrieved from http://dosoaftor.ru/2018/11/28/naskalnaya-zhivopis-altamiryi/

Drobyshevsky, S. V. (2017). Non-missing link. Book 1, 2. Monkeys and all-all-all.- Corpus, 2017.

Dunbabin, K. M. D. (1999). Mosaics of the Greek and Roman World. Cambridge, UK: Cambridge University Press.

Eco, U. (2004). *The evolution of medieval aesthetics*. St. Petersburg, Russia: Azbuka-klassika.

Edelman, G. M. (1989). The Remembered Present: A Biological Theory of Consciousness. New York: Basic Books.

Edelman, G. M. (2001). Building a Picture of the Brain. In G. M. Edelman, & J.-P. Changeux (Eds.), *The Brain*. London, UK: Transaction Publishers.

Edelman, G. M., & Mountcastle, V. B. (1978). *The Mindful Brain: Cortical Organization and the Group-Selective Theory of Higher Brain Function*. Cambridge, MA: MIT Press.

Edgerton, S. Y. Jr. (1977). The Renaissance Rediscovery of Linear Perspective. *The Art Bulletin*, *59*(2), 281–282.

Egyptopedia. Arkhitektura-srednego-tsarstva. (n.d.). Retrieved from http://egyptopedia.info/a/1432-arkhitektura-srednego-tsarstva

Egyptopedia. Iskusstvo Drevnego tsarstva. (n.d.). Retrieved from http://egyptopedia.info/i/715-iskusstvo-drevnego-tsarstva

Egyptopedia. Iskusstvo Novigo tsarstva. (n.d.). Retrieved from http://egyptopedia.info/i/1593-iskusstvo-novogo-tsarstva

Egyptopedia. Kultura Srednego tsarstva. (n.d.). Retrieved from http://egyptopedia.info/i/716-kultura-srednego-tsarstva

Egyptopedia. Selskoe-khozyajstvo. (n.d.). Retrieved from http://egyptopedia.info/s/866-selskoe-khozyajstvo

Egyptopedia. Voennaya politika. (n.d.). Retrieved from http://egyptopedia.info/v/857-vneshnyaya-politika

Ellard, C. (2017). *Habitat: how architecture affects our behavior and well-being*. Moscow, Russia: Alpina Publisher.

Ellis, L. (1984). The Cucuteni–Trypillia culture: study in technology and the origins of complex society. Volume 217 of the BAR international series; British Archaeological Reports (B.A.R): Oxford, 1984.

Employees. Minoan_Mycenaen. (n.d.). Retrieved from http://employees.oneonta.edu/farberas/arth/Images/ARTH209images/minoan/octopus_vase.jpg

E-reading Club. Ehrman. (n.d.). Retrieved from https://www.e-reading.club/chapter.php/1005287/4/Erman_-_Zhizn_v_Drevnem_Egipte.html

E-reading_club. Ehrman. (n.d.). Retrieved from https://www.e-reading.club/chapter.php/1005286/13/Erman_Adolf_Gosudarstvo_armiya_i_obschestvo_Drevnego_Egipta.html

Erikh, T. (2007). Lunar God: Terra.

Estateline. Dosug. Pictures. (n.d.). Retrieved from http://estateline.ru/dosug/pictures/12102/

European steppe-eneolithic. (n.d.). Retrieved from https://indo-european.eu/tag/steppe-eneolithic/

Evpatori Pechat Mohendzo. (n.d.). Retrieved from http://www.evpatori.ru/pechat-iz-moxendzho-daro.html

Fakty-ob-osminogah. (n.d.). Retrieved from https://www.voprosy-kak-i-pochemu.ru/interesnye-fakty-ob-osminogax/

Falk, D. (2011). The Fossil Chronicles: How Two Controversial Discoveries Changed Our View of Human Evolution. University of California Press Fb. Article 410425. (n.d.). Retrieved from http://fb.ru/article/410425/muraveynik-ustroystvo-etapyi-postroyki-foto-muraveynik-iznutri-delenie-na-kastyi-i-interesnyie-faktyi-iz-jizni-muravev

Faqindecor. (n.d.). *Frank Lloyd Wright.* Retrieved from http://faqindecor.com/frjenk-llojd-rajt-istorija-arhitektury-napisannaja-zolotymi-bukvami/

Farmer, S., Sproat, R., & Witzel, M. (2004). The collapse of the Indus-script thesis: The myth of a literate Harappan civilization. *Electronic journal of Vedic studies, 11*(2), 19-57.

Fechner, G. T. (1882). *Revision der Hauptpuncte der Psychophysik, Leipzig Aristotle. (1976-1983). Works in 4 Volumes.* Moscow, Russia: Misl.

Fedorov-Davydov, A. A. (1957). Picture A. Rylov "In the blue space". *Art.* 1957. No. 8. Fedotovartist. Avtoportret. (n.d.). Retrieved from http://fedotovartist.beon.ru/photos/1944-629-sergei-fedotov-avtoportret-jekspressionizm.html

Ferguson, W. K. (1951). The Interpretation of the Renaissance: Suggestions for a Synthesis. *Journal of the History of Ideas*, 483–495.

Ferrebeekeeper. Etruscan painting. (2014). Retrieved from https://ferrebeekeeper.wordpress.com/2014/06/13/etruscan-painting/

Ferring, R., Oms, O., Agustí, J., Berna, F., Nioradze, M., Shelia, T., ... & Lordkipanidze, D. (2011). Earliest human occupations at Dmanisi (Georgian Caucasus) dated to 1.85–1.78 Ma. Proceedings of the National Academy of Sciences, 108(26), 10432-10436.

Festinger, L., & Carlsmith, J. M. (1959). Cognitive consequences of forced compliance. *Journal of Abnormal and Social Psychology, 58*(2), 203–211. doi:10.1037/h0041593 PMID:13640824

Field, J. V. (1997). *The Invention of Infinity, mathematics, and arts in the Renaissance.* London, UK: Oxford University Press.

Finlayson, S. (n.d.). A sample of the Neanderthals' rock pattern. Graham Cave, Gibraltar, not later than 39 thousand years ago. Retrieved from Gibraltar Museum.

Fleming, F., & Lothian, A. (1997). *The Way to Eternity: Egyptian Myth.* Amsterdam, The Netherlands: Duncan Baird.

Flickr-photos-cavorite. (n.d.). Retrieved from https://www.flickr.com/photos/cavorite/98591365/in/set-1011009/

Forum. Gallery-original-view. (n.d.). Retrieved from http://www.vvv.ru/forum_gallery/original_view.php?id=92007&tid=1

François, V. (1965). *The Poems* (G. Kinnell, Trans.). Houghton Mifflin.

Fromm, E. (2012). *Anatomy of human destructiveness.* Big Press.

Fuster, J. M. (2003). *Cortex and Mind. Unifying Cognition.* Oxford, UK: Oxford University Press.

Gagge, A. P., Stolwijk, J. A. J., & Hardy, J. D. (1967). Comfort and thermal sensations and associated physiological responses at various ambient temperatures. *Environmental Research, 1*(1), 1–20.

Gardens. Istoriua parkovigo. (n.d.). Retrieved from http://www.best-gardens.ru/landshaftnyiy-dizayn/istoriya-sadovo-parkovogo-iskusstva/sadyi-drevnego-egipta-krita/

Gardiner, A. (1964). *Egypt of the Pharaohs*. Oxford, UK: University Press.

Gat, A. (2006). *War in Human Civilization*. New York: Oxford University Press.

Gegenfurtner, K. R., & Sharpe, L. T. (Eds.). (1999). *Color Vision. From Genes to Perception*. Cambridge, UK: Cambridge University Press.

Genofond. (n.d.). Retrieved from http://генофонд.рф/?page_id=3362

Genofond. (n.d.). Retrieved from http://генофонд.рф/?page_id=7277

Geological origin of Melitopol sandstone Kamyana Mohyla. (n.d.). Retrieved from http://www.arabatka.in.ua/?p=1441

Georgieva, S. A., Belikina, N. V., Prokofjeva, L. I., Korshunov, G. V., Kirichuk, V. F., Golovchenko, V. M., & Tokaeva, L. K. (1981). Human physiology. Moscow, Russia: Meditsyna.

Georgieva, S. A., Belikina, N. V., Prokofyeva, L. I., Korshunov, G. V., Kirichuk, V. F., Golovchenko, V. M., & Tokayeva, L. K. (1981). Human Physiology. Moscow, Russia: Medicine.

Gerasimov, M. M. (1960). *The appearance of a person from Nezvisko. Materials and research in the field of archeology of the Southwestern USSR and the Romanian People's Republic* (pp. 83–90). Kishinev.

Gergen, K. J. (1973). Social Psychology as History. *Journal of Personality and Social Psychology*, *26*(2), 309–320. doi:10.1037/h0034436

Gershenzon-Chegodaeva, N. M. (1983). *Bruegel*. Moscow, Russia: Isskustvo.

Gilbert, D. (1948). Foundations of Geometry. Moscow, Russia: OGIZ.

Gilbert, D., & Berneis, P. (1982). Foundations of Mathematics, vol. 1-2. Moscow, Russia: Nauka.

Gillings, R. J. (1972). *Mathematics in the Time of the Pharaohs*. New York: Dover.

Glamurchic. (n/d). *Pugacheva, Galkin*. Retrieved from https://glamurchik.tochka.net/197883-pugacheva-povezla-galkina-kupatsya-v-livadiyu/

Glazychev, V. (2017). *Urbanistics*. Europe Publishing House.

Global History of Arts: in 6 Vol. (1956). Academy of Fine Arts of the USSR. Institute of Theory and History of Fine Arts. Moscow, Russia: State Publ. House "Iskusstvo".

Gödel, K. (1931). Über formal unentscheidbare Sätze der Principia Mathematica und verwandter Systeme, I. *Monatshefte für Mathematik und Physik*, *38*(1), 173–198. doi:10.1007/BF01700692

Godish, T. (2001). *Indoor Environmental Quality*. Boca Raton, FL: CRC Press.

Golan, A. (1993). Myth and Symbol. Moscow, Russia: Russlit.

Golan, A. (1993). *Myth and Symbol. Moscow, Russia:* Russlit.

Golos. Andrea del Verocchio. (n.d.). Retrieved from https://golos.io/vp-painting/@vp-painting/andrea-del-verrokko-velikii-master-epokhi-vozrozhdeniya

Gombrikh, E. (1988). History of Arts: OOO Izdatelstvo AST.

Google. Leonardo. Tajnaya-vecera. (n.d.). Retrieved from https://sites.google.com/site/leonardo123vinchi/kartinnaa-galerea/tajnaa-vecera

Gorelik, G. S. (1959). *Oscillations and Waves. Introduction to Acoustics, Radiophysics, and Optics*. Moscow, Russia: Fizmatlit.

Gower, P. (2019) Groundbreaking study reveals southern Africa as birthplace of modern Homo sapiens, who migrated due to climate shifts. Retrieved from up.ac.za. University of Pretoria.

Grant, E. (1994). *Planets, Stars, & Orbs: The Medieval Cosmos, 1200–1687*. Cambridge, UK: Cambridge University Press.

Grant, E. (2001). *God and Reason in the Middle Ages*. Cambridge, UK: Cambridge University Press. doi:10.1017/CBO9780511512155

Grant, H., Halvorson, T., & Higgins, T. (2014). *Psychology of motivation. How deep installations affect our desires and actions*. AST.

Grashchenkov, V. N. (1988). "Judgment of the Eye" in the theory and practice of the art of the Italian Renaissance. [Moscow, Russia.]. *Soviet Art Studies*, *24*, 112.

Greatpicture. test-na-intuitsiyu. (n.d.). Retrieved from http://www.greatpicture.ru/test-na-intuitsiyu-naskolko-u-vas-razvita-intuitsiya/

Gregory, R. J. (2003). *"The History of Psychological Testing" (PDF). Psychological Testing: History, Principles, and Applications*. Allyn & Bacon.

Griaznov, M. P. (1999). Afanasievo Culture along the Yenisei. St. Petersburg, Russia: Academic Press.

Grimal, N. (1992). *A History of Ancient Egypt*. Blackwell Books.

Gromov, A. A. (1997). *Origin and Links of the Okunev Culture Population*. St. Petersburg, Russia: Petro-RIF.

Guner. Bibliotec Culture. (n.d.). Retrieved from https://www.gumer.info/bibliotek_Buks/Culture/gombr/02.php

Haaland, M. (2011). Presumably, the oldest human painting of a modern type. Blombos Cave, 2011. Retrieved from https://vokrugsveta.ua/science/najden-drevnejshij-v-mire-risunok-13-09-2018

Haken, G. (1985). Synergetics: a hierarchy of instabilities in self-organizing systems and devices. Moscow, Russia: Mir.

Hameroff, S. (2006). Consciousness, Neurobiology and Quantum Mechanics: The Case for a Connection. In J. Tuszynski (Ed.), *The Emerging Physics of Consciousness*. Berlin, Germany: Springer-Verlag. doi:10.1007/3-540-36723-3_6

Hameroff, S., & Penrose, R. (1996). Orquestrated Reduction of Quantum Coherence in Brain Microtubules: A model for Consciousness. *Journal of Consciousness Studies*, *3*, 36–53.

Hands, J. (2015). *Cosmosapiens: Human Evolution from Origin of the Universe*. New York, NY: Overlook Duckworth.

Harari, F. (1973). Graph Theory. Moscow, Russia: Mir.

Harris, L., & Jenkin, M. (Eds.). (2007). *Computational Vision in Neural and Machine Systems*. Cambridge, UK: Cambridge University Press.

Harvard, H. (1998). *History of Modern Art: Painting, Sculpture, Architecture, Photography*. New York: Harry N. Abrams.

Hause, S., & Maltby, W. (2001). A History of European Society. [Belmont, CA: Thomson Learning.]. *Essentials of Western Civilization*, 2, 245–246.

Heaven, P., & Furnham, A. (1998). *Personality and Social Behavior*. Routledge.

Hegel. (1913). Phenomenology of the Spirit. Saint Petersburg.

Hegel. (1916). The science of Logic. Saint Petersburg, 1916.

Henderson, G. (1977). *Early Medieval Art* (Rev. ed.). Harmondsworth, UK: Penguin.

Herman, H. (1943). *Das Glasperlenspiel*. Henry Holt and Company.

Heyes, C., & Huber, L. (Eds.). (2000). *The Evolution of Cognition. Cambridge, MA: Bradford Book.*

Hieronymus, B. (1974). *Reproduction album / Text author and compiled by Fomin, G. I.* Moscow, Russia: Isskustvo.

Higgins, A. C., & Wilson, E. O. (1975). Sociobiology: The New Synthesis: Harvard University Press.

Hill, M. (2007). *Gifts for the gods: images from Egyptian temples*. New York: The Metropolitan Museum of Art.

Historic Cultural Heritage MuseumP. L. A. T. A. R. (n.d.). Retrieved from https://antikvar.ua/tripolskie-ornamenty/

History of Mankind. World History: in 9 vol. Vol. 3. Western Asia and Africa (1903). Saint Petersburg, Russia: Publishing partnership "Prosvescheniye".

History, W. Vol.1 Stone Age. (1955). Moscow, Russia: Gospolitizdat.

History. Egipetskie ieroglify. (n.d.). Retrieved from http://history-doc.ru/egipetskie-ieroglify/

Hockney, D. (2001). *Secret Knowledge: Rediscovering the Lost Techniques of the Old Masters*. London, UK: Viking Press.

Hockney, D., & Falco, M. (2003). Optics of the Dawn of Renaissance. *Technical Digest of the Optical Society, 87th Annual Meeting*, 852-857.

Hollis, J. (2013). Pass in the middle of the road. How to overcome the midlife crisis and find a new meaning of life: Kogito-Center.

Homans, G. C. (1961). *Social behavior: its elementary forms*. New York: Harcourt, Brace and World, Inc.

Horek. (n.d.). Retrieved from http://horek.spb.ru/skelet-delfina-foto.html

Huizinga, J. (1988). The Autumn of the Middle Ages. A study of the forms of life and thought in the XIV and XV centuries in France and the Netherlands. Moscow, Russia: Science.

Human Brain Project. (n.d.). Retrieved from https://www.humanbrainproject.eu/en/

Humphreys, M. (2016). *Adaptive thermal comfort: foundations and analysis*. Abingdon, U.K.: Routledge.

Hunter, G. R. (1934). *The Script of Harappa and Mohenjodaro and Its Connection with Other Scripts*. London, UK: Kegan Paul.

Igordiksa. (n.d.). Retrieved from http://igordiksa.com/?p=7871&lang=ru

Ikonnikov, A. V. (1985). Art, environment, time: Aesthetic organization of the urban environment. Moscow, Russia: Soviet artist.

Iovlev, V. I. (1996). *Ecopsychology for architects: process and form*. Ekaterinburg, Russia: Architecton.

Isaeva, K. (2016, Aug. 16). *Russian fairy tales in the works of Ivan Bilibin*. Retrieved from www.rbth.com

Isatorius. Livejournal. (n.d.). Retrieved from https://isartorius.livejournal.com/photo/album/1107/?page=28

Iskusstvoved. Kniznaya miniatyra. (n.d.). Retrieved from http://iskusstvoed.ru/2018/09/01/книжная-миниатюра-эпохи-каролингов-к

Istorii zemli-tripolskaya. (n.d.). Retrieved from http://историиземли.рф/tripolskaya-kultura.html

Isweet. Risunki detej. (n.d.). Retrieved from http://isweet.com.ru/risunki-detej-5-let-foto.html

Ivanitskiy, M. F. (2008). Human Anatomy (with foundations of dynamic and sporting morphology): Guide for Institute of Physical Culture. Moscow, Russia: Olimpiya.

Ivanitskiy, M. F. (2008). *Human Anatomy (with foundations of dynamic and sporting morphology): Guide for Institute of Physical Culture. Moscow, Russia*: Olimpiya.

Japinga-Original-Art. (2012). Retrieved from https://japingkaaboriginalart.com/articles/contemporary-aboriginal-art-in-australia/

Johanson, D. C., & Wong, K. (2009). *Lucy's Legacy: The Quest for Human Origins*. Crown.

Joiner, A. D. (1985). The origin of art / A. D. Joiner. Moscow, Russia: Art, 1985.

Kant, I. (2007). *Criticism of Pure Reason. Moscow, Russia*: Eksmo. doi:10.1007/978-1-137-10016-0

Kaplya-sveta-posle. (n.d.). Retrieved from http://kaplyasveta.ru/posle-altamiry/altamira-impressionizm-15-000-let-do-n-e-chast-2-2.html

Kauffman, S. (1993). *The Origins of Order: Self-Organization and Selection in Evolution*. Oxford, UK: Oxford University Press.

Kauffman, S. (1995). *At Home in the Universe: The Search for the Laws of Self-Organization and Complexity*. New York: Oxford University Press.

Kaufman, A. S. (2009). *IQ testing 101*. Springer Pub.

Kavtaradze, S. (2017). *Anatomy of architecture. Seven books on logic, form, and meaning*. Higher School of Economics.

Kaznacheev, V. P., & Spirin, E. A. (1991). *Cosmoplanetary Phenomenon of Human: The Problem of Integrated Study*. Novosibirsk, Russia: Nauka.

Kelly, R. C. (2000). *Warless societies and the origin of war*. Ann Arbor: University of Michigan Press. doi:10.3998/mpub.11589

Kemclub. Photo 3674067. (n.d.). Retrieved from http://kemclub.ru/photo/3674067/

Kemp, M. (1989). *The Science of Art*. New Haven, CT: Yale University Press.

Kemp, M. (2000). *Visualizations: the nature book of art and science*. Berkeley, CA: The University of California Press.

Kevin, N. L. (2017). *Darwin's Unfinished Symphony: How Culture Made the Human Mind*. Princeton University Press.

Khrisanfova, E. N., & Perevozchikov, I. V. (2005). Anthropology, 4th ed.: Moscow, Russia: Publishing house Moscow University, Science, 2005.

Kifishyn, A. G. (2001). The ancient sanctuary of Kamyana Mohyla. Deciphering experience of a proto-Sumerian archive of XII-III millennium BC. In 7 volumes. Kiev, Ukraine: Aratta.

King, Margaret L. (1991). Women of the Renaissance. 1991.

Kleene, S. (1957). Introduction to metamathematics. Moscow, Russia: Ed. John lit.

Kleene, S. (1973). Mathematical Logic. Moscow, Russia: Nauka.

Kleene, S., & Wesley, R. (1978). Foundations of Intuitionistic Mathematics. Moscow, Russia: Nauka.

Klein, F. (1956). A comparative review of the latest geometric research ("Erlangen program"). In *On the foundations of geometry. A collection of classical works on Lobachevsky geometry and the development of its ideas,* Moscow, Russia, pp. 399-434.

Klyosov, A. A., & Tyunyaev, A. A. (2009). *The origin of man according to archeology, anthropology and DNA genealogy*. Boston, MA.

Knews. Mnenie. Religia. (n.d.). Retrieved from https://knews.kg/2018/03/13/mnenie-religiya-okazyvaet-bolshoe-vliyanie-na-molodezh-kyrgyzstana/

Knorr, W. R. (1992). When circles don't look like circles: An optical theorem in Euclid and Pappus. [Springer Berlin/Heidelberg.]. *Archive for History of Exact Sciences, 44*(4), 287–329.

Kobylina, M. M. (1939). *Art of Old Rome*. Moscow, Russia.

Koehler, R. (2010, June 10). The Painter Sam Francis. Variety. Retrieved June 22, 2010.

Koffka, K. (2014). Principles of Gestalt Psychology: Mimesis International.

Kolcaba, K. (2003). Comfort Theory and Practice: A Vision for Holistic Health Care and Research.

Kolotov, S. M. (1998). To the history of imaging methods. *Applied Geometry and Engineering Graphics, 63,* 239–255.

Kolpinskiy, Yu. D. (1970). Monuments of world art: 9 vol. Russian Academy of Arts. Institute of Theory and History of Fine Arts. Moscow, Russia: Art, V. 3: The Art of the Aegean World and Ancient Greece.

Kolpinskiy, Yu. D., & Britova, N. N. (1983). Monuments of world art: 9 vol. Russian Academy of Arts. Institute of Theory and History of Fine Arts. Moscow, Russia: Art, V. 7: The Art of Etruscans and Ancient Rome.

Korobko, V. I. (1997). Golden proportion and harmony issues of systems. Moscow, Russia: Publisher. Association of Building Universities of CIS states.

Korobko, V. I. (1997). The Golden Proportion and Problems of Harmony Systems. Moscow, Russia: Ed. Association of Universities of Civil Engineering of the CIS countries.

Kovalyev, Yu. N., & Nitsyn, A. Yu. (2009). The connection of the visual perception of space with the spiritual currents of the Renaissance in the South and North of Europe. Geometrical and computer modeling, 24. pp. 81-91. Kharkiv, 2009.Kubovy, M. (1988). The Arrow in the Eye: The Psychology of Perspective and Renaissance Art. Cambridge, UK: Cambridge University Press.

Kovalyov, N. (2009). The Image of Space in the Primeval Epoch. Technical aesthetics and design, 6. pp. 27–33.

Kovalyov, Y. N., & Grebenyuk, M. F. (2007). C-space description languages. Applied geometry and eng. Graphics, 77, pp. 29-35.

Kovalyov, Y. N., & Nitsyn, A. Y. (2010). Space in the Painting of Ancient Egypt. Technical aesthetics and design. 2010, 7. pp. 23–35.

Kovalyov, Y. N., & Nitsyn, A. Y. (2012). Development of Space Representations in Medieval Painting. *Proceedings of the Tauride State Agrotechnological University. 14 International Scientific-Practical Conference "Contemporary Problems of Geometric Modeling of SPGM 14"*. Melitopol, Ukraine: TDATU. 2012. pp. 63-69.

Kovalyov, Y. N., & Vereschaga, V. M. (2012). Applied Geometry: Descriptive Geometry, Engineering, and Computer Graphics, Modern Trends. Kiev, Ukraine: Omega-L.

Kovalyov, Y. N., Nitsyn, A. Y., & Matveeva, E. L. (2011). The Development of Space Depictions in Ancient Painting. Technical Aesthetics and Design. 2011, 9. pp. 10–19.

Kovalyov, Yu. N., & Nitsyn, A. Yu. (2012). Depicting the Subjective Space in Painting. *Technical Aesthetics and Design, 8*. 49–56.

Kovalyov, Y. N. (1996). *Geometric modeling of human-machine systems: the development of apparatus. Kiev, Ukraine*: KMUGA.

Kozlov, C. (n.d.). Retrieved from https://www.e-reading.club/chapter.php/102033/2/Filip%27echev%2C_Kozlov_-_Krokodily.html

Krause, J., Fu, Q., Good, J. M., Viola, B., Shunkov, M. V., Derevianko, A. P., & Pääbo, S. (2010). The complete mitochondrial genome of an unknown hominin from southern Siberia. *Nature*, *464*(7290), 894–897. doi:10.1038/nature08976 PMID:20336068

Krskstate. Okunevskaya kultura. (n.d.). Retrieved from http://my.krskstate.ru/docs/early_history/okunevskaya-kultura/

Kuzhekina, Y. I., Voronina, M. A., & Lukoyanova, A. S. (2016). The anthropic principle in cosmology. *Research and methodological online journal. Concept*, *2*, 531–535.

Kuzmin, Y. F. (2017). *Geoarchaeology: natural science methods in archaeological research*. Tomsk, Russia: TSU Publishing House.

Kwang Hyun, K. (2015). Origins of Bipedalism. *Brazilian Archives of Biology and Technology*, *58*(6), 929–934. doi:10.1590/S1516-89132015060399

Lahr, C. D. (1997). *Mathematics and Knowledge: Models of Reality*. Hanover, NH: Dartmouth College.

Lascaux-culture. (n.d.). Retrieved from http://www.lascaux.culture.fr/#/fr/02_00.xml

Lawrence, H. K. (1996). War before Civilization. Oxford University Press edition.

Leakey, M. D. (1979). *Olduvai Gorge: My search for early man*. Collins.

Leibniz, G. V. (1982-1989). Works in 4 Volumes. *Moscow, Russia*: Misl.

Leonard, W. R., Snodgrass, J. J., & Robertson, M. L. (2007). Effects of brain evolution on human nutrition and metabolism. Annu Rev Nutr. 2007. – Vol. 27. – pp. 311-327. doi:10.1146/annurev.nutr.27.061406.093659

Leontiev, N. V., Kapelko, V. F., & Yesin, Yu. N. (2006). *Sculptures and Stelas of the Okunev Culture*. Abakan, Russia: Khakassia Research Institute of Language, Literature, and History.

Leroi-Gourhan, A. (1978). Signs in Prehistoric Art: Maseno.

Leroi-Gourhan, A. (1978). Signs in Prehistoric Art: Mazeno.

Leroi-Gourhan, A. (1967). *Treasures of Prehistoric*. New York: Harry N. Abrams.

Leroi-Gourhan, A. (1982). *The Dawn of European Art: An Introduction to Palaeolithic Cave Painting*. Cambridge, UK: Cambridge University Press.

Leroux R. (1970). Human Ecology - the science of housing. Moscow, Russia: Izd. construction literature.

Levins, R. (1968). Evolution in Changing Environments: Some Theoretical Explorations. Princeton University Press.

Levi-Strauss, C. (1994). Primitive Thinking. Moscow, Russia: Republic.

Levi-Strauss, C. (2001). Structural Anthropology. Moscow, Russia.

Lewis-William, D. (2002). *The Mind in the Cave*. London, UK: Thames and Hudson.

Lhote, H. (1962). The Search for the Tassilli Frescoes. Moscow, Russia: Publisher of Oriental literature, p. 140.

Libet, B. (2004). *Mind Time: The temporal Factor in Consciousness*. Cambridge, MA: Harvard University Press.

Lin, Mike V. (2010). Modern Design. Walkthrough. Moscow, Russia: Astrel, AST.

Linden, A. (2007). Reclassification of Homo floresiensis as Homo species indeterminate to accurately reflect a lack of existing knowledge. *Totem: The University of Western Ontario Journal of Anthropology*, *15*(1), 5–20.

Litos-perigord. (n.d.). Retrieved from http://lithos-perigord.org/spip.php?rubrique25

Locke, D. (1985-1988). Works in 3 Volumes. *Moscow, Russia*: Misl.

Lomov, B. F. (1979). Systematicity as a principle of mathematical modeling in psychology. *Matters of cybernetics*, 50, 3-18.

Losev, A. F. (1978). *Aesthetics of the Renaissance*. Moscow, Russia: Misl.

Losskiy, N. O. (1992). *Intuitivism. Moscow, Russia*: Progress.

Lossky, N. O. (1992). *Intuitionism*. Moscow, Russia: Progress.

Lsvsx-livejornal. (n.d.). Retrieved from https://lsvsx.livejournal.com/482239.html

Luria, A. R. (1985). *Sensation and Perception*. Moscow, Russia: Publisher. MGU.

Lvov, S. L. (1971). *Pieter Bruegel the Elder*. Moscow, Russia: Isskustvo.

Maat-object. (n.d.). Retrieved from http://maat.org.ru/object/2004/2004- 02.shtml

Machotka, P. (1996). *Cézanne: Landscape into Art*. United States: Yale University Press.

Maddox, T. (2018). Smart cities: 6 essential technologies. Tech Republic; Retrieved April 1, 2018.

Magazindomov. Gorodskoj dom v yaponii. (n.d.). Retrieved from http://www.magazindomov.ru/2014/07/30/gorodskoj-dom-v-yaponii-67/

Maggiani, A. (1985). *Artistic crafts: Northern Etruria in Hellenistic Rome*. Italy: Electra.

Mail. Community-catastrof. (2007). Retrieved from https://my.mail.ru/community/catastrof/14C6C1BFB14FA238.html

Mail. Community-catastrof. (n.d.). Retrieved from https://m.my.mail.ru/community/catastrof/130BB97B091C716B.html

Mail. Question 39179570. (n/d). Retrieved from https://otvet.mail.ru/question/39179570

Mallén, E. (2003). *The Visual Grammar of Pablo Picasso*. New York: Peter Lang.

Mango, C., & Hawkins, E. J. W. (1964). The Apse Mosaics of St. Sophia in Istanbul. *Report on Work Carried out.*

Mantu, C.-M. (2000). Cucuteni–Trypillia cultural complex: relations and synchronisms with other contemporaneous cultures from the Black Sea area. Studia Antiqua et Archaeologica. Iaşi, Romania: Iaşi University. VII: 267.

Maria, P. (1979). *Sixteen Paintings in the National Museum of Ukrainian Folk Art*. Leningrad: Shown on A Set of Sixteen Color Postcards Published by Aurora Art Publishers.

Markov, A. V. (2011-2012). Human evolution. Book 1, 2. Moscow, Russia. *Astrel: Corpus*, *2011*, 2012.

Marysrosaries. Saint Peter. (n.d.). Retrieved from http://www.marysrosaries.com/collaboration/index.php?title=File:Saint_Peter_Icon_from_Saint_Catherine%27s_Monastery,_Mount_Sinai_001.jpg

Maslow, A. (2003). Motivation and Personality: Peter.

Matie, M. E. (1956). Ancient Egyptian Myths Moscow, Russia: Publ. House of the Academy of Arts of the USSR.

Matie, M. E. (1966). Art of Ancient Egypt. Moscow, Russia: Iskusstvo.

Matveev, A. N. (1985). Optics: *Moscow, Russia*: Higher School. My-3-sotki. (n.d.). Retrieved from http://my-3sotki.ru/?p=2042

Maugh, I. I., & Thomas, H. (2009). Venus figurine sheds light on origins of art by early humans. *Los Angeles Times*. Retrieved May 14, 2009.

Mbradio. (n.d.). Retrieved from https://mbradio.ru/publication/1356/

McKenna, T. (1995). Food of Gods. Moscow, Russia: Publishers of Transpersonal Institute.

McPherron, S. P. (2017). The Age of the Homo sapiens fossils from Jebel Irhoud (Morocco) and the origins of the Middle Stone Age. *Nature*, *8*(June), doi:10.1038/nature22335

Meade, M. (1977). *Eleanor of Aquitaine: A Biography*. Hawthorn.

Megabook-media. Fivy. (n.d.). Retrieved from https://megabook.ru/media/Фивы%20(фрески%20гробницы%20Нефертари)

Megalitica. Naskalnaya zhivopis. (n.d.). Retrieved from https://megalithica.ru/naskalnaya-zhivopis-nourlangie-kakadu-rock-art.html

Mendez, F. L., Poznik, G. D., Castellano, S., & Bustamante, C. D. (2016). The Divergence of Neandertal and Modern Human Y Chromosomes. *The American Journal of Human Genetics*, 2016. Gazeta. Science. (n.d.). Retrieved from https://www.gazeta.ru/science/2007/10/11_a_2232959.shtml?updated

Menotti, F., & Korvin-Piotrovsky, A. (Eds.). (2012). *The Trypillia Culture giant-settlements in Ukraine. Formation, Development, and Decline*. Oxford, UK: Oxbow Books.

Mesarovic, M., & Takahara, Y. (1978). General system theory: mathematical foundation. Moscow, Russia: Mir.

Mesarovic, M., Mano, D., & Takahara, Y. (1973). Theory of hierarchic multi-level systems. Moscow, Russia: Mir.

Mets, F. I. (2001). Samuskiy burial ground. Nations and cultures of Tomsk-Narym Priobye: Materials for the encyclopedia of Tomsk region. Tomsk, 129-130.

Mif-Meduza. (2016). Retrieved from https://mif-medyza.ru/2016/10/28/древнегреческие-вазы-геометрические/ Minoanatlantis.

Miller, D. (2009). The Comfort of Things.

Miller, E. K., & Cohen, J. D. (2001). An Integrative Theory of Prefrontal Cortex Function. *Annual Review of Neuroscience, 24*(1), 167–202. doi:10.1146/annurev.neuro.24.1.167 PMID:11283309

Miller, N. (1993). Euclid Redivivus: A Hypothesis on a Proposition. *Sighttte des Beaux-Arts, 121*, 213–226.

Milner, J. (1992). *Mondrian*. London, UK: Phaidon.

Minoan_Flotilla. (n.d.). Retrieved from https://www.minoanatlantis.com/pix/Minoan_Flotilla_Fresco_Three_Panels_800px.jpg

Minoan_Mycenaen. (n.d.). Retrieved from http://employees.oneonta.edu/farberas/arth/ARTH209/minoan_mycenaean.html

Mircea, E. (2002). History of faith and religious ideas. Vol. 1. From the Stone Age to the Eleusinian Mysteries, Vol. 2. From Gautama Buddha to the triumph of Christianity: Criterion Employees.

Mircea, E. (1998). *A myth of the Eternal Return.*

Mkhitaryan, N. M. (2016). Man, and Dwelling. Gangemi Editore International Publishing.

Mkhitaryan, N. M. (2017). *Man and Dwelling*. Gangemi Editore International Publishing.

Mkhitaryan, N. M. (2018). *Habitat: images of comfort*. Kiev, Ukraine: Naukova Dumka.

Mkhitaryan, N. M., Badeyan, G. V., & Kovalyov, Y. N. (2004). *Ergonomic aspects of complex systems*. Kiev, Ukraine: Naukova Dumka.

Mkhitaryan, N. M., Badeyan, G. V., & Kovalyov, Y. N. (2004). *Ergonomic Aspects of Complex Systems. Kiev, Ukraine*: Naukova Dumka.

Monfasani, J. (2016). *Renaissance Humanism, from the Middle Ages to Modern Times*. Taylor & Francis.

Monserrat, J. (2007), Neural Networks and Quantum Neurology: A Speculative Heuristic Towards the Architecture of Psyquism. In IWINAC 2007, 1, 1-20. Springer. doi:10.1007/978-3-540-73053-8_1

Monserrat, J. (2016). How the brain and neural networks explain human reality. *Pensamiento, 72*(273), pp. 1043-1070.

Monserrat, J. (2001). Engramas neuronales y teoría de la mente. *Pensamiento, 57*(2001), 176-211.

Monuments of World Art: in 9 vol. – Vol. 2: Art of Ancient East. (1968). Academy of Arts of the USSR. Research Institute of Theory and History of Fine Arts. Moscow, Russia: Iskusstvo.

Monuments of world art: in 9 vol. (1987). Russian Academy of Arts. Institute of Theory and History of Fine Arts. Moscow, Russia: Isskustvo, 8. Italian art of late 13th - 15th centuries / I. A. Smirnova. (1987).

Monuments of world art: in 9 vol. / Russian Academy of Arts. Institute of Theory and History of Fine Arts. Moscow, Russia: Isskustvo, 1. Art of Italy of the XVI century / E. I. Rotenberg (1967).

Monuments of world art: in 9 vol. / Russian Academy of Arts. Institute of Theory and History of Fine Arts. Moscow, Russia: Isskustvo, 9: Art of Renaissance in The Netherlands, France, England / T. S. Voronina, N. L. Maltseva, V. V. Starodubova. (1994).

Moon, W. G. (1983). *Ancient Greek Art and Iconography*. Madison, WI: University of Wisconsin Press.

Moor, P. (2007). Archived 21 December 2007 at the Wayback Machine.

Morris, M. E., Adair, B., Miller, K., Ozanne, E., Hansen, R., Pearce, A. J., ... & Said, C. M. (2013). Smart-home technologies to assist older people to live well at home. *Journal of aging science, 1*(1), 1-9.

Morz, F. (1949). Fluctuations and Sound. M.L.: GITTL.

Mullins, C. (2006). *Painting People: Figure Painting Today*. New York: D. A. P.

Museo-Archeologic-Napoli. (n.d.). Retrieved from https://www.museoarcheologiconapoli.it/en/room-and-sections-of-the-exhibition/frescoes/

Mykhailov, B. D. (2005). Petroglyphs of Kamyana Mohyla: Semantics. Chronology. Interpretation. Kiev, Ukraine: MAUP.

Narod. USSR. Laktionov. (n.d.). Retrieved from http://webstarco.narod.ru/ussr/laktionov.html

National Geographic. Bull-leaping. (n.d.). Retrieved from https://www.nationalgeographic.org/media/bull-leaping/

Néret, G. (2003). Kazimir Malevich and Suprematism 1878-1935, Taschen, 2003.

Nesselstrauss, C. G. (2000). The art of the Early Middle Ages. St. Petersburg, Moscow: Azbuka-Klassika.

Ngasanova. Livejournal. (n.d.). Retrieved from https://ngasanova.livejournal.com/627043.html

Ngorod-doistoricheskie-pamyatniki. (n.d.). Retrieved from http://ngorod.net/doistoricheskie-pamyatniki/peshhera-lasko/photo-4/

Niemitz, C. A. (2002). Theory on the Evolution of the Habitual Orthograde Human Bipedalism - The "Amphibisce Generalistheorie". *Anthropologischer Anzeiger*, *60*(1), 3–66. doi:10.1127/anthranz/60/2002/3 PMID:12058577

Nikulin, N. N. (1999). *The Golden Age of the Dutch painting. XV century*. Moscow, Russia: Publishing House AST.

Nitsyn, A. Yu. (1999). Results of the experiment on the study of the dependence of the visually perceived size of an object on the distance to the observer. Transactions of the Tauride States. Agricultural Academy. Melitopol, Ukraine: TGATA, 1999, 4. Applied geometry and engineering graphics, 7. pp. 68–70.

Nitsyn, A. Yu. (2006). Results of the experiment to determine the distance from the center of projection to the picture plane. *Geometrical computer modeling*. Kharkiv, Ukraine: KhDUKhT, 2006. *VIP*, *15*, 91–94.

Nosov, K. S. (n.d.). *Siege machinery of the Assyrians*. Retrieved from http://xlegio.ru/siege-machinery/poliorketika-and-paraskeuastika/assyrian-siege-machinery/

Olsen, S. (2006). *The Golden Section, Nature's Greatest Secret.* Wooden Books, Ltd., 2006.

Olsen, S. (2006). The Golden Section, Nature's Greatest Secret", Wooden Books, Ltd.

Openlibrary. Occulta philosiphia. (n.d.). Retrieved from https://openlibrary.org/books/OL24342362M/Occvlta_philosophia

Oringo. Tripolskie Ornamenty. (n.d.). Retrieved from https://oringo.com.ua/forum/tripolskie-ornamenty

Orlando, D. G. A. (2015). An introduction to universal, quantum-galactic mathematics. Retrieved from https://www.researchgate.net/profile/Giovanni_Orlando/publication/331928401_Orlando_Intro_Galactic_Math/links/5c935ae745851506d71ff7ba/Orlando-Intro-Galactic-Math.pdf

Osorina, M. (2016). Secret Children's World of in the Space of the Adult's World: Peter.

Ostrovsky, G. (1967). Joseph Bokshay. Moscow, Soviet artist, 1967.

Our civilization. Mohenjo. (n.d.). Retrieved from https://our-civilization.com/civilizacia/arhitektura-drevnih/mohendzo-daro

Pal, D., Funilkul, S., Charoenkitkarn, N., & Kanthamanon, P. (2018). Internet-of-Things and Smart Homes for Elderly Healthcare: An End User Perspective.

Pallasweb. Virgin and Child. (n.d.). Retrieved from https://www.pallasweb.com/deesis/virgin-and-child-hagia-sophia.html

Pally, R. (2018). *The mind-brain relationship* (n.d.). London, UK: Routledge. doi:10.4324/9780429482465

Palmer, D. (2007). *The Origins of Man: an illustrated history of human evolution.* New Holland Publishers.

Panero, J., & Zelnik, M. (2006). *Fundamentals of ergonomics. Man, space, interior. Handbook of project standards.* Astrel.

Panofsky, E. (2004). Perspective as a "Symbolic Form". Gothic Architecture and Scholasticism. St. Petersburg, Russia: Azbuka-Klassika.

Panofsky, E. (2004). *Perspective as a "symbolic form". Gothic architecture and scholasticism: Trans. from German – St.* Petersburg, Russia: Azbuka-klassika.

Panofsky, E. (2004). *Perspective as a "symbolic form". Gothic architecture and scholastics.* Saint Petersburg, Russia: Azbuka-Klassika.

Panofsky, E. (2006). *Renaissance and the "Renaissances" in the art of the West.* St. Petersburg, Russia: Azbuka-klassika.

Paoletti, J. T., & Radke, G. M. (2011). *Art in Renaissance Italy* (4th ed.).

Park, M. S., Nguyen, A. D., Aryan, H. E., U, H. S., Levy, M. L., & Semendeferi, K. (2007). Evolution of the human brain: Changing brain size and the fossil record. *Neurosurgery, 60*(3), 555–562. doi:10.1227/01.NEU.0000249284.54137.32 PMID:17327801

Parpola, A. (1987-2010). Corpus of Indus seals and inscriptions, Helsinki, Finland: Suomalainen Tiedeakatemia (Academia scientiarum Fennica).

Parpola, A. (2010). Dravidian solution to the Indus script problem. Central Institute of Classical Tamil, 2010.

Parygin, B. D. (2010). Social Psychology. Origins and prospects. Saint Petersburg, Russia: SPbGUP, 2010.

Passek, T. (1961). Early agricultural (Tripoli) tribes of Transnistria, MIA, n. 84. Moscow, Russia, 1961.

Patrick, J. A. (Ed.). (2007). Renaissance and Reformation (5 vol 2007).

Pavlov, I. P. (1949). *Selected works. Moscow, Russia:* Academy of Sciences of USSR.

Penrose, R. (1994). *Shadows of the Mind: A Search for the Missing Science of Consciousness.* Oxford, UK: Oxford University Press.

Penrose, R., & Hameroff, S. (2011). Consciousness in the Universe: Neuroscience, Quantum Space-Time Geometry and Orch or Theory. In R. Penrose & ... (Eds.), *Consciousness and the Universe: Quantum Physics, Evolution, Brain, and Mind. Cambridge* , MA: Cosmology Science Publishers.

Peter Summers. Archeology. (n.d.). Retrieved from https://www.petersommer.com/blog/archaeology-history/minoan-frescoes

Piaget, J. (1951). *The Psychology of Intelligence.* London, UK: Routledge and Kegan Paul.

Pikabu. Gibel Harappskoy. (n.d.). Retrieved from https://pikabu.ru/story/gibel_kharappskoy_tsivilizatsii_5265336

Pikabu-Story-Sentyabr. (n.d.). Retrieved from https://pikabu.ru/story/12_sentyabrya_1940_goda_byila_naydena_peshchera_lasko__quotsikstinskaya_kapella_naskalnoy_zhivopisi_5335111

Pinch, G. (1995). *Magic in Ancient Egypt*. University of Texas Press.

Plato. (2006). Works in 4 Volumes - Saint Petersburg: Publishing House of the University of Saint Petersburg. Publishing House of Oleg Abyshko, Popmech-Ddesign-news. (n.d.). Retrieved from https://www.popmech.ru/design/news-359292-potryasayushche-krasivye-snimki-zvezdnogo-neba-iz-kabiny-pilota/

Polesn_inters_anatomy. (n.d.). Retrieved from http://www.polezen.ru/interes/anatomy.php

Pomerantseva, N. A. (2006). Primitive Art. From Ancient Cultures to Early Civilizations: [album]. Moscow, Russia: White City.

Popmech. Science News-437992-7. (n.d.). Retrieved from https://www.popmech.ru/science/news-437992-7-faktov-o-gibride-neandertalca-i-denisovskogo-cheloveka/#part0

*Popol-Vuh*. (1993). St. Petersburg, Russia: Nauka.

Potaev, I. (2008). *Urban planning and territorial planning*. Phoenix.

Potts, R. (1998). Environmental Hypotheses of Hominin Evolution. *American Journal of Physical Anthropology*, 107, 93–136. doi:10.1002/(SICI)1096-8644(1998)107:27

Pravoslavie. (n.d.). Mount Athos. Retrieved from https://pravoslavie.ru/100213.html

Proistoria. Assets-galleries. (n.d.). Retrieved from http://www.proistoria.org/assets/galleries/184/cueva-de-las-manos-4.jpg

Prokhorov, M. I. (1946). *City Square*. Moscow, Russia: State Architectural Publishing House.

Prokofiev, V. N. (1985). *On art and art history. Articles from different years*. Moscow, Russia: Sov. Khudozhnik.

Propp, V. Ya. (2000). Historical Backgrounds of the Fairy Tale. Moscow, Russia: Labyrinth.

Psicabi. Test na Chuvstvitelnost. (n.d.). Retrieved from https://psycabi.net/testy/507-test-na-chuvstvitelnost-diagnostika-emotsionalnosti-metodika-v-v-suvorova

Psicabi. Test Ravena. (n.d.). Retrieved from https://psycabi.net/testy/717-test-ravena-progressivnye-matritsy-raven-progressiv-matrices-metodiki-dlya-diagnostiki-intellekta-vzroslykh

Psylist. Prakticum-00465. (n.d.). Retrieved from https://psylist.net/praktikum/00465.htm

Pulson. Isrusstvo-risunki. (n.d.). Retrieved from https://pulson.ru/iskusstvo/risunki-shizofrenika-12-foto.html

Quesabesde-notitias-nomada (n.d.). Retrieved from http://www.quesabesde.com/noticias/nomada-altamira-museo,1_5259

Radhakrishnan, S. (1956-1957). Indian Philosophy. V. 1-2. *Moscow, Russia*: Nauka.

Radhakrishnan, S. (1956-1957). Indian Philosophy. Vol. 1-2. Moscow, Russia.

Ragain, C. J. (2015) A Review of Color Science in Dentistry: The Process of Color Vision. Retrieved from https://symbiosisonlinepublishing.com/dentistry-oraldisorders-therapy/dentistry-oraldisorders-therapy34.php

Ragozina, Z. A. (1902). History of Assyria. From the Rise of the Assyrian State to the Fall of Nineveh. St. Petersburg, Russia: Publ. House of A.F. Marks.

Ragozina, Z. A. (1902). History of Chaldaea. From the oldest times to the rise of Assyria. St. Petersburg, Russia: Publ. House of A.F. Marks.

Ramage, N. H., & Ramage, A. (2009). Roman Art: Romulus to Constantine. New Jersey: Pearson Prentice-Hall.

Ramirez, T., & Herbig, R. (2016) Optimising hearing aid processing for music appreciation. Retrieved from https://www.entandaudiologynews.com/features/ent-features/post/optimising-hearing-aid-processing-for-music-appreciation

Rao, R. P. N. (2010). Probabilistic Analysis of an Ancient Undeciphered Script (PDF). *IEEE Computer.*, *43*(4), 76–80. doi:10.1109/MC.2010.112

Ratsel, F. (1902). Ethnic Science: in 2 vol. St. Petersburg, Russia: Publishing Partnership “Prosvescheniye”

Rauschenbach, B. V. (1986). Systems of Perspective in Visual Arts. General Theory of Perspective. Moscow, Russia: Science.

Rauschenbach, B. V. (2001). The geometry of Picture and Visual Perception. Saint Petersburg, Russia: Azbuka-Classic.

Rauschenbach, B. V. (2001). The geometry of Picture and Visual Perception. St. Petersburg, Russia: Azbuka-Classic.

Rauschenbach, B. V. (2001). *The geometry of the picture and visual perception*. Saint Petersburg, Russia: Azbuka-Klassika.

Raushenbakh, B. V. (1980). Spatial Constructions in Pictorial Art. Outline of Basic Methods. Moscow, Russia: Nauka.

Raushenbakh, B. V. (1986). Systems of Perspective in the Pictorial Art. General Theory of Perspective. Moscow, Russia: Nauka.

Raushenbakh, B. V. (2001). The geometry of a Picture and Visual Perception. St. Petersburg, Russia: Azbuka-Klassika.

Realhistory. Etruscan tomb. (n.d.). Retrieved from http://realhistoryww.com/world_history/ancient/Misc/Med/Etruscan_tomb_painting1.htm

Reich, D., Green, R. E., Kircher, M., Krause, J., Patterson, N., Durand, E. Y., ... Pääbo, S. (2010). Genetic history of an archaic hominin group from Denisova Cave in Siberia. *Nature*, *468*(7327), 1053–1060. doi:10.1038/nature09710 PMID:21179161

Reimers, N. F. (1994). Ecology. Laws, principles, hypotheses. Moscow, Russia.

Religion-peace-pervobytnaya. (n.d.). Retrieved from https://religionpeace.ru/pervobytnoe/pervobytnye-pecshery.html

Religions-union. (n.d.). Afon. Retrieved from https://religions.unian.net/afon/1488789-7-tayn-i-samyih-bolshih-zagadok-svyatoy-goryi-afon.html

Reznikova, Z., & Ryabko, B. (2011). Numerical competence in animals, with insight from ants. *Behaviour*, *148*(4), 405–434. doi:10.1163/000579511X568562

Rice, D. T. (1968). *Byzantine Art* (3rd ed.). Penguin Books.

Robins, G. (2008). *The Art of Ancient Egypt* (Rev. ed.). Cambridge, MA: Harvard University Press.

Rose, K. D. (1994). The earliest primates. Evolutionary Anthropology: Issues. *News Review (Melbourne)*, *3*(5), 159–173.

Rudynskyi, M. Ya. (1961). *Kamyana Mohyla (framework of rock paintings)*. Kyiv, Ukraine: Publishing House of Academy of Sciences of USSR.

Ruggiero, G. (2015). *The Renaissance in Italy: A Social and Cultural History of the Rinascimento*. Cambridge, UK: Cambridge University Press.

Rutravel. The region of Tassili. (n.d.). Retrieved from https://www.rutraveller.ru/place/13545

Ryabinin, I. (2015). *Origin of man*. Baltic Book Company.

Rybakov, B. A. (1965). Cosmogony and mythology of the Eneolithic farmers. Soviet Archeology, 1965, No. 1-2.

Sacred Texts. (n.d.). Retrieved from https://www.sacred-texts.com/egy/)

Sadiq-Khan, D. (2017). *Battle of the city: how to change our streets. Revolutionary ideas in urban planning*. Moscow, Russia: Olympus Business publishing house.

Schmidt, K. (2015). *Premier temple. Göbekli tepe (Le)*. Göbelki Tepe. CNRS Editions.

Schwartz, J. M., Stapp, H. P., & Beauregard, M. (2005). Quantum Physics in Neuroscience and Psychology: A Neurophysical Model of Mind/Brain Interaction. *Philosophical Transaction of the Royal Society of London, Series B*, *360*(1458), 1309–1327. doi:10.1098/rstb.2004.1598 PMID:16147524

Selye, H. (1982). *Stress without Distress*. Moscow, Russia: Progress.

Serbia-Lepensky-Vir. (n.d.). Retrieved from http://44serbia.ru/lepenski-vir-nahodki-i-zagadki/

Shafer, B. E. (Ed.). (1991). *Religion in Ancient Egypt: Gods, Myths, and Personal Practice*. Cornell University Press.

Sherif, M. (1954). Experiments in group conflict. *Scientific American*, *195*(5), 54–58. doi:10.1038cientificamerican1156-54

Shimko, V. T. (1990). *Architectural Formation of the Urban Environment: A Manual for Architectural Specialties of Universities*. Moscow, Russia: Higher. Sc.

Shlahova, A. (2000). Problems in the Perception of Perspective in Drawing. [Oxford, UK: The Alden Press.]. *International Journal of Art & Design Education*, *19*(1), 102–109.

Shlain, L. (1991). *Art & Physics: Parallel Visions in Space, Time, & Light*. New York: William Morrow and Company.

Shtoer, R. (1992). Multi-criteria optimization: theory, calculations, and applications. Moscow, Russia: Radio and communication.

Silenciollama-blogpost. Adam Kadmon. (n.d.). Retrieved from https://silenciollama.blogspot.com/2010/11/adam-kadmon.html

Silnaya mplitva Bogu. (n.d.). Retrieved from https://сильная-молитва.рф/molitva-k-bogu-totu

Sinelnikov, R. D., & Sinelnikov, Ya. R. (1996). *Atlas of Human Anatomy in 4 Volumes. Moscow, Russia*: Medicine.

Sinisgalli, R. (2011). *On Painting. A New Translation and Critical Edition*. Cambridge, UK: Cambridge University Press.

Skinner, B. F. (1965). *Science and Human Behavior*. Free Press.

Skinner, B. F. (2002). *Beyond Freedom and Dignity*. Hackett Publishing.

Smith, W. S., & Simpson, W. K. (1998). *The Art and Architecture of Ancient Egypt* (3rd ed. 1998). Yale University Press.

Soroko, E. M. (1984). *Structural Harmony of Systems. Moscow, Russia*: Nauka.

Spengler, O. (1993). The Decline of the West. Essays on the morphology of world history. M.

Sperry, R. W. (2018). Consciousness, free will, and personal identity 1. In D. A. Oakley, & H. C. Plotkin (Eds.), *Brain, behavior, and evolution* (pp. 219–228). London, UK: Routledge. doi:10.4324/9781315149523-10

Spinoza, B. (1999). Works in 2 Volumes. Saint Petersburg, Russia: Science. Symbiosis-online-publishing. (n.d.). Retrieved from https://symbiosisonlinepublishing.com/dentistry-oraldisorders-therapy/dentistry-oraldisorders-therapy34.php

Spivey, N. (1997). *Etruscan Art*. London, UK: Thames and Hudson.

Stapp, H. (2001). Quantum Theory and the Role of Mind in Nature. *Foundations of Physics, 31*(10), 1465–1499. doi:10.1023/A:1012682413597

Steingräber, S. (2006). *The abundance of Life: Etruscan Wall Painting*. Los Angeles, CA: J. Paul Getty Museum.

Stephens, J. (1990). *The Italian Renaissance: The Origins of Intellectual and Artistic Change Before the Reformation*. Routledge.

*Stierlin, H.* (2004). Greece: From Mycenae to the Parthenon. TASCHEN.

Stolyar, A. (1985). The origin of the visual arts. Moscow, Russia: Art.

Storinka-Kiev. Maria Primachenko. (n.d.). Retrieved from https://storinka-m.kiev.ua/product.php?p_id=2646

Stuard, S. M. (1987). *Women in Medieval History and Historiography*. Philadelphia, PA: University of Pennsylvania Press. doi:10.9783/9781512807295

Studblue. Notes-midterm. (n.d.). Retrieved from http://www.studyblue.com/notes/note/n/art-midterm/deck/1189362

Studfiles, preview-5254880. (n.d.). Retrieved from https://studfiles.net/preview/5254880/

Summit, J. (2011). Renaissance Humanism and the Future of the Humanities. *Literature Compass* (2012), 9-10, pp: 665–678.

Symbol. Shumerskaya pismennost. (n.d.). Retrieved from http://the-symbol.ru/shumerskaya_pismenost.htm

Tajfel, H., & Turner, J. C. (1986). The social identity theory of intergroup behavior. In S. Worchel, & W. G. Austin (Eds.), *Psychology of Intergroup Relations*. Chicago, IL: Nelson-Hall.

Talbot, R. (2003). Speculations on the Origin of Linear Perspective. *Nexus Network Journal, 5*(1), 64–98.

Tarski, A. (1931). Sur les ensembles définissables de nombres réels I, *Fundamenta Mathematicae, 17,* 210-239 in Tarski (1983), 110-142.

Tarski, A. (1936). Grundlegung der wissenschaftlichen Semantik, Actes du Congrès international de philosophie scientifique, Sorbonne, Paris 1935, vol. III, Language et pseudo-problèmes, Paris, Hermann, 1936, pp. 1–8 in Tarski (1983), 401-408.

Taylor, E. B. (1989). Primitive culture. Moscow, Russia: Science.

Taylor, J. C. (1981). *Learning to Look: A Handbook for the Visual Arts*. Chicago, IL: University of Chicago Press.

Teilhard de Chardin, P. (1956). *Oeuvres, t.1: Le phénomène humain*. Paris, France: Les Éditions du Seuil.

Telegraf. Aziatskie pyramidy. (n.d.). Retrieved from https://telegraf.com.ua/ekologiya/450729-aziatskie-piramidyi-fotogalereya.html

Tetior, A. N. (2008). *Urban ecology*. Moscow, Russia.

*The Gilgamesh epic ("On the one who sees all")*. (2015). St. Petersburg, Russia: Nauka.

The-history-blog. Kalpis. (n/d). Retrieved from http://www.thehistoryblog.com/wp-content/uploads/2012/06/kalpis.jpg

Thiébaut, P. (2018). *Mucha et l'Art Nouveau*. Paris, France: Éditions du Chêne. (in French)

Thomas, W., & Coolidge, F. L. (2012). *How to Think Like a Neandertal*. Oxford University Press.

Thorndike, E., Watson, D., & Karelin, A. (1988). Behaviorism.

Tiazhelov, V. N. (1981). The Art of the Middle Ages in Western and Central Europe. Moscow, Russia: Iskusstvo.

Todeschini, M. (1949). La Teoria delle Apparenze. Istituto Italiano D'Arti Grafiche.

Tononi, G., Boley, M., Marcello, M., & Koch, C. (2016). Integrated Information Theory: From Consciousness to the Physical Substrate. *Nature Reviews. Neuroscience*, *17*(7), 450–461. doi:10.1038/nrn.2016.44 PMID:27225071

Traditionalist-livejournal. (n.d.). Retrieved from https://ltraditionalist.livejournal.com/617658.html

Treegene. (n.d.). Retrieved from https://treegene.kz/uslugi/ychromosome-heritage/attachment/567

Triandis, H. C. (1994). Culture and Social Behavior: McGraw-Hill College.

Trinixy. Risunki devushki. (n.d.). Retrieved from https://trinixy.ru/143795-risunki-devushki-stradayuschey-shizofreniey-14-foto.html

Trollno. (n.d.). *Zhizn v zamre.* Retrieved from https://trollno.com/zhizn-v-zamke-kak-vyglyadit-dom-ally-pugachevoj-i-maksima-galkina/

Truisheva. Kitayskaya zhivopis. (n.d.). Retrieved from http://turisheva.ru/2015/05/08/kitayskaya-zhivopis-iv-vi-veka-n-e/

Tuck, S. L. A. (2015). *History of Roman Art.* Malden, MA: Wiley Blackwell.

Turaev, B. A. (1922). Ancient Egypt. Moscow, Russia: Publ. House "Ogni".

Turner, R. V. (2009). *Eleanor of Aquitaine: Queen of France, Queen of England.* Yale University Press.

Ufa-mama-forum. (n.d.). Retrieved from https://www.ufamama.ru/forum/messages.aspx?topicid=9392

Universal History of Architecture. (2008). *Ancient World Architecture* (Vol. I). Moscow, Russia: Arhitektura-S.

Urbina, S., & Anastasi, A. (1997). *Psychological testing.* (7th ed.). Upper Saddle River, NJ: Prentice Hall.

Urmantsev, Y. A. (1974). The symmetry of Nature and Nature of Symmetry: Philosophic and scientific aspects. Moscow, Russia: Mysl.

Urmantsev, Y. A. (1988). Evolution study or General system development of nature, society, and mind. Puschino: Institute of phytophysiology Academy of Sciences of USSR.

Urmantsev, Y. A. (1974). *The symmetry of Nature and Nature of Symmetry Philosophical and Natural Scientific Aspects. Moscow, Russia*: Misl.

Urmantsev, Y. A. (1988). *Evolutionary or General Theory of Development of Systems of Nature, Society, and Thinking.* Pushchino, Russia: Institute of Plant Physiology, USSR Academy of Sciences.

Urochishe Kalbak-Tash. (n.d.). Retrieved from http://kp74.ru/urochishhe-kalbak-tash.html

Usenkomaxim. Laurent Parcelier. (n.d.). Retrieved from http://usenkomaxim.ru/2013/01/03/laurent-parcelier-sovremennyiy-frantsuzskiy-impressionizm/

Uzgorod-novosti. Bokshai. Panoramic view of Uzhgorod. (n.d.). Retrieved from http://uzhgorod.in/novosti/2015/dekabr/uzhgorod_v_iskusstve_naslazhdaemsya

Uzgorod-novosti. Erdely. View of Uzhgorod. (n.d.). Retrieved from http://uzhgorod.in/novosti/2015/dekabr/uzhgorod_v_iskusstve_naslazhdaemsya

Vadetskaya, E. B. Sculptures of the Okunev Culture. (1980). London, UK: Science.

Vadetskaya, E. B. (1986). *Archeological Monuments in the Middle Yenisei Steppes*. Okunev Culture. London, UK.

van den Broecke, M. P. R. (1996). Ortelius Atlas Maps. An illustrated Guide. HES Publishers, 1996.

van Mander, K. (2007). *The book about artists*. St. Petersburg, Russia: Azbuka-klassika.

Vasyutinskiy, N. A. (1990). *Golden Proportion*. Moscow, Russia: Molodaya Gvardiya.

Vektork-naskalnaya-zhivopis. (n.d.). Retrieved from http://vectork.org/naskalnaya-zhivopis-peshher-lasko/

Velchinskaya, I. L. (1977). *Albrecht Altdorfer*. Moscow, Russia: Isskustvo.

Verman, K. (1904) History of Arts of all Times and Peoples: in 3 vol. / K. Verman. St. Petersburg, Russia: Publishing Partnership "Prosvescheniye"

Veselovskiy, N. I. (1893). Report of Imperial Archaeological commission for the year 1890, Saint Petersburg.

Videiko, M. (2010). *Ukraine: from Trypillia to Rus*. Kyiv, Ukraine: Krion.

Vishnyatsky, L. B. (2010). Neanderthals: the story of a failed humanity. Publishing house "Nestor-Istoria".

Visloguzov, A. (2016). *Features of Modern Design of Heating, Ventilation, and Air Conditioning Systems in Public, Multi-Storey, and High-Rise Buildings: A Tutorial*. SKFU.

Vitruvius. (1936). *Ten books about architecture*. Moscow, Russia: Publishing House of the Academy of Architecture.

Vityuk, E. (2015). *Search of an Ideal City*. Architecton.

von Bertalanffy, L. (1950). The Theory of Open Systems in Physics and Biology. *Science*, *111*(2872), 23–29. doi:10.1126cience.111.2872.23 PMID:15398815

Voronin, A. N., Ziatdinov, Y. K., Charchenko, A. V., & Ostashevkiy, V. V. (1995). *Complex technical and Human-machine systems: research methods. Germany:* Fakt.

Vselova. Zodchiy Hesira. (n.d.). Retrieved from http://vseslova.com.ua/word/«Зодчий_Хесира»._Рельеф_из_гробницы_%28Древний_Египет%29_%28изображение%29-38640

Vygotsky, L., Köhler, F., & Koffka, K. (1988). Gestalt psychology: ozon.ru.

Warspot. Pervobytnaya-voyna. (n.d.). Retrieved from https://warspot.ru/5116-pervobytnaya-voyna-masshtab-tragedii

Watson, J. (1913). Psychology from the Behavioral Perspective.

*Wattenmaker , R. J. (1993).* Great French Paintings from the Barnes Foundation: Impressionist, Post-impressionist, and Early Modern. New York: Alfred A. Knopf.

Waude, A. (2017). *Social Influence: Why We Conform in Groups*. Retrieved from www.psychologistworld.com

Webcitation. (n.d.). Retrieved from https://www.webcitation.org/63ZdUemZU

Weekend. (n.d.). Retrieved from https://weekend.rambler.ru/read/38910641-skolko-kaloriy-szhigaet-mozg-i-kak-eto-rasschitat/

Weitzmann, K., & Chatzidakis, M. (1982). *The Icon*. London, UK: Evans Brothers.

Westminster-Abbey. About Abbey. (n.d.). Retrieved from https://www.westminster-abbey.org/ru/about-the-abbey/history/retable

Weyl, G. (1968). *Symmetry.* Moscow, Russia: Nauka.

Whitton, D. (1998). The Society of Northern Europe in the High Middle Ages, 900–1200. In G. Holmes (Ed.), The Oxford Illustrated History of Medieval Europe. Oxford, UK: Oxford University Press. pp. 115–174.

Wikiart. Maria Primachenko. (n.d.). Retrieved from https://www.wikiart.org/ru/mariya-primachenko/black-beast-1936

Wikiart. Sam Francis. (2012). Retrieved from https://www.wikiart.org/en/sam-francis/untitled-yellow-1961

Wikimedia. (n/d). The_Universe. Retrieved from https://commons.wikimedia.org/wiki/File:The_universe.jpg

Wikimedia. Battle of Crecy. (n.d.). Retrieved from https://commons.wikimedia.org/wiki/File:Battle_of_crecy_froissart.jpg?uselang=ru

Wikimedia. Bilibin. Vasilisa the Beautiful. (n.d). Retrieved from https://commons.wikimedia.org/wiki/File:Vasilisa.jpg

Wikimedia. Bosh. Judgment. (n.d.). Retrieved from https://commons.wikimedia.org/wiki/File:Hieronymus_Bosch_-The_Last_Judgement.jpg

Wikimedia. Bosh. The Garden. (n.d.). Retrieved from https://commons.wikimedia.org/wiki/File:Hieronymus_Bosch_-_The_Garden_of_Earthly_Delights_-_The_exterior_(shutters).jpg

Wikimedia. Bosh. Vision of Hereafter. (n.d.). Retrieved from https://commons.wikimedia.org/wiki/File:Poliptyk_Hieronima_Bocha_Visions_of_the_Hereafter.JPG

Wikimedia. Botticelli. Simonetta. (n.d.). Retrieved from https://commons.wikimedia.org/wiki/File:Sandro_Botticelli_066.jpg

Wikimedia. Botticelli. The Cestello Annunciation. (n.d.). Retrieved from https://commons.wikimedia.org/wiki/File:Botticelli,_annunciazione_di_cestello_02.jpg

Wikimedia. Bruegel the Elder. Tower. (n.d.). Retrieved from https://commons.wikimedia.org/wiki/File:Pieter_Bruegel_the_Elder_-_The_Tower_of_Babel_(Vienna)_-_Google_Art_Project_-_edited.jpg

Wikimedia. Conquest of Constantinople. (n.d.). Retrieved from https://commons.wikimedia.org/wiki/File:ConquestOfConstantinopleByTheCrusadersIn1204.jpg

Wikimedia. Cubiculum. Villa Fannius. (n.d.). Retrieved from https://commons.wikimedia.org/wiki/File:Cubiculum_(bedroom)_from_the_Villa_of_P._Fannius_Synistor_at_Boscoreale_MET_DP170950.jpg

Wikimedia. Dong Yan. Wintry Groves. (n.d.). Retrieved from https://commons.wikimedia.org/wiki/File:Dong_Yuan._Wintry_Groves_and_Layered_Banks_ca._950_(181,5x116,5cm)_Kurokava_Inst._Hyogo.jpg?uselang=ru

Wikimedia. Etruscan civilization. (1988). Retrieved from https://en.wikipedia.org/wiki/File:Etruscan_civilization_map.png

Wikimedia. Etruscan mural. (n.d.). Retrieved from https://commons.wikimedia.org/wiki/File:Etruscan_mural_achilles_Troilus.gif

Wikimedia. Fourth Style. (n.d.). Retrieved from https://commons.wikimedia.org/wiki/File:Fourth_Style_wall_painting_with_architectural_landscape,_a_detail_from_a_panel_from_the_portico_of_the_Temple_of_Isis_in_Pompeii,_Naples_National_Archaeological_Museum_(14584129434).jpg

Wikimedia. Franqis Tomb painting. (n.a.). Retrieved from https://commons.wikimedia.org/wiki/File:François_Tomb_Painting_02.jpg

Wikimedia. Gans Memling. (n.d.). Retrieved from https://commons.wikimedia.org/wiki/File:Hans_Memling_Passione.jpg

Wikimedia. Gu Kaychdji. Fea. (n.d.). Retrieved from https://commons.wikimedia.org/wiki/File:Гу_Кайчжи._Фея_реки_Ло._Деталь_-_Фея._Гугун,_Пекин.jpg?uselang=ru

Wikimedia. Interior of Hagia Sophia. (n.d.). Retrieved from https://commons.wikimedia.org/wiki/File:Interior_of_Hagia_Sophia.jpg

Wikimedia. Iphigenia. Taurus. (n.d.). Retrieved from https://commons.wikimedia.org/wiki/File:Iphigenia_in_Tauris_MAN_Napoli_Inv111439.jpg

Wikimedia. Jan van Eyck. (n.d.). Retrieved from https://commons.wikimedia.org/wiki/File:Jan_van_Eyck_070.jpg

Wikimedia. Jan van Eyck. Diptych. (n.d.). Retrieved from https://commons.wikimedia.org/wiki/File:Jan_van_Eyck_-_Diptych_-_WGA07587.jpg?uselang=ru

Wikimedia. Jan van Eyck. Madonna. (n.d.). Retrieved from https://commons.wikimedia.org/wiki/File:Jan_van_Eyck_-_The_Madonna_in_the_Church_-_Google_Art_Project.jpg

Wikimedia. Joan of Arc. (n.d.). Retrieved from https://commons.wikimedia.org/wiki/File:Joan_of_Arc_miniature_graded.jpg

Wikimedia. Karl III_monks/ (n.d.). Retrieved from https://commons.wikimedia.org/wiki/File:KarlII_monks.jpg

Wikimedia. Klimt. Portrait of Adele Bloch-Bauer II. (n.d.). Retrieved from https://commons.wikimedia.org/wiki/File:Gustav_Klimt_047.jpg

Wikimedia. Lascaux. Megaloceros. (2009). Retrieved from https://commons.wikimedia.org/wiki/File:Lascaux,_Megaloceros.jpg?uselang=ru

Wikimedia. Last geometric pyxis. (n.d.). Retrieved from https://commons.wikimedia.org/wiki/File:Late_geometric_pyxis_Ancient_Agora_Museum.jpg?uselang=ru

Wikimedia. Leonardo. Madonna Litta. (n.d.). Retrieved from https://commons.wikimedia.org/wiki/File:Leonardo_da_Vinci_attributed_-_Madonna_Litta.jpg

Wikimedia. Leonardo. Mona Lisa. (n.d.). Retrieved from https://commons.wikimedia.org/wiki/File:Mona_Lisa,_by_Leonardo_da_Vinci,_from_C2RMF_retouched.jpg

Wikimedia. Li Cheng. A Solitary Temple. (n.d.). Retrieved from https://commons.wikimedia.org/wiki/File:Li_Cheng_-_A_Solitary_Temple_Amid_Clearing_Peaks_-_Google_Art_Project.jpg?uselang=ru

Wikimedia. Lucas Cranah. Judgment. (n.d.). Retrieved from https://commons.wikimedia.org/wiki/File:Lucas_Cranach_d._Ä._-_The_Judgment_of_Paris_-_WGA05629.jpg?uselang=ru

Wikimedia. Madonna. Child. (1994). Retrieved from https://commons.wikimedia.org/wiki/File:KellsFol007vMadonnaChild.jpg

Wikimedia. Malevich. Black suprematic square. (n.d.). Retrieved from https://commons.wikimedia.org/wiki/File:Чёрный_супрематический_квадрат._1915._ГТГ.png

Wikimedia. Massccio. Trinitia. (2002). Retrieved from https://commons.wikimedia.org/wiki/File:Masaccio,_trinità.jpg

Wikimedia. Miniature hydria. Thebes. (n.a.). Retrieved from https://commons.wikimedia.org/wiki/File:Miniature_hydria_Thebes_Louvre_A566.jpg?uselang=ru

Wikimedia. Mondrian. Composition. (n.d.). Retrieved from https://commons.wikimedia.org/wiki/File:Mondrian,_Composition_with_color_planes_and_gray_lines,_1918.jpg

Wikimedia. Mosaic of Theodora. (n.d.). Retrieved from https://commons.wikimedia.org/wiki/File:Mosaic_of_Theodora_-_San_Vitale_-_Ravenna_2016_(4).jpg

Wikimedia. Neolithique-001. (n.d.). Retrieved from https://commons.wikimedia.org/wiki/File:Néolithique_0001.jpg?uselang=ru

Wikimedia. Oranta. Kyiv. (n.d.) Retrieved from https://commons.wikimedia.org/wiki/File:Oranta-Kyiv.jpg?uselang=ru

Wikimedia. Petus Cristus. (n.d.). Retrieved from https://commons.wikimedia.org/wiki/File:Petrus_Christus,_Portrait_of_a_young_girl.jpg

Wikimedia. Phaistos Disk. (n.d.). Retrieved from https://en.wikipedia.org/wiki/File:Δίσκος_της_Φαιστού_πλευρά_B_6381.JPG

Wikimedia. Phaistos Disk. (n.d.). Retrieved from https://en.wikipedia.org/wiki/Phaistos_Disc#/media/File:Δίσκος_της_Φαιστού_πλευρά_A_6380.JPG

Wikimedia. Pierro della Francesko. (n.d.). Retrieved from https://commons.wikimedia.org/wiki/File:Piero_della_Francesca_044.jpg

Wikimedia. Quenn-megaron. (n.d.). Retrieved from https://commons.wikimedia.org/wiki/File:Queen%27s_Megaron_with_dolphins_Knossos_Palace.jpg

Wikimedia. Raffael-058. (n.d.). Retrieved from https://commons.wikimedia.org/wiki/File:Raffael_058.jpg

Wikimedia. Rylov. In blue space. (n.d.). Retrieved from https://commons.wikimedia.org/wiki/File:Rylov_Blue_Expanse.jpg?uselang=ru

Wikimedia. Savanorola. Rerrara. (n.d.). Retrieved from https://commons.wikimedia.org/wiki/File:Savonarola_(Ferrara).jpg?uselang=ru

Wikimedia. Tablette-lineare. (n.d.). Retrieved from https://commons.wikimedia.org/wiki/File:Tablette_lineaire-B.jpg?uselang=ru

Wikimedia. The Turee Gracies. (n.d.). Retrieved from https://commons.wikimedia.org/wiki/File:The_Three_Graces,_from_Pompeii_(fresco).jpg

Wikimedia. Trinity. Rublev. (n.d.). Retrieved from https://commons.wikimedia.org/wiki/File:Angelsatmamre-trinity-rublev-1410.jpg

Wikimedia. Waldorf Hotel. (n.d.). Retrieved from https://commons.wikimedia.org/wiki/File:Demolished_Astoria_hotel,_NYC.jpg?uselang=ru

Wikimedia. Xiao and Xiang rivers. (n.d.). Retrieved from https://commons.wikimedia.org/wiki/File:Xiao_and_Xiang_rivers.jpg?uselang=ru

Wikipedia Encaustic Virgin. (n.d.). Retrieved from https://en.wikipedia.org/wiki/File:Encaustic_Virgin.jpg

Wikipedia. (n.d.). Astoria Hotel. Retrieved from https://en.wikipedia.org/wiki/File:Astoria_Hotel_-_First_Floor_Plan.png

Wikipedia. (n.d.). *Waldorf-Astoria.* Retrieved from https://en.wikipedia.org/wiki/Waldorf–Astoria_(1893–1929)

Wikipedia. Aboriginal rock art. (n.a.). Retrieved from https://en.wikipedia.org/wiki/File:Aboriginal_rock_art_on_the_Barnett_River,_Mount_ElizabethStation.jpg

Wikipedia. Altdorfer. The Battle of Alexander, (n.d.). Retrieved from https://ru.wikipedia.org/wiki/Битва_Александра#/media/File:Albrecht_Altdorfer_-_Schlacht_bei_Issus_(Alte_Pinakothek,_München).jpg

Wikipedia. Ancient Egypt. (n.d.). Retrieved from https://en.m.wikipedia.org/wiki/File:Ancient_Egypt_map-hiero.svg

Wikipedia. Bilibin. Alconost. (n.d.). Retrieved from https://ru.wikipedia.org/wiki/Файл:Bilibin_Alkonost.JPG

Wikipedia. Bosch. The Cart of Hay. (n.d.) Retrieved from https://ru.wikipedia.org/wiki/Воз_сена_(триптих_Босха)#/media/File:Haywain_central_panel_of_the_triptych_WGA.jpg

Wikipedia. Bosch. The Garden of Earthly Delights. (n.d.). https://ru.wikipedia.org/wiki/Сад_земных_наслаждений_(триптих_Босха)#/media/File:El_jardín_de_las_Delicias,_de_El_Bosco.jpg

Wikipedia. Bruegel the Elder. Hunters in the Snow. (n.d.). Retrieved from https://ru.wikipedia.org/wiki/Охотники_на_снегу#/media/File:Pieter_Bruegel_d._Ä._106b.jpg

Wikipedia. Bruegel the Elder. Soroka. (n.d.). Retrieved from https://ru.wikipedia.org/wiki/Сорока_на_виселице#/media/File:Die_Elster_auf_dem_Galgen.jpg

Wikipedia. Bruegel the Elder. The Fall of Icarus. (n.d.). Retrieved from https://ru.wikipedia.org/wiki/Падение_Икара#/media/File:Pieter_Bruegel_de_Oude_-_De_val_van_Icarus.jpg

Wikipedia. Bruegel the Elder. The Fight Between Carnival and Lent. (n.d.). Retrieved from https://ru.wikipedia.org/wiki/Битва_Масленицы_и_Поста_(картина)#/media/File:Pieter_Bruegel_d._Ä._066.jpg

Wikipedia. Bruegel the Elder. Triumph of Death. (n.d.). Retrieved from https://ru.wikipedia.org/wiki/Триумф_смерти_(картина)#/media/File:Thetriumphofdeath.jpg

Wikipedia. Bruegel the Elder-001. (n.d.). Retrieved from https://ru.wikipedia.org/wiki/Список_картин_Питера_Брейгеля_Старшего#/media/File:Pieter_Bruegel_d._Ä._001.jpg

Wikipedia. Cezanne, Mont Sainte-Victoire. (n.d.). Retrieved from https://en.wikipedia.org/wiki/File:Paul_Cézanne_210.jpg

Wikipedia. Cezanne. Banks of the Marne. (n.d.). Retrieved from https://en.wikipedia.org/wiki/File:Paul_Cézanne_104.jpg

Wikipedia. Chatal-Hyyk. (n.d.). Retrieved from https://ru.wikipedia.org/wiki/Чатал-Хююк

Wikipedia. Dalí. The Persistence of Memory. (n.d.). Retrieved from https://ru.wikipedia.org/wiki/Файл:The_Persistence_of_Memory.jpg

Wikipedia. Empress Zoe. (n.d.). Retrieved from https://en.wikipedia.org/wiki/File:Empress_Zoe_mosaic_Hagia_Sophia.jpg

Wikipedia. Huan Gunvan. (n.d.). Retrieved from https://ru.wikipedia.org/wiki/Хуан_Гунван#/media/File:Huang_Gongwang._Clearing_After_Sudden_Snow._ca._1340._104,6_cm_long.Palace_Museum,_Beijing.jpg

Wikipedia. Kompen Rober. (n.d.). Retrieved from https://ru.wikipedia.org/wiki/Кампен,_Робер#/media/File:RobertCampinWorkshop-Man-1314.jpg

Wikipedia. Kvadrat Aristoeles. (n.d.). Retrieved from https://ru.wikipedia.org/wiki/Файл:Quadrat_Aristoteles.jpg

Wikipedia. Mirovoe derevo. (n.d.). Retrieved from https://ru.wikipedia.org/wiki/Мировое_древо#/media/File:Yggdrasil_AM_738_4to.jpg

Wikipedia. Mosaic Empereor Justinian. (n.d.). Retrieved from https://en.wikipedia.org/wiki/File:The_mosaic_of_Emperor_Justinian_and_his_retinue.jpg

Wikipedia. Ni Tsan. (n.d.). Retrieved from https://ru.wikipedia.org/wiki/Файл:Ни_Цзань.Воды_и_бамбуковый_домик._1343_Музей_кит._истории,Пекин.jpg#/media/File:Ни_Цзань.Воды_и_бамбуковый_домик._1343_Музей_кит._истории,_Пекин.jpg

Wikipedia. Nurenberg. Dance of Death. (n.d.). Retrieved from https://en.wikipedia.org/wiki/File:Nuremberg_chronicles_-_Dance_of_Death_(CCLXIIIIv).jpg

Wikipedia. Persidskaya Zhivopis. (n.d.). Retrieved from https://ru.wikipedia.org/wiki/Персидская_живопись#/media/File:Demotte_Shahname_001.jpg

Wikipedia. Persidskaya Zhivopis. (n.d.). Retrieved from https://ru.wikipedia.org/wiki/Персидская_живопись#/media/File:Junaid_001.jpg

Wikipedia. Persidskaya Zhivopis. (n.d.). Retrieved from https://ru.wikipedia.org/wiki/Персидская_живопись#/media/File:Yusef_Zuleykha.jpg

Wikipedia. Picasso_Gernica. (n.d.). Retrieved from https://ru.wikipedia.org/wiki/Файл:PicassoGuernica.jpg

Wikipedia. Soulages. Honolulu. Painting. (n.d.). Retrieved from https://en.wikipedia.org/wiki/File:%2717_December_1966%27_by_Pierre_Soulages,_Honolulu_Museum_of_Art.JPG

Wikipedia. Teorema Neter. (n.d.). Retrieved from https://ru.wikipedia.org/wiki/Теорема_Нетер

Wikipedia. Teoria Kalutsi. (n.d.). Retrieved from https://ru.wikipedia.org/wiki/Теория_Калуцы_Клейна.

Wikipedia. Tree of Life. (n.d.). Retrieved from https://en.wikipedia.org/wiki/File:Tree_of_life_bahir_Hebrew.svg

Wikipedia. Tsan Syan. (n.d.). Retrieved from https://ru.wikipedia.org/wiki/Цянь_Сюань#/media/File:2_Qian_Xuan_Wang_Xizhi_Watching_Geese._(23,2x92,7cm)_Metropolitan_Museum_of_Art,_N-Y.jpg

Wikipedia. Turinskiy avtoportret. (n.d.). Retrieved from https://ru.wikipedia.org/wiki/Туринский_автопортрет#/media/File:Leonardo_self.jpg

Wikipedia. wiki. Complex. (n.d.). Retrieved from https://ru.wikipedia.org/wiki/Системный_подход

Wikipedia. wiki. Complex. (n.d.). Retrieved from https://ru.wikipedia.org/wiki/Сложная_система

Wikiway. Bolgaria-peshera. (n.d.). Retrieved from https://wikiway.com/bolgaria/peshchera-magura/

Wilde, C. (1994). Painting, Alberty and the Wisdom of Minerva. *British Journal of Aesthetics*, *34*, 48–59.

Wilkinson, T. (2010). *The Rise and Fall of Ancient Egypt: The History of a Civilisation from 3000BC to Cleopatra*. London, UK: Bloomsbury.

Williams, D. (1999). *Masterpieces of Classical Art, 2009*. British Museum Press.

Williams, N. D. (2006). *Design of familiar things*. Moscow, Russia: Publishing House.

Winkler, G., Niebuhr, K., & Schurz, G. (1903). The History of Mankind. World History: in 9 vol.: Vol. 3. *West Asia and Africa*. St. Petersburg, Russia: Publishing Partnership.

Wolff, N., & Hameroff, S. (2001, November). A Quantum Approach to Visual Consciousness. *Trends in Cognitive Sciences*, *5*(11), 472–478. doi:10.1016/S1364-6613(00)01774-5 PMID:11684479

Wong, K. (2014). Tiny Genetic Differences between Humans and Other Primates Pervade the Genome. *Scientific American*.

World History: in 13 vol. (1955-1970). Academy of Sciences of the USSR. Moscow, Russia: State Publ. House of Polit. Literature.

World Rock Art. (n.d.). Retrieved from http://worldrockart.ru/risunki/petroglify-altaya/

World-rock-art. (n.d.). Retrieved from http://worldrockart.ru/risunki/laas-gaal-somaliland/

World-rock-art. (n.d.). Retrieved from http://worldrockart.ru/risunki/serra-da-kapivara-braziliya/

World-rock-art. (n.d.). Retrieved from http://worldrockart.ru/risunki/tadrart-akakus-liviya/

Wright, L. (1983). *Perspective in Perspective*. London, UK: Routledge & Kegan Paul.

Xlegio. Siege machinery. (n.d.). Retrieved from http://xlegio.ru/siege-machinery/poliorketika-and-paraskeuastika/assyrian-siege-machinery/

Yandex-collection-cart. (n.d.). Retrieved from https://yandex.ru/collections/card/5a07c8eb0c1ed2002fed3a3e/

Yaroslav. (n.d.). Hrudchevka. Retrieved from https://yaroslavl.n1.ru/articles/yaroslavskaya_hruschevka_carstvo_ili_myshelovka-50334181/

Yatromanolakis, D. (2016). *Epigraphy of Art: Ancient Greek Vase-Inscriptions and Vase-Paintings*. Oxford, UK: Archaeopress.

Yoga Sutras by Patanjali. (2000). Moscow, Russia: Janus books.

Zidermane, A. A. (1988). Some issues in chronobiology and chronomedicine: Zinatne.

Zidermane, A. A. (1988). *Certain questions of chronobiology and chronomedicine*. Zinatne.

Znamerovskaya, T. P. (1975). Quattrocento problems and Masaccio's work: *Abstract of a dissertation for the degree of Doctor of Art History*. Leningrad, Russia. State un-t them.

# About the Authors

**Yury N. Kovalyov,** Doctor of Technical Science, Professor Head of Industrial Design and Computer Technologies Department of the Mykhailo Boichuk Kyiv State Academy of Decorative Applied Arts and Design, Ukraine. Author of 233 scientific publications (5 monographs, 4 textbooks) in the fields of geometry foundations, mathematical modeling and multicriteria optimization of complex systems and human-machine systems, habitat design, sociometry, psychology. Hobbies: poetry (author of 2 books), philosophy, yoga, sport, travel.

**Nver. M. Mkhitaryan,** Doctor of Technical Science, Professor, Corresponding Member of the National Academy of Science of Ukraine; Foreign Member of the National Academy of Sciences of Armenia; Full Member of the American Society of Civil Engineers (ASCE); Honorable director of the Institute of Renewable Energy (National Academy of Science of Ukraine), Owner of the holding company "Emerald Palace Group" (Dubai, United Arab Emirates). Author of more than 250 scientific publications (14 monographs) in the fields of building construction and technology, energy efficiency, alternative and renewable energy, complex systems optimization. Hobbies: travel, philosophy and psychology, collecting and studying ancient books and ancient weapons.

**Alexander Yu. Nitsyn,** Doctor of Technical Science, Professor of Department of Geometry Modeling and Computer Graphics at National Technical University "Kharkiv Polytechnic Institute". Author of more than 70 scientific publications (1 monograph, 2 textbooks) in the fields of geometry, perspective, computer graphics. Hobbies: books, fine art.

# Index

## A



## B



## C



## D



## E

# F



# G



# H



# I



# J



# K



# M

## T



## U



## W

www.ingramcontent.com/pod-product-compliance
Lightning Source LLC
LaVergne TN
LVHW080532300926
849162LV00007B/1589
* 9 7 8 1 7 9 9 8 1 7 0 6 2 *